Lecture Notes in Physics

The Editorial Policy for Proceedings

The series Lecture Notes in Physics reports new developments in physical research and teaching – quickly, informally, and at a high level. The proceedings to be considered for publication in this series should be limited to only a few areas of research, and these should be closely related to each other. The contributions should be of a high standard and should avoid lengthy redraftings of papers already published or about to be published elsewhere. As a whole, the proceedings should aim for a balanced presentation of the theme of the conference including a description of the techniques used and enough motivation for a broad readership. It should not be assumed that the published proceedings must reflect the conference in its entirety. (A listing or abstracts of papers presented at the meeting but not included in the proceedings could be added as an appendix.)
When applying for publication in the series Lecture Notes in Physics the volume's editor(s) should submit sufficient material to enable the series editors and their referees to make a fairly accurate evaluation (e.g. a complete list of speakers and titles of papers to be presented and abstracts). If, based on this information, the proceedings are (tentatively) accepted, the volume's editor(s), whose name(s) will appear on the title pages, should select the papers suitable for publication and have them refereed (as for a journal) when appropriate. As a rule discussions will not be accepted. The series editors and Springer-Verlag will normally not interfere with the detailed editing except in fairly obvious cases or on technical matters.
Final acceptance is expressed by the series editor in charge, in consultation with Springer-Verlag only after receiving the complete manuscript. It might help to send a copy of the authors' manuscripts in advance to the editor in charge to discuss possible revisions with him. As a general rule, the series editor will confirm his tentative acceptance if the final manuscript corresponds to the original concept discussed, if the quality of the contribution meets the requirements of the series, and if the final size of the manuscript does not greatly exceed the number of pages originally agreed upon. The manuscript should be forwarded to Springer-Verlag shortly after the meeting. In cases of extreme delay (more than six months after the conference) the series editors will check once more the timeliness of the papers. Therefore, the volume's editor(s) should establish strict deadlines, or collect the articles during the conference and have them revised on the spot. If a delay is unavoidable, one should encourage the authors to update their contributions if appropriate. The editors of proceedings are strongly advised to inform contributors about these points at an early stage.
The final manuscript should contain a table of contents and an informative introduction accessible also to readers not particularly familiar with the topic of the conference. The contributions should be in English. The volume's editor(s) should check the contributions for the correct use of language. At Springer-Verlag only the prefaces will be checked by a copy-editor for language and style. Grave linguistic or technical shortcomings may lead to the rejection of contributions by the series editors. A conference report should not exceed a total of 500 pages. Keeping the size within this bound should be achieved by a stricter selection of articles and not by imposing an upper limit to the length of the individual papers. Editors receive jointly 30 complimentary copies of their book. They are entitled to purchase further copies of their book at a reduced rate. As a rule no reprints of individual contributions can be supplied. No royalty is paid on Lecture Notes in Physics volumes. Commitment to publish is made by letter of interest rather than by signing a formal contract. Springer-Verlag secures the copyright for each volume.

The Production Process

The books are hardbound, and the publisher will select quality paper appropriate to the needs of the author(s). Publication time is about ten weeks. More than twenty years of experience guarantee authors the best possible service. To reach the goal of rapid publication at a low price the technique of photographic reproduction from a camera-ready manuscript was chosen. This process shifts the main responsibility for the technical quality considerably from the publisher to the authors. We therefore urge all authors and editors of proceedings to observe very carefully the essentials for the preparation of camera-ready manuscripts, which we will supply on request. This applies especially to the quality of figures and halftones submitted for publication. In addition, it might be useful to look at some of the volumes already published. As a special service, we offer free of charge LATEX and TEX macro packages to format the text according to Springer-Verlag's quality requirements. We strongly recommend that you make use of this offer, since the result will be a book of considerably improved technical quality. To avoid mistakes and time-consuming correspondence during the production period the conference editors should request special instructions from the publisher well before the beginning of the conference. Manuscripts not meeting the technical standard of the series will have to be returned for improvement.

For further information please contact Springer-Verlag, Physics Editorial Department V, Tiergartenstrasse 17, D-69121 Heidelberg, FRG

H.-J. Röser K. Meisenheimer (Eds.)

Jets in Extragalactic Radio Sources

Proceedings of a Workshop
Held at Ringberg Castle, Tegernsee, FRG
September 22-28, 1991

Springer-Verlag Berlin Heidelberg GmbH

Editors

Hermann-Josef Röser
Klaus Meisenheimer
Max-Planck-Institut für Astronomie
Königstuhl 17, D-69117 Heidelberg, FRG

ISBN 978-3-662-13950-9 ISBN 978-3-540-47925-3 (eBook)
DOI 10.1007/978-3-540-47925-3

Originally published by Springer-Verlag Berlin Heidelberg New York in 1993
Softcover reprint of the hardcover 1st edition 1993

Typesetting: Camera ready by author/editor using the TEX macropackage from Springer-Verlag Berlin Heidelberg GmbH.

58/3140-543210 - Printed on acid-free paper

Preface

In late September 1991, a workshop was held at Ringberg castle (Bavaria), the conference center of the Max-Planck Society, on the topic "Jets in Extragalactic Radio Sources". Three and a half years had elapsed since our first meeting at Ringberg at which we had discussed the very special subject of "Hot Spots in Extragalactic Radio Sources". Inevitably, jets had also been discussed and we wanted to come back to this topic a few years later, when new observational tools such as the Hubble Space Telescope, ROSAT and the VLBA were about to shed new light on radio sources in general and on their jets in particular. Since quite a few of the participants had attended both workshops, a forum was set up to return to some of the ideas that had been brought up a few years ago. We were able both to discuss the progress achieved and to talk about the problems encountered. So these proceedings should illustrate our current understanding of radio sources in general and the jet phenomenon in particular.

The meeting was organized by K. Meisenheimer (Max-Planck-Institut für Astronomie), R.A. Perley (National Radio Astronomy Observatory), H.-J. Röser (Max-Planck-Institut für Astronomie) and P.A.G. Scheuer (Mullard Radio Astronomy Observatory). The final organization of the sessions was, however, a collaborative effort by the review speakers and Frazer Owen, Sam Falle for the Thursday morning and Guy Pooley and Mike Norman for the Friday morning session. We would like to thank them all for their help and for spending several evenings in the "Library" instead of in the "Hexenzimmer".

A generous grant from the Max-Planck Society made it possible for us to support invited speakers. The MPS also made the Ringberg facilities available to us. The organizers would like to thank Dr. D. Nickel from the Generalverwaltung of the MPS and Mr. A. Hörmann and his experienced staff at the castle for their excellent support which made the stay at Ringberg so enjoyable.

The editors would like to sincerely thank Miss Rachel Blythe who very carefully read through all of the final manuscripts. She helped to eliminate a lot of typing errors and in addition made many suggestions for improvements of the English. Our thanks also go to Mrs. Karin Dorn and Mrs. Martina Weckauf for their skilled preparation of the camera-ready manuscripts.

Heidelberg,
April 1993

Klaus Meisenheimer and Hermann-Josef Röser
on behalf of the Scientific Organizing Committee

Contents

List of Participants

Appl Stefan	Landessternwarte	Königstuhl D6900 Heidelberg 1	Germany
Begelman Mitchell C.	JILA	University of Colorado Bolder, CO 80309-0440	USA
Bicknell Geoffrey V.	Mt. Stromlo and Siding Spring Observatories	Private Bag 2606 Woden, ACT	Australia
Biermann Peter L.	MPI für Radioastronomie	Auf dem Hügel 69 D5300 Bonn 1	Germany
Biretta John A.	NRAO	P.O. Box 0 Socorro, NM 87801	USA
Black Adam R.S.	Mullard Radio Astron. Obs. Cavendish Laboratory	Madingley Road Cambridge CB3 0HE	U.K.
Bregman Joel N.	Astronomy Department University of Michigan	Ann Arbor, MI 48109	USA
Brinkmann Wolfgang P.	MPI für Physik und Astrophysik Institut für Extraterr. Physik	Giessenbachstraße D 8046 Garching	Germany
Browne Ian W.A.	University of Manchester Nuffield Radio Astron. Lab.	Jodrell Bank, Macclesfield, Cheshire SK11 9DL	U.K.
Burns Jack O.	Department of Astronomy New Mexico State University	Las Cruces, NM 88003	USA
Camenzind Max	Landessternwarte	Königstuhl D6900 Heidelberg 1	Germany
Clarke David A.	National Center for Super- computing Applications	University of Illinois Urbana-Champaign, IL 61801	USA
Conway Robin G.	University of Manchester Nuffield Radio Astron. Lab.	Jodrell Bank, Macclesfield, Cheshire SK11 9DL	U.K.
Crane Philippe	E.S.O.	Karl-Schwarzschild-Straße 2 D-8046 Garching	Germany
de Ruiter Hans Rudolf	Istituto Di Radioastronomia	Via Irnerio 46 I-40126 Bologna	Italy
Eilek Jean A.	NRAO	P.O. Box 0 Socorro, NM 87801	USA
Falle Sam A.E.G.	Department of Applied Mathematics	Leeds University Leeds LS2 9JT	U.K.
Fanti Carla	Istituto Di Radioastronomia	Via Irnerio 46 I-40126 Bologna	Italy

Fanti Roberto	Istituto Di Radioastronomia	Via Irnerio 46 I-40126 Bologna	Italy
Giovannini Gabriele	Istituto Di Radioastronomia	Via Irnerio 46 I-40126 Bologna	Italy
Hardee Philip E.	Department of Physics & Astronomy	University of Alabama Tuscaloosa, AL 35487	USA
Hes Ronald	E. S. O.	Karl-Schwarzschild-Straße 2 D-8046 Garching	Germany
Kirk John G.	MPI für Kernphysik	Saupfercheckweg 1 D-6900 Heidelberg 1	Germany
Komissarov Sergey S.	Astro-Space Center Lebedev Physical Institute	Leminsky prospect 53 117924 Moscow	Russia
Krichbaum Thomas P.	MPI für Radioastronomie	Auf dem Hügel 69 D-5300 Bonn 1	Germany
Leahy J. Patrick	University of Manchester Nuffield Radio Astron. Lab.	Jodrell Bank, Macclesfield, Cheshire SK11 9DL	U.K.
Litchfield Simon	Department of Physics and Astronomy	Lancashire Polytechnic Preston, PR1 2TQ	U.K.
Meisenheimer Klaus	MPI für Astronomie	Königstuhl 17 D-6900 Heidelberg 1	Germany
Norman Michael L.	National Center for Super- computing Applications	University of Illinois Urbana-Champaign, IL 61801	USA
Owen Frazer N.	NRAO	P.O. Box 0 Socorro, NM 87801	USA
Pearson Thimothy J.	California Institute of Technology	Pasadena, CA 91125	USA
Perley Richard A.	NRAO	P.O. Box 0 Socorro, NM 87801	USA
Pooley Guy	Mullard Radio Astron. Obs. Cavendish Laboratory	Madingley Road Cambridge CB3 0HE	U.K.
Robson Ian	Department of Physics and Astronomy	Lancashire Polytechnic Preston, PR1 2TQ	U.K.
Scheuer Peter A.G.	Mullard Radio Astron. Obs. Cavendish Laboratory	Madingley Road Cambridge CB3 0HE	U.K.
van Breugel Wil	Institute of Geophysics and Planetary Physics	Lawrence Livermore Nat. Lab. Livermore, CA 94551-9900	USA
Vestergaard Marianne	Kapteyn Laboratorium	P.O. Box 800 NL-9700 AV Groningen	The Nether lands
Wagner Stefan J.	Landessternwarte	Königstuhl D6900 Heidelberg 1	Germany
Wardle John	Physics Department Brandeis University	South Street Waltham, MA 02254	USA

Yates Mark G.	MPI für Astronomie	Königstuhl 17 D-6900 Heidelberg 1	Germany
Zensus J. Anton	NRAO	P.O. Box 0 Socorro, NM 87801	USA

Introduction

> ... A curious straight ray lies in a gap in the nebulosity in p.a. 20°, apparently connected with the nucleus by a thin line of matter. The ray is brightest at its inner end, which is 11″ from the nucleus. ...
>
> *Heber Doust Curtis*
> *Publications of the Lick Observatory, Vol. XIII — 1918*

Curtis' description of M 87 is the first reference to what we nowadays call an "extragalactic jet". In recent years jets have become a central issue not only in the study of radio galaxies — like M 87 — but also in the study of young stellar objects and other astronomical sources. The term *jet* itself was first introduced in Baade's and Minkowski's seminal paper (1954b) "On the identification of radio sources", where they write on NGC 4486 (M 87):

> *... NGC 4486, in the center of the field, is a member of the Virgo cluster of nebulae. At first sight, it is not obvious why this object should be a relatively strong radio source; it does not differ essentially in size or brightness from other bright members of the cluster. ... a nebula of type E0 with a rich envelope of globular clusters; this is not a rare type of nebula. However, NGC 4486 has a unique peculiarity which has been known for a long time. In the center of the nebula ... is a straight jet, extending from the nucleus in position angle 290°. ... No possibility exists at this time of forming any hypothesis on the formation of the jet, the physical state of its material, and the mechanism which connects the existence of the jet with the observed radio emission.*

Almost 40 years later, our knowledge of radio sources has increased tremendously and our physical picture of these features has advanced considerably. In this introduction, we wish to briefly recall the highlights of research on extragalactic radio sources to illustrate how the central role of the jet phenomenon emerged in this field. More extensive reviews and conference proceedings are summarized at the end.

In 1946 Hey and collaborators discovered the first discrete cosmic radio source, Cygnus A (although a local maximum at this position was already evident in Reber's maps of 1944). But it was not until 1954 that Baade and Minkowski (1954a), on the basis of considerably improved radio positions provided in 1951 by F.G. Smith, could demonstrate that the optical counterpart was a distant galaxy. Previously, Bolton, Stanley and Slee (1949) had already proposed that the strong radio emission catalogued as Virgo A and Centaurus A originates from the relatively nearby galaxies M 87 and NGC 5128 respectively. Thus, a small fraction of galaxies seemed to be strong radio emitters.

With improved angular resolution due to interferometric setups the spatial structure of these extragalactic sources in the radio domain was revealed step by step. Jennison and Das Gupta (1953) discovered that Cygnus A is an extended source with two lobes of radio emission. These lobes straddle the central optical galaxy identified by Baade and Minkowski (1954a) far beyond its visible boundaries. This picture soon proved to be typical for many of these objects, especially for the most powerful sources. The next step in resolution was provided by the NRAO radio link interferometer. Miley and Wade (1971) showed that bright hot spots of subarcsecond extent are imbedded in the outermost rim of the lobes at both ends. This picture was further refined by radio maps obtained with the Cambridge-5km-interferometer also of Cygnus A (Hargrave, Ryle 1974 and 1975). At a resolution of $2'' \times 3''$, the contrast in surface brightness between the bright hot spots and the general lobe emission was at least 150:1, making the analysis of structural details in the lobes a question of dynamic range.

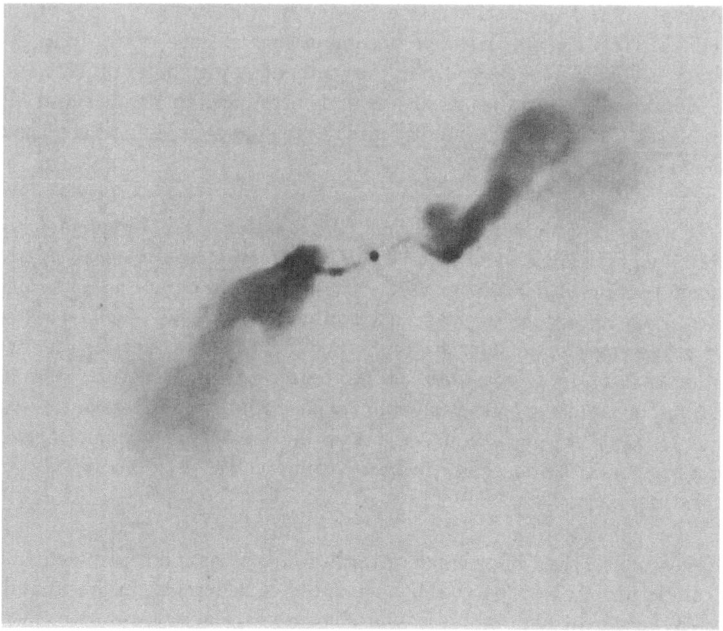

Fig. 1. The radio source Hydra A is of type FR I and shows symmetrical jets starting bright near the core and then fading away. This map was obtained at 6 cm with the VLA in all its configurations (Taylor *et al.* 1990). North is left, East is at bottom.

Not all radio sources show their brightest features in the far outskirts of the source. As with Hydra A in Fig. 1, in many sources the brightest component is close to the nucleus, and the source gradually fades away towards the outskirts. In their classic paper, Fanaroff and Riley (1974) pointed out that this morphological division is also a division in total luminosity: classifying radio sources by the ratio R of the distance between the regions of highest brightness on opposite sides of the central core to the total extent of the source measured from the lowest contour, they found that nearly all sources with R less than 0.5 (class I) have luminosities at 178 MHz below $\approx 2 \times 10^{25}$ W Hz^{-1} sr^{-1} Those above $R = 0.5$ (class II) nearly all have luminosities above that value. This coined the term Fanaroff Riley II sources (FR II) for powerful sources with the characteristic double lobe structure and hot spots at the outer ends. The corresponding weaker sources, where the central emission dominates, are referred to as Fanaroff Riley I sources (FR I).

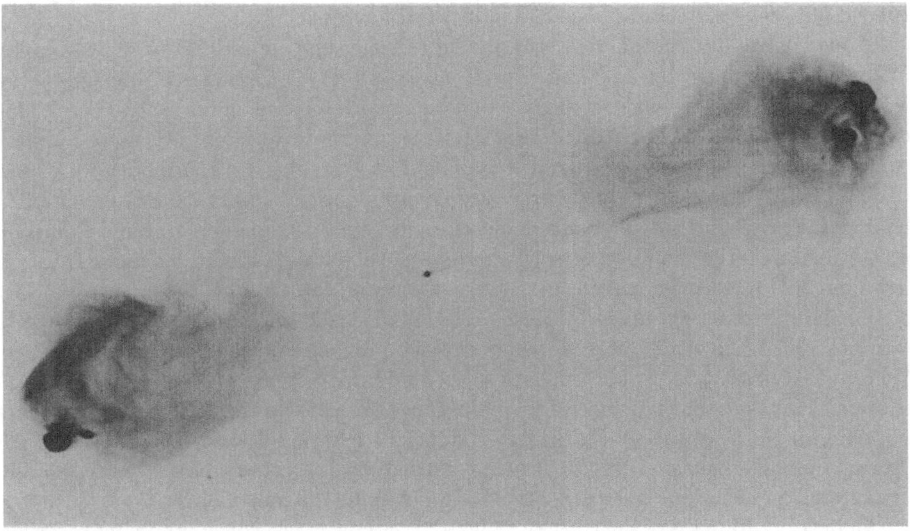

Fig. 2. **Cygnus A is a classical double radio source of type FR II. This map was obtained at 6 cm with the VLA in all its configurations (Perley, *priv.comm.*; North up and East to the left).**

The discovery of bright hot spots in FR II sources provided a challenge for any model of the double lobed radio sources. As Longair, Ryle and Scheuer (1973) discuss in detail, the source of the energy supply seems to lie in the central source. This energy must then somehow be symmetrically injected into the lobes. In many models, however, in which fields and relativistic particles are supplied from the nucleus, most of the energy is lost by adiabatic expansion before it can reach the radio hot spots. Furthermore, in view of the large energy requirements, it seems neccessary to contain fields and energetic particles within the lobes. All these considerations finally led to the development of the "twin-exhaust" model of double radio galaxies (Scheuer 1974, Blandford and Rees 1974): a twin beam of plasma carrying mass, energy and momentum out towards the giant lobes is produced in the central nucleus. This beam is decelerated by the ambient medium in a shock, where part of its energy is converted into highly relativistic particles, giving rise to the intense, compact hot spot emission. From here the energy is deposited into the lobe volume.

This picture was finally confirmed with the dramatic improvement both in sensitivity and dynamic range achieved with the Very Large Array (VLA) near Soccoro (New Mexico, see Heeschen 1980). It enabled Perley, Dreher and Cowan (1984) to detect a faint jet in Cygnus A several tens of kpc long connecting the nucleus (within the optical galaxy) with the western lobe and hot spots. Furthermore, the maps with a dynamic range of 3000:1 (at 6 cm) revealed a filamentary structure of the lobe emission. All these structural details — nucleus, lobes with filaments, jet and hot spots — are nicely visible in Fig. 2. Further examples of classical radio double sources may be found in a review by Perley (1989). More than 100 radio sources with jets are known (see list in the review by Bridle and Perley 1984). The one-sidedness of the kpc-scale jets observed in Cygnus A turned out to be a general phenomenon in FR II sources.

Thus far, we have considered only the large scale (kpc) properties of radio sources. When interferometry by intercontinental networks (very long baseline interferometry, VLBI) became feasible in about 1970, attention became focused on the central cores. One of the principle difficulties in the late sixties was the rapid variability of some compact central sources. It posed considerable problems for the synchrotron models (*e.g.* van der Laan 1966), as source diameters deduced from the variability time-scales implied higher energy densities than could be accommodated by the models (inverse Compton scattering limits the radiation density and thus the observable brightness temperature is less than 10^{12} K, an effect called the "inverse Compton catastrophe"). One possible way out of the dilemma was given by Rees (1966, 1967), who pointed out that the inverse Compton problem would be reduced if the cores expand with relativistic velocities. When VLBI observations of 3C 279 and 3C 273 (Whitney *et al.* 1971, Cohen *et al.* 1971) showed changes in the structure of the core, these were most easily interpreted as being due to an increase in the source diameter with an expansion velocity of about $3c$. Such "superluminal" expansion was subsequently also found in other sources and is readily interpreted as being due to relativistic motion close to the line of sight.

Modern VLBI maps of the cores of radio sources now reveal considerably more detail. On the parsec scale we see jets whose knots seem to move out at relativistic speeds. These inner jets are always on the same side as the large, kpc-scale jets and they are also always one-sided. Therefore, it is tempting to assume that they represent the start of the large-scale jet. Besides mapping these sources at high spatial resolution, further clues are obtained from the variability studies at all frequencies but one has to bear in mind that time-scales and geometries may be severely affected by relativistic effects. Thus, VLBI and variability studies of the cores of radio sources are vital contributions to our understanding of the jet phenomenon in extragalactic radio sources.

Much progress has been made in recent years towards understanding radio galaxies, and with the advent of new observational tools further progress will certainly follow. But among other questions we still ask ourselves

- how are the jets formed and how are they kept collimated?
- why are jets in FR II sources always one-sided? Is this due to Doppler beaming, as suggested from the nature of jets on the pc-scale?
- what is the physical difference between FR I and FR II sources?
- can the different types of *observed* radio source properties be reconciled with a basic underlying object viewed from different angles as proposed by *e.g.* Orr and Browne (1982) and Barthel (1989) (unified schemes)?

At our second Ringberg workshop we addressed these aspects from both observational and theoretical points of views. The session topics ranged from phenomenology and models of radio sources to the influence of the environment on radio galaxies. However, despite all the progress made in recent years, in the end we are still asking the same basic questions as Baade and Minkowski did.

Selected review papers and conference proceedings:

Begelman, M.C., Blandford, R.D., Rees, M.J.: 1984, *Theory of Extragalactic Radio Sources*, *Rev. Mod. Phys.* **56**, 255.

Bridle, A.H., Eilek, J.A. (eds.): 1984, *Physics of Energy Transport in Extragalactic Radio Sources*, NRAO Workshop No. 9, NRAO, Green Bank.

Bridle, A.H., Perley, R.A.: 1984, *Extragalactic Radio Jets*, *Ann. Rev. Astron. Astrophys.* **22**, 319.

Ferrari, A., Pacholczyk, A.G. (eds.): 1983, *Astrophysical Jets*, Proceedings of a workshop held at Torino, Italy 1983. Reidel, Dordrecht.

Heeschen, D.S., Wade, C.M. (eds.): 1982, *Extragalactic Radio Sources* (IAU Symposium No. 97), Reidel, Dordrecht

Hughes, P.A. (ed.): 1991, *Beams and Jets in Astrophysics*, Cambridge University Press, Cambridge.

Meisenheimer, K., Röser, H.-J. (eds.): 1989, *Hot Spots in Extragalactic Radio Sources*, Proceedings of a workshop held at Ringberg Castle 1988, Lecture Notes in Physics No. 327, Springer Verlag, Berlin Heidelberg.

Miley, G.K.: 1980, *The Structure of Extended Extragalactic Radio Sources*, *Ann. Rev. Astron. Astrophys.* **18**, 165.

Roland, J., Sol, H., Pelletier, G. (eds.): 1992, *Extragalactic Radio Sources — From Beams to Jets*, Proceedings of a workshop held at Paris 1991. Cambridge University Press, Cambridge.

Saikia, D.J., Salter, C.J.: 1988, *Polarization Properties of Extragalactic Radio Sources*, *Ann. Rev. Astron. Astrophys.* **26**, 93.

Zensus, J.A., Pearson, T.J. (eds.): 1987, *Superluminal Radio Sources*, Cambridge University Press, Cambridge.

———: 1990, *Parsec-Scale Radio Jets*, Cambridge University Press, Cambridge.

Literature cited:

Baade, W., Minkowski, R.: 1954a, *Astrophys. J.* **119**, 206.

———: 1954b, *Astrophys. J.* **119**, 215.

Barthel, P.D.: 1989, *Astrophys. J.* **336**, 606.

Bolton, J.G., Stanley, G.J., Slee, O.B.: 1949, *Nature* **164**, 101.

Hargrave, P.J., Ryle, M.: 1974, *Mon. Not. R. astr. Soc.* **166**, 305.

———: 1975, *Mon. Not. R. astr. Soc.* **175**, 481.

Hey, J.S., Parsons, S.J., and Phillips, J.W.: 1946, *Nature* **158**, 234.

Heeschen, D.S.: 1981, in *"Telescopes for the 1980s"*, ed. G. Burbidge and A. Hewitt, Annual Reviews Monograph, Palo Alto, CA

Jennison, R.C., Das Gupta, M.S.: 1953, *Nature* **172**, 996,

Miley, G.K., Wade, C.M.: 1971, *Astrophysical Letters* **8**, 11.

Orr, M.J.L., Browne, I.W.A.: 1982, *Mon. Not. R. astr. Soc.* **200**, 1067.

Perley, R.A., Dreher, J.W., Cowan, J.J.: 1984, *Astrophys. J. (Lett.)* **285**, L35.
Perley, R.A.: 1989, in *Hot Spots in Extragalactic Radio Sources*, ed. K. Meisenheimer and H.-J. Röser, Lecture Notes in Physics No. 327, Springer Verlag, Berlin Heidelberg.
Reber, G.: 1944, *Astrophys. J.* **100**, 279.
Rees, M.J.: 1966, *Nature* **211**, 468.
——: 1967, *Mon. Not. R. astr. Soc.* **135**, 345.
Taylor, G.B., Perley, R.A., Inoue, M., Kato, T., Tabara, H. and Aizu, K.: 1990, *Astrophys. J.* **360**, 41.
van der Laan, H.: 1966, *Nature* **211**, 1131.

DRAGNs

J.P. Leahy

University of Manchester
Nuffield Radio Astronomy Laboratories
Jodrell Bank
Macclesfield, Cheshire SK11 9DL
England

Abstract: DRAGN, from 'double radio source associated with a galactic nucleus', is my proposed name for the hitherto anonymous phenomenon with which we are concerned at this meeting. This includes the radio sources associated with radio galaxies, radio-loud quasars and some Seyfert galaxies. Two aspects of this phenomenon are reviewed. One is recent evidence that the host galaxies of such radiosources are gas-rich in an absolute sense. The second is the relationship between the detailed structure of radio jets and the overall structure of the radio source, with particular emphasis on the structures at the end of jets.

1 Introduction

I want to make three points, all suggested by my assigned title 'Phenomenology of Radio Galaxies'. Curiously, the editors did not actually want me to discuss this topic, at least in its strict sense. The problem is that we lack a name for the phenomenon which this meeting has been convened to discuss, so we have made do by misusing other terms (such as 'radio galaxy') to fill the gap. In Sect. 2 I solve this problem by proposing a brand new name; in view of rapid and uncontrolled drift in the meaning of many astronomical terms, I provide the new name with an extensive operational definition and make explicit its underlying theoretical presuppositions.

In fact, the study of radio galaxies *per se* is going through a particularly interesting period. Section 3 reviews some highlights and underlines their relevance for our subject. Finally, in Sect. 4 I review the interconnections between jets and lobes and use this as a basis for a rationalised morphological classification scheme.

I assume $H_0 = 100\,\mathrm{km}^{-1}\mathrm{Mpc}^{-1}$ throughout this article.

2 An Anonymous Phenomenon

We have known for years that the majority, but not all, of catalogued extragalactic radio sources are examples of the same general phenomenon. There is little practical difficulty in distinguishing this phenomenon from unrelated ones such as the radio emission from normal spirals, from starburst galaxies, and from 'cluster haloes', but it is not so easy to put the difference into words.

Examples of our phenomenon cover a huge range of structure, linear size, and luminosity. Moreover, they are associated with a diverse collection of hosts: active spirals (*i.e.* Seyferts), S0s, active ellipticals (*i.e.* radio galaxies) and quasars. Amongst this variety it is hard to pin down a single feature shared by all, except for the synchrotron radiation process which is also characteristic of starbursts and many other kinds of object. From a phenomenological point of view, we are dealing with a *syndrome*, *i.e.* a collection of traits which occur together in various different combinations in different manifestations of the phenomenon. The key traits are all features of the radio structure (my terminology follows Muxlow & Garrington 1991):

Jets, hot spots, and **lobes:** The large-scale structure can be loosely described as made up of these three distinct kinds of components; *our phenomenon is defined by the presence of at least one of these.* Although some sources appear to contain only one kind of extended structure (*e.g.* the 'naked jets', de Ruiter, this volume), the vast majority contain at least two, and some contain all three.

Double structure: Some kind of duality in the radio structure is present in at least 98% of cases. The remaining 2% include objects which seem to be genuinely one-sided (Saikia *et al.* 1989) and amorphous sources such as the head-tail 3C 264 (Gavazzi *et al.* 1981).

Cores: i.e. high-brightness compact regions co-incident with the active nucleus. Cores occur in more than 90% of cases.

In common usage 'lobe' is a cover-all term for components more diffuse than 'jets' and 'hot spots', so it may seem risky to make it a definitive feature for our phenomenon. Fortunately, lobes have more in common than mere low surface brightness: they show billowy or filamentary substructure, often a high degree of polarization with B-fields parallel to the filaments or billows, and relatively steep spectra ($\alpha \gtrsim 0.7$). Where lobes can be observed with good signal-to-noise they have well-defined perimeters where the intensity drops to zero, with very high polarization and the B-field closely wrapping around the edge (Leahy & Perley 1991). These properties distinguish lobes from diffuse emission in other types of radio source.

The definition explicitly excludes sources in which only a core has been detected. Luminous examples of such 'compact' sources usually turn out to have either the faint extended structure of a fairly conventional type, or else to consist of a very small but otherwise normal structure (*i.e.* the 'compact double' sources). However, a very large proportion of galaxies (including our own) contain extremely weak compact radio cores and it would be risky to assume that these are necessarily examples of the phenomenon I have been trying to isolate.

Since nobody has yet proposed a name for this phenomenon, I shall do so now: we should call them DRAGNs, which is an acronym for '**D**ouble **R**adio source **A**ssociated with **G**alactic **N**ucleus'.

Let me be clear about what a DRAGN is. It is the 'object' revealed by the radio emission (usually assumed to consist of very low-density plasma containing substantial energy density in relativistic electrons and magnetic fields). It is not to be confused with the active nucleus or with the host galaxy: the DRAGN is physically distinct from both. For instance, DRAGNs can be either much smaller or much larger than their host galaxies; only rarely and probably by co-incidence are they similar in size. It is surprisingly difficult to maintain this obvious distinction because we usually say that a DRAGN is 'identified' with a galaxy or quasar, and use the name of the DRAGN and the host galaxy interchangeably, as in Cen A / NGC 5128.

Finally, the term DRAGN is a *name*, and not a *definition*. I choose it because it is short, pronouncible, unencumbered with alternative meanings, and, I hope, catchy. The expanded version is a reasonable description of the vast majority of DRAGNs, but a non-double source (*e.g.* 3C 273), or one not obviously associated with a galactic nucleus (see later), may still be a DRAGN.

As noted above, DRAGNs form a superficially heterogeneous population. What binds them all together, phenomenologically at least, is that all the various classes merge into one another: *there are sources whose structure is intermediate between any two classes of DRAGN.* We can only divide DRAGNs into distinct sub-classes by imposing arbitrary boundaries on a continuous distribution in structure, luminosity, and size.[1]

But we know there is a more fundamental reason for identifing all DRAGNs as a single phenomenon. There *is* a single, magic, morphological feature shared by them all. Following Bridle's (1986) definition of an extragalactic jet as "a feature which *looks* as if it *might* be a collimated stream or flow emanating from the centre of an AGN", we can say that:

> *DRAGNs are radiosources which* look *as if they* might *be produced by the interaction between extragalactic jets and their environment.*

The word 'interaction' is an essential part of this definition. The behaviour of a jet (or a discrete plasma cloud) ejected into a perfect vacuum is particularly simple and differs in almost every way from the observed structure of DRAGNs. This was recognised well over 20 years ago as the 'adiabatic expansion problem', and in the failure of radio lobes to be spherical or jets to be purely conical. Every plausible theory proposed since then has fundamentally involved interaction with the external medium, whether to explain the bending of jets in narrow-angle tails, the existence of hot spots and elongated lobes in classical doubles, or the variable cone angle of jets. Therefore, it is depressing to find papers proudly announcing that this or that peculiarly distorted DRAGN is 'evidence for interaction with the environment'. The structure of *all* DRAGNs constitutes such evidence. What is different about the highly distorted DRAGNs (*e.g.* 3C 171, Heckman *et al.* 1984) is that the environment is extremely inhomogeneous, containing dense clouds of 10^4 K gas embedded in the hotter ISM. What is peculiarly important about such DRAGNs is not the fact of the interaction, but the opportunity to observe a component

[1] In particular this applies to the Fanaroff-Riley (1974) division, as can be confirmed by a close look at the images of the B2 galaxy sample (Fanti *et al.* 1987 and references therein).

of the interacting environment, via optical line emission, as well as the DRAGN. These results give us just a small taste of the improved understanding we can expect when XMM and AXAF enable us to see the hot gas which is, of course, the main component of the environment around DRAGNs.

3 Phenomenology of Radio Galaxies

We can conveniently define a radio galaxy as an elliptical or S0 galaxy containing an active nucleus which produces a DRAGN. (This preserves the tradition that spirals with active nuclei are always known as Seyferts, whether a DRAGN is present or not). The key question about radio galaxies is, of course, *why* they produce DRAGNs. We cannot yet answer this, but important clues are emerging from work on the related observational question, 'what is the difference between radio galaxies and other galaxies (apart from the DRAGN)?'

To date, most studies of radio galaxies have focussed on their stellar structure and kinematics. Wrobel & Heeschen (1991) review the current situation, which is quite confusing. One problem is that DRAGNs occur across nine orders of magnitude in radio power and there are important differences between the 'typical' hosts of DRAGNs at different power levels.

Radio galaxies are mostly very similar to large ellipticals (*e.g.* Owen & Laing 1989); the probability of producing a DRAGN increases with optical luminosity and decreases with DRAGN power (*e.g.* Sadler *et al.* 1989). At a given total luminosity S0s and spirals are much less likely to produce a DRAGN. Sadler *et al.* find that S0s produce DRAGNs about as often as small Es with the same luminosity as the S0 bulge; such DRAGNs are extremely weak ($P_{1.4\,\mathrm{GHz}} \lesssim 10^{21}\,\mathrm{W\,Hz^{-1}}$) and rarely occur in radio-selected samples.

Apart from the luminosity bias, radio galaxies differ in a few more subtle ways from typical ellipticals. Most notably Bender *et al.* (1989) find that even a weak disc component dramatically reduces the probability of producing a weak DRAGN ($P_{1.4\,\mathrm{GHz}} \sim 10^{22}\,\mathrm{W\,Hz^{-1}}$); in contrast, Heckman and co-workers have shown that a significant fraction of radio galaxies at powers around the FR break ($P_{1.4\,\mathrm{GHz}} \sim 10^{25}\,\mathrm{W\,Hz^{-1}}$) are disturbed ellipticals which they believe may have gained a disc component from a recent merger (Smith *et al.* 1990). There are gross differences between high-z radio galaxies ($P_{1.4\,\mathrm{GHz}} \sim 10^{28}\,\mathrm{W\,Hz^{-1}}$) and nearby ellipticals (e.g. McCarthy *et al.* 1987); but since these differences are seen as the *result* of nuclear activity they seem irrelevent to our key question.

Intriguing as they are these results do not point in any clear direction. It might help to look instead at the interstellar medium, which is the component of the galaxy that must be directly responsible for both the confinement of the DRAGN and the fuelling of the central engine.

Our ideas about the interstellar medium of elliptical galaxies have been revolutionised in the last ten years. Massive ellipticals certainly contain $O(10^9 h^{-2.5}\mathrm{M_\odot})$ of 10^7 K gas (*e.g.* Canizares *et al.* 1987), and it seems increasingly likely that many of them contain similar amounts of cold (*i.e.* molecular) gas, as will be discussed below.

It must be crucial to the physics of jets that they propagate down a strong pressure gradient defined by the hot gas. For a 'generic' large elliptical, in the central parsec, the modern interpretation of the broad-line region (*e.g.* Perry & Dyson 1985) suggests a pressure of a few $\times 10^{-4}$ Pa on fairly general principles. The pressure must drop steeply from this region to reach that in the general galaxy core which *Einstein* data puts at $\sim 10^{-11}$ Pa. Beyond the core radius of perhaps a few hundred parsecs the pressure falls again and, in a few cases, *Einstein* could track it down to $\lesssim 10^{-13}$ Pa at ~ 50 kpc (e.g. Trinchieri *et al.* 1986). This pressure gradient is large enough to keep a jet travelling more or less in the radial direction, but usually less than the fall in pressure in an adiabatic conical expansion, so the general trend is continually to re-collimate the jet. This is the solution to the puzzle raised by Icke (1991) as to how jets can maintain their coherence over so many orders of magnitude in scale size.

The difference between the distribution of gas in ellipticals and spirals is very relevalent to the production of DRAGNs. Because they lack the hot coronae, the pressure in spirals at radii of $O(10\,\mathrm{kpc})$ is much lower than in ellipticals (except in the disc plane). Furthermore, the IGM usually has a higher pressure around ellipticals because they tend to occur in clusters. This weaker confinement around spirals will lead to rapid quenching of any DRAGN whose jets escape the disc. However, this cannot be the main reason for the lack of powerful DRAGNs in Seyferts since even their parsec-scale radio cores are far less luminous than in radio galaxies (normalised, for instance, to the same narrow-line luminosity).

Explanations of the spiral–elliptical difference usually invoke the denser cold ISM in spiral discs. This is claimed to provide generally more fuel in spirals (to explain why 'most' AGN are Seyferts rather than radio galaxies), while at the same time 'smothering' the jets in Seyferts (*e.g.* Osterbrock 1991). The extra hot gas in ellipticals is deemed irrelevant here as most of it is far from the nucleus. These explanations fall apart when examined in detail. The 'rarity' of radio galaxies at least partly reflects the rarity of ellipticals: for instance, about 14% of galaxies in the *Revised Shaply-Ames Catalog* are classified as E or E/S0 (Sandage & Tammann 1981), which can be compared with the "15 – 20%" of of radio-loud AGNs in the PG sample (Kellermann *et al.* 1989). The concept of 'smothering' appears to be free of any physical motivation; everything we know about jets suggests that, for a given energy and momentum flux, short jets will be much *brighter* because the same energy would be deposited in a much smaller volume.

Fortunately, we should probably reject the original premise that the environs of spiral nuclei are richer in cold gas. Sadler & Gerhard (1985) observe dust lanes (mostly at radius ~ 1 kpc) in 20% of ellipticals and, allowing for projection effects, infer them in $\sim$ 40%. Ebneter *et al.* (1988) find nuclear dust patches in $\gtrsim 40\%$. Phillips *et al.* (1986) find LINER-type emission lines from a kpc-scale component in more than 50% of ellipticals, suggesting that substantial amounts of cold gas may be available for ionization. Knapp *et al.* (1989) detected 50% of a sample of E+S0s, at 60 and/or 100 μm in co-added IRAS data, and found a significant, roughly linear, correlation (albeit with very large scatter) with L_B (Knapp *et al.* 1990). For massive ellipticals ($L_B > 3\times 10^{10}\,L_\odot$), $L_{FIR} \gtrsim 10^9\,L_\odot$, in the same range as for typical spirals ($L_B \approx 10^{10}\,L_\odot$). The FIR spectrum is clearly discontinuous with any nuclear radio emission (Knapp & Patten 1991) and is presumed to be due to dust. The 60/100 μm colours and hence the inferred dust masses are also similar to those of typical quiescent spirals.

In normal spirals there is a good correlation between FIR and the mass in H_2 (*e.g.* Young & Scoville 1991). Lees *et al.* (1991) have found CO in a number of (mostly low-luminosity) ellipticals and find a similar ratio of FIR to H_2 mass. If this also applies to massive ellipticals then they contain up to $\sim 10^{10} M_\odot$ of H_2. H I *is* severely depleted in ellipticals relative to spirals, but even so, van Gorkom *et al.* (1989) detect H I in absorption towards a number of radio cores (and find it redshifted relative to the galaxy's systemic velocity in every case!).

All of these suggest that there is a substantial 'cold' component in many ellipticals, and furthermore that *the cold gas in ellipticals is very centrally concentrated*, *i.e.* all these components are probably associated with the 1 kpc discs.[2] If these size and mass estimates are correct, then massive ellipticals have a central density of cold gas at least as high and perhaps higher than do spirals.

How do radio galaxies fit into this picture? An old result (Kotanyi & Ekers 1979) is that radio jets in FR Is tend to be perpendicular to the central dust-lanes. Is this because the central engine itself is aligned, or because the jets bend away from the dust-lane as they propagate out? VLBI shows that these kpc-scale jets are usually well aligned with the parsec-scale jets (for FR Is), so apparently the central engine is aligned with, and hence related to, the kpc-scale disc.

There have been some claims that radio galaxies have unusually large amounts of cold gas compared to other ellipticals. Knapp *et al.* (1990) find that the FIR-to-blue flux ratio of radio galaxies is 2–3 times higher than 'normal' ellipticals. The effect is not very obvious on a luminosity–luminosity plot, however, and could be simulated by a small non-linearity in the correlation since the radio galaxies occur mainly at the bright end. Unpublished data by Wardle & Knapp also suggests that H I is more common in radio galaxies than in 'normal' ellipticals.

In any case, it is clear that a massive, centrally concentrated cold component is much too common to be the sole cause of a DRAGN. Only in special circumstances can some, perhaps a small fraction, of this gas be fed to the central object. There are several possibilities, but interactions with other galaxies are an attractive option, given the disturbed structures of many radio galaxies. It is sometimes suggested that such interactions must be full-blown mergers in order to transfer cold gas into elliptical nuclei (*e.g.* Knapp *et al.* 1990), but because the fuel is apparently already in place in many large ellipticals, much weaker interactions will probably suffice.

4 How Jets End

Jets in DRAGNs come in two or three flavours, already distinguished by Bridle & Perley (1984). The following descriptions are for archetypes; as with DRAGN structure generally, jets with intermediate structure are easy to find.

Type I jets: large opening angle ($5° - 30°$); initial 'gap' of $O(1\,\mathrm{kpc})$ between core and rapid brightening at 'base knot', followed by a smooth brightness decline; projected B-field mostly perpendicular to jet (sometimes parallel at the edges and the base); brightness asymmetry between the two jets typically a factor of 2; cross-section centre-brightened. DRAGNs containing archetypical type I jets are always FR Is (the converse is not true).

[2] With the probable exception of the H I.

Type II jets: small or zero opening angle, irregular ('knotty') structure; *B*-field parallel to jet except for deviations to align with sharp edges in the knots; large asymmetry in brightness between jet and counter-jet (if any); in a few cases (*e.g.* Cygnus A, M 87) clearly limb-brightened in places (very few type II jets have been observed with sufficient transverse resolution to detect limb-brightening). Type II jets are found in FR II DRAGNs and in at least one case (PKS 1333-33, Killeen *et al.* 1986) in the 'gap' before the start of type I jets; there are hints of this in other cases and it may be the general rule.

Tails or **plumes** constitute a third type of jet-like feature. They have an irregular, often wiggly structure which can show regions of rapid widening, roughly constant width and occasionally clear narrowing. There is usually a close symmetry in radio power between the tails on each side of the DRAGN. The term 'plume' is mostly used when the overall structure is nearly straight, as in 3C 31 (Strom *et al.* 1983). Tails are fed through jets of either of the previous types. Tails and plumes are usually described as 'lobes' rather than 'jets', but most models involve continued, although subsonic, outward flow in these components. Their detailed structure is typical of lobes in general; because their internal filaments are roughly aligned along the tail, their projected *B*-fields are also roughly parallel to the tail (cf. the high-quality images of tails in O'Donoghue *et al.* 1990 [in the following ODOE]).

Another major type of lobe is the **bridge**. Bridges were first defined as diffuse emission in FR II DRAGNs filling the gap between the core and the hot spots, and therefore surrounding the presumed path of the jet. In modern images we often see the jet running through the bridge (Black, this volume), and it turns out that similar bridges usually surround those type I jets which do not produce tails.

The transitions from type II to type I jet and from jet to lobe provide some important clues to the nature of these flows. Type II jets generally end in bright features which outshine the jet itself: the hot spots in FR IIs (feeding a bridge), and apparently the base knots in FR Is (feeding a type I jet). There are also a number of DRAGNs with tails which begin with bright knots (or hot spots) which in turn are fed by large-scale ($O(20\,\mathrm{kpc})$), faint, narrow jets which are often knotty and asymmetric. Good examples are 3C 171, Hydra A and the prototype 'wide-angle-tail' 3C 465 (Leahy 1984). In fact, many of the DRAGNs showing this structure have been classified as WATs, even when their tails are essentially straight (*e.g.* 0043+201, see ODOE).[3] Although *B*-field configurations have not been reliably determined in any of these inner jets, I predict that most will turn out to have parallel magnetic fields and hence to be type II jets.

Type I jets end quietly. In fact, they usually outshine any downstream structure, which gives DRAGNs with such jets their FR I classification. De Ruiter (this volume) discusses cases where the emission stops completely at the end of weak type I jets. If the jet feeds a bridge as in B2 1321+31 (Fanti *et al.* 1982) or 3C 296 (Leahy & Perley 1991) it either fades out near the outer edge of the lobe or curves around and flares to merge smoothly into the bridge. If the jet feeds a plume as in 3C 31, the transition

[3] Not all DRAGNs classified as WATs show this structure: unfortunately different authors use the term WAT in such a variety of ways that it is now essentially meaningless. Most would agree that the mild bending which originally defined the class is not a significant unifying property, as it can occur in otherwise very dissimilar DRAGNs, and apparently has no common origin.

is more interesting: there is a characteristic pattern seen in a number of well-studied cases including 3C 449 (Perley *et al.* 1979), 3C 66B (Leahy *et al.* 1986), and 2236−176 (see ODOE). What happens is that the jet curves sharply and simultaneously fades, then curves back towards the original direction and flares into an oblong or triangular component. The B-field in the first half of this component forms a ring; in the centre of the ring the field stays perpendicular. The second half of the component is very highly polarized, especially along the edge opposite the jet entry point. The plume emerges from this component at a fainter brightness level, with a parallel field, and often in a new direction.

There are a number of DRAGNs which consist entirely of lobes with virtually no trace of large-scale jets or hot spots. Some of these (*e.g.* M 87) show a 'steep-spectrum core', essentially a small-scale FR I embedded in the larger structure. Others (*e.g.* 3C 310, van Breugel & Fomalont 1984) have a prominent flat-spectrum core, while others again (*e.g.* 3C 314.1, Leahy & Perley 1991) are among the small fraction of DRAGNs which lack any sign of current nuclear activity. In most cases, the lobe structure is similar to that of low-luminosity bridges, in which case these jetless DRAGNs are called 'relaxed doubles' or 'core-haloes', depending partly on whim and partly on core prominence. Presumably there are also tailed DRAGNs in which nuclear outflow has ceased, but these seem to be hard to detect: I christen this mostly hypothetical class 'dead tails'. Candidates for such objects include tailed DRAGNs in clusters in which jets have not yet been found (but perhaps we just have not looked hard enough), and 1253+257, an extremely large and faint twin-plume-like structure at the edge of the Coma cluster (Giovannini *et al.* 1985). If dead tails mostly look like 1253+257 it is no surprise that we have found very few of them.

These connections between jets and lobes allow us to organize the various classes of DRAGN in a simple way (Fig. 1). Classification in this scheme depends on whether hot spots are visible or not, since this can be determined from maps of modest quality; but a closer look shows that this usually corresponds to the distinction between type II and I jets. The term "WAT" is used for DRAGNs with hot spots and tails only for want of a better term (see footnote 3).

The classes in Fig. 1 are not rigid or well-separated. Instead they represent archetypes which function as marker poles in a multi-dimensional morphological continuum. The classification is not completely arbitary because these archetypes are at the centres of clusters in this morphological space. The B2 sample (de Ruiter *et al.* 1990) provides a good test of the scheme as its 100 DRAGNs cover a range of powers ($10^{22} < P_{1.4\,\mathrm{GHz}} < 10^{26}\,\mathrm{W\,Hz^{-1}}$) in which all classes occur. I find that about two-thirds of the well-resolved cases are close enough to an archetype to be classified unambiguously. A few of the remainder might be easily classified from better images, but more often the structure is truely intermediate between the archetypes: some have moderately compact components which could be called either weak hot spots or merely a 'lump' in the lobe, or the lobe may be entirely on one side of the jet, in which case it might be called either a very distorted tail or a bridge.

The Fanaroff-Riley division corresponds roughly to the distinction between classical doubles (FR II), and the other classes in Fig. 1 (FR I), although strict application of FR's original criterion would lead to some relaxed doubles and "WATs" being classified as marginal FR IIs. Note that the various classes included under the FR I umbrella differ

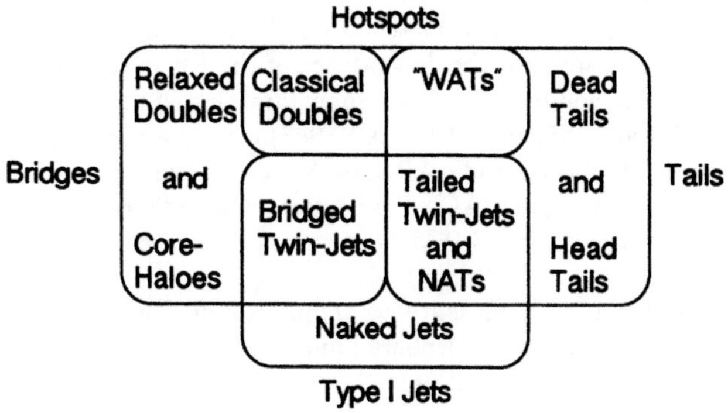

Fig. 1. Morphological scheme for DRAGNs

from each other as much as they differ from FR IIs. The notion of a 'typical' FR I is therefore something of a myth, especially as no one class occurs much more commonly than the rest.

The various classes have different luminosity distributions, although there is considerable overlap. Classical doubles are generally the most powerful and the naked jets apparently the weakest. The "WATs" and relaxed doubles both resemble FR IIs (in different ways) and it is interesting that at higher powers both become more common relative to DRAGNs with type I jets; in fact, the rare FR Is with powers above the FR break are mostly of one of these classes.

What light does this scheme shed on the physics of DRAGNs? For one thing I think it highlights the peculiarity of the type I jets. Although these are the jets which have been studied most intensely, I am not satisfied that the best current models, those of Geoff Bicknell, really capture their essence. According to Geoff, type I jets are regions of turbulent deceleration from moderately supersonic to roughly transonic flow; they are fed by supersonic type II jets and lead, in tailed twin-jets (TTJs), to subsonic buoyant plumes. But apparently this same supersonic-to-subsonic transition can occur in one step, at the hotspot, in the "WATs". Why are *two* well defined and morphologically distinctive transition regions (base knot, jet-plume structure) needed in the TTJs? And how could the unique perpendicular B-field structure of type I jets survive in the strong shear which is supposed to provide the decelerating turbulent stress? At the least, type I jets must be quite different from the laboratory turbulent jets to which they are often compared. Elsewhere (Leahy 1991) I have suggested that magnetic stresses may play an important role in suppressing shear in these jets; as an added bonus magnetically-active jets come complete with multiple sonic points (Alvén, fast and slow magnetosonic) which it is tempting to associate with the multiple flow transitions in twin-jet sources.

More generally, correlations between structure and radio power are intriguing because there is no simple reason for a direct connection: 'shape' is a dimensionless parameter. Therefore, the rather good connection between power and FR type, in particular, is telling us that many of the parameters which might be thought to influence structure are either constant from DRAGN to DRAGN or else cancel out in some way. To unravel this we need to find some parameters apart from radio power which do influence structure; an exciting example is the dependence of the FR break power on host galaxy luminosity (Owen, this volume). Another important clue is that DRAGNs in Seyferts (*e.g.* 3C 71 / NGC 1068, Wilson & Ulvestad 1987) often have hot spots but are *much* fainter than classical doubles in radio galaxies. We could learn a lot from looking at the structural differences between DRAGNs in Seyferts and radio galaxies in the context of the differences between the central ISM in spirals and ellipticals.

Acknowledgements: It is a pleasure to thank Robin Conway, Simon Garrington, Jaqueline van Gorkom, Jill Knapp, Robert Laing and Ralph Spencer for useful discussions and suggestions.

References

Bender, R., Surma, P., Döbereiner, S., Möllenhoff, C. & Madejsky, R.: 1989, *Astron. Astrophys.* **217**, 35.
Bridle, A.H.: 1986, *Can.J.Phys.* **64**, 353.
Bridle, A.H., Perley, R.A.: 1984, *Ann. Rev. Astron. Astrophys.* **22**, 319.
Canizares, C.R., Fabbiano, G., Trinchieri, G.: 1987, *Astrophys. J.* **312**, 503.
de Ruiter, H.R., Parma, P., Fanti, C., Fanti, R.: 1990, *Astron. Astrophys.* **227**, 351.
Ebneter, K., Djorgovski, S., Davis, M.: 1988, *Astron. J.* **95**, 422.
Fanaroff, B.L., Riley, J.M.: 1974, *Mon. Not. R. astr. Soc.* **167**, 31P.
Fanti, C., Fanti, R., de Ruiter, H.R., Parma, P.: 1987, *Astron. Astrophys. Suppl. Ser.* **69**, 57.
Fanti, R., Lari, C., Parma, P., Bridle, A.H., Ekers, R.D., Fomalont, E.B.: 1982, *Astron. Astrophys.* **110**, 169.
Gavazzi, G., Perola, G.C., Jaffe, W.: 1981, *Astron. Astrophys.* **103**, 35.
Giovannini, G., Feretti, L., Andernach, H.: 1985, *Astron. Astrophys.* **150**, 302.
Heckman, T.M., van Breugel, W.J.M., Miley, G.K.: 1984, *Astrophys. J.* **286**, 509.
Icke, V.: 1991, in *Beams & Jets in Astrophysics*, ed. P.A. Hughes, Cambridge: Cambridge University Press, p. 232.
Kellermann, K.I., Sramek, R., Schmidt, M., Shaffer, D.B., Green, R.: 1989, *Astron. J.* **98**, 1195.
Killeen, N.E.B., Bicknell, G.V., Ekers, R.D.: 1986, *Astrophys. J.* **302**, 306.
Knapp, G.R., Bies, W.E., van Gorkom, J.H.: 1990, *Astron. J.* **99**, 476.
Knapp, G.R., Guhathakurta, P., Kim, D.-W., Jura, M.: 1989, *Astrophys. J. Suppl.* **70**, 329.
Knapp, G.R., Patten, B.M.: 1991, *Astron. J.* **101**, 1609.
Kotanyi, C.G., Ekers, R.D.: 1979, *Astron. Astrophys. (Lett.)* **73**, L1.
Leahy, J.P.: 1984, *Mon. Not. R. astr. Soc.* **208**, 323.
——: 1991, in *Beams & Jets in Astrophysics*, ed. P.A. Hughes, Cambridge: Cambridge University Press, p. 100.
Leahy, J.P., Jägers, W.J., Pooley, G.G.: 1986, *Astron. Astrophys.* **156**, 234.
Leahy, J.P., Perley, R.A.: 1991, *Astron. J.* **102**, 537.
Lees, J.F., Knapp, G.R., Rupen, M.P., Phillips, T.G.: 1991, *Astrophys. J.* **379**, 177.
McCarthy, P.J., van Breugel, W., Spinrad, H., Djorgovski, S.: 1987, *Astrophys. J. (Lett.)* **321**, L29.
Muxlow, T.W.B., Garrington, S.T.: 1991, in *Beams & Jets in Astrophysics*, ed. P.A. Hughes, Cambridge: Cambridge University Press, p. 52.

O'Donoghue, A.A., Owen, F.N., Eilek, J.A.: 1990, *Astrophys. J. Suppl.* **72**, 75.
Osterbrock, D.E.: 1991, *Rep.Prog.Phys.* **54**, 579.
Owen, F.N., Laing, R.A.: 1989, *Mon. Not. R. astr. Soc.* **238**, 357.
Perley, R.A., Willis, A.G., Scott, J.S.: 1979, *Nature* **281**, 437.
Perry, J.J., Dyson, J.E.: 1985, *Mon. Not. R. astr. Soc.* **213**, 665.
Phillips, M.M., Jenkins, C.R., Dopita, M.A., Sadler, E.M., Binette, L.: 1986, *Astron. J.* **91**, 1062.
Sadler, E.M., Gerhard, O.E.: 1985, *Mon. Not. R. astr. Soc.* **214**, 177.
Sadler, E.M., Jenkins, C.R., Kotanyi, C.G.: 1989, *Mon. Not. R. astr. Soc.* **240**, 591.
Saikia, D.J., Junor, W., Muxlow, T.W.B., Tzioumis, A.K.: 1989, *Nature* **339**, 286.
Sandage, A., Tammann, G.A.: 1981, *A Revised Shapley-Ames Catalog of Bright Galaxies*, Carnagie Institute of Washington.
Smith, E.P., Heckman, T.M., Illingworth, G.D.: 1990, *Astrophys. J.* **356**, 399.
Strom R.G., Fanti, R., Parma, P., Ekers, R.D.: 1983, *Astron. Astrophys.* **122**, 305.
Trinchieri, G., Fabbiano, G., Canizares, C.R.: 1986, *Astrophys. J.* **310**, 637.
van Breugel, W., Fomalont, E.B.: 1984, *Astrophys. J. (Lett.)* **282**, L55.
van Gorkom, J.H., Knapp, G.R., Ekers, R.D., Ekers, D.D., Laing, R.A., Polk, K.S.: 1989, *Astron. J.* **97**, 708.
Wilson, A.S., Ulvestad, J.S.: 1987, *Astrophys. J.* **319**, 105.
Wrobel, J.M., Heeschen, D.S.: 1991, *Astron. J.* **101**, 148.
Young, J.S., Scoville, N.Z.: 1991, *Ann. Rev. Astron. Astrophys.* **29**, 581.

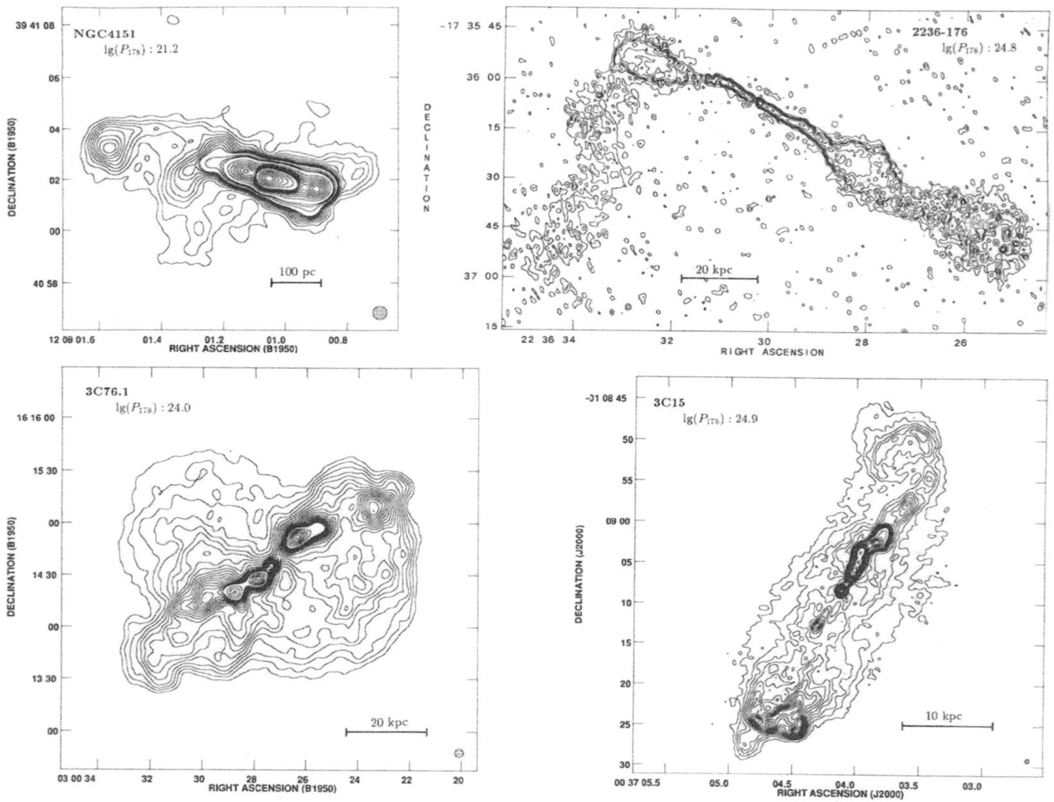

Fig. 2. Maps of typical DRAGNs. The contours are initially linear in each case. The log of the radio power at 178 MHz in W Hz^{-1} sr^{-1} is given in each map. **NGC 4151:** Tailed(?) Twin-Jet (Sab / Sy 1.5) (courtesy M. Kukula, D. Longley & A. Pedlar). **2236–176:** Tailed Twin-Jet/WAT (cD in cluster) (courtesy A. O'Donoghue, F. Owen & J. Eilek). **3C 76.1:** Bridged Twin-jet (E galaxy). **3C 15:** Bridged Twin-Jet/Classical Double (E).

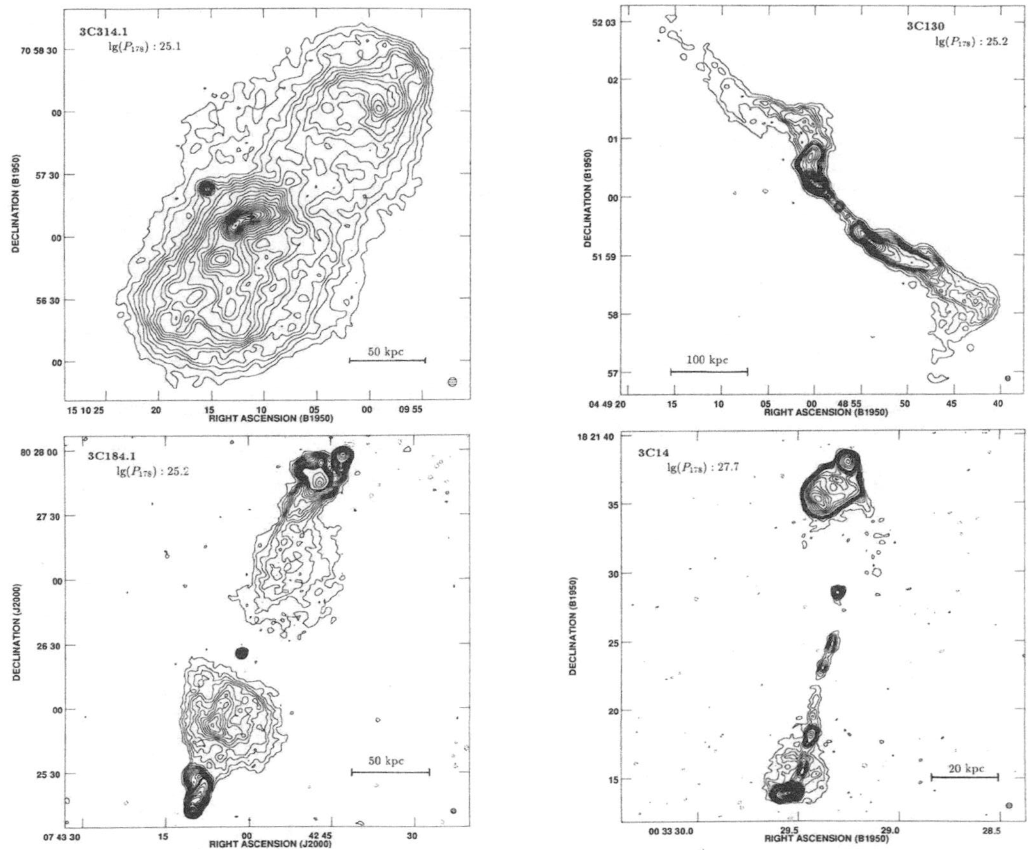

Fig. 3. More maps of typical DRAGNs. **3C 314.1:** Relaxed Double (E in cluster). **3C 130:** WAT (cD in cluster). **3C 184.1:** Classical Double (E). **3C 14:** Classical Double (QSO). (courtesy S. Garrington and E. Lüdke).

Jets and Related Issues: Observations of Nearby FR II Radio Galaxies.

Adam R.S. Black

Mullard Radio Astronomy Observatory
Cavendish Laboratory
Madingley Road
Cambridge CB3 0HE
U.K.

Abstract: An overview is given of the morphological properties of jets and hot spots in a sample of 30 nearby FR II radio galaxies, following extensive observations at 8 GHz using the VLA. Jets are found to be common (in more than 57%) and predominantly 1-sided; their detection requires high dynamic-range imaging. The hot spots are strikingly different in detail; it is argued that this dissimilarity reflects their transience.

1 Observations

A sample of 30 nearby FR II extragalactic radio sources was selected from the 3CR catalogue with $z < 0.15$ and $P_{178} > 1.5\times10^{25}$ W Hz^{-1} sr^{-1}. With high-resolution ($0''.25$) and high-sensitivity imaging the goal was to detect and trace the path of any faint jet into the hot spots and to follow the post-shock flow. Previous observations of jets in FR II sources (*e.g.* 3C 405 — Carilli, Dreher and Perley 1989) resolved widths of $\sim$ 1 kpc: sub-kpc spatial resolution (of $\sim$ 0.8 kpc) was therefore considered necessary. Trends in and relationships between the hot spots and their 'feeding' jets were thus to be identified through analysis of this sample.

The VLA, with its new, high-sensitivity 8-GHz receiver system, is well suited to observe such jets and hot spots. Most sources were observed at A-, B-, C- and D-configurations to obtain exceptional *u-v* coverage and allow imaging of structures over a wide range of scales (from $0''.25$ to $190''$).

A value for Hubble's constant of H_0=50 km s^{-1} Mpc^{-1} and for the deceleration parameter of q_0=0 have been assumed throughout.

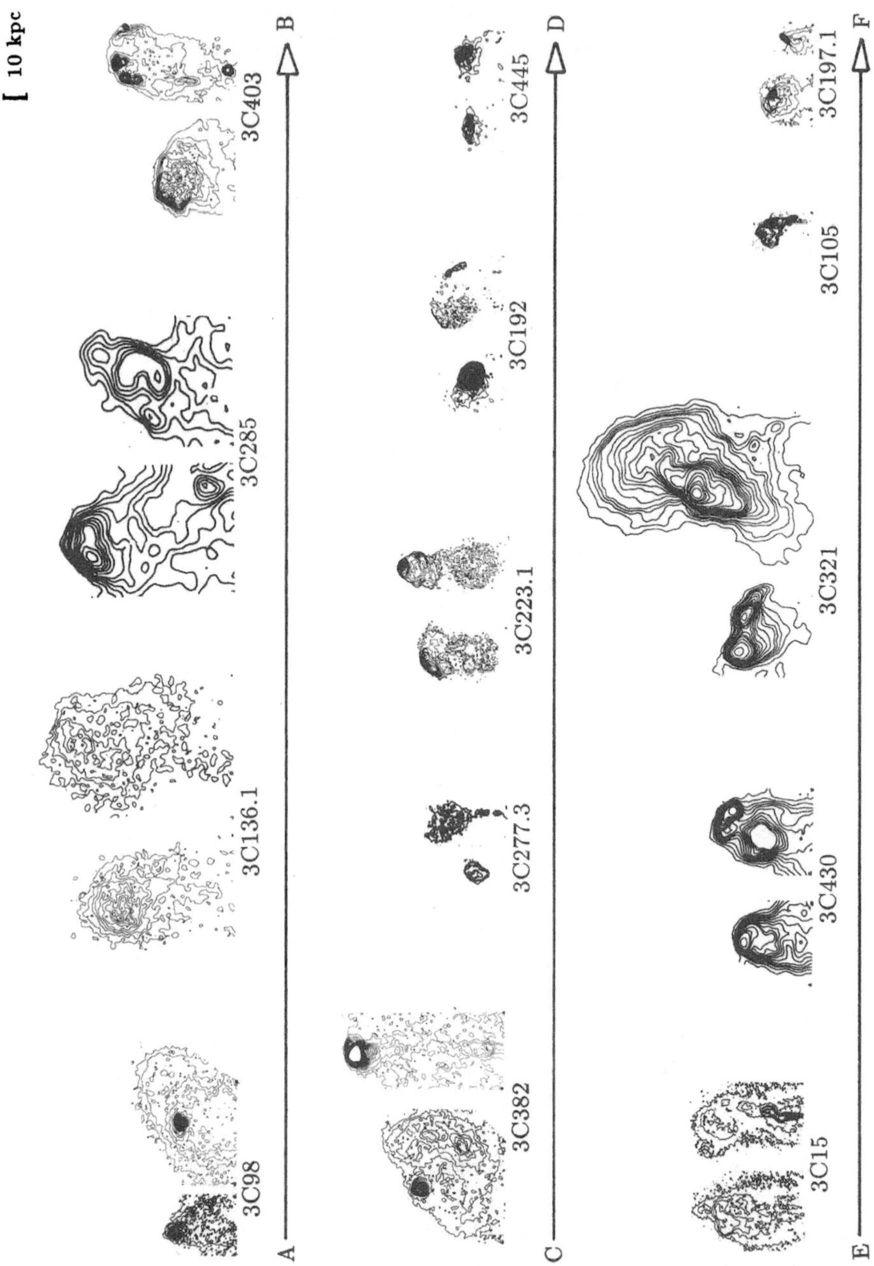

Fig. 1a. Hot spots from 28 nearby FR II radio sources. The hot spots from opposite lobes of each source are re-orientated with the central core (and anticipated jet) vertically below and re-sized to the same absolute, physical scale (given by the linear scale-line). They are ordered in increasing P_{178} (source luminosity measured at 178 MHz) — from label A (Fig. 1a, top right-hand corner) through to label L (Fig. 1b, bottom left-hand corner).

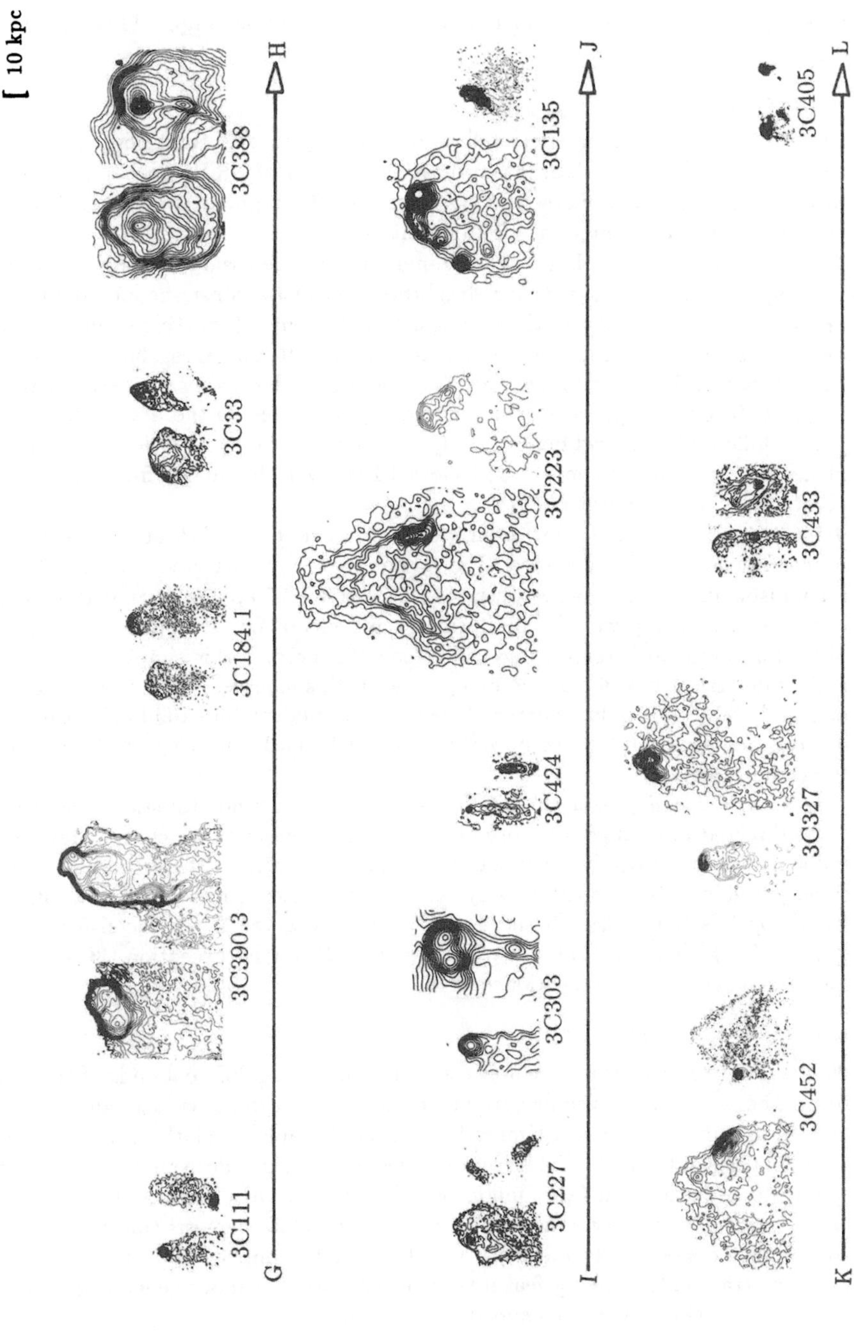

Fig. 1b. Hot spots from nearby FR II radio sources, continued.

2 Review

A review is given of the most striking trends and features in this sample of low-luminosity FR II radio sources.

Properties of the jets:

i. *Jets are common.* Jets are detected unambiguously in 57% (and possibly 70%) of the sources – far more than in any previous studies of FR II sources: in a recent observational review, Muxlow and Garrington (1991) report a detection rate of less than 10% for FR II sources at all luminosities.

ii. *Jets are faint and weak.* High signal-to-noise ratio is necessary to distinguish these jets; typically they show a surface-brightness contrast against the lobe of only ~5 and contribute only ~3% of the luminosity of the source (omitting several striking exceptions). If these properties are common to all FR II sources, the low detection rate of jets until now may be understood as being due to insufficiently sensitive imaging. In our sample, better imaging has steadily led to the detection of more jets, as follows: Bridle and Perley (1984) reported jets in only 7 (23%); a compilation of the best available images in 1988 showed 10 (33%); the images from this project (Black *et al.* 1991) reveal 17 (57%).

iii. *Jets are 1-sided.* Of the 17 sources in which jets are detected, 76% are 1-sided in that no counter-jet *at all* is detected. The remaining 4 have detected counter-jets and are unusual in the following respects: 3C 452 and 3C 405 appear as archetypal FR II radio sources having well-defined 'primary and secondary' hot spots at the heads of trailing lobes with remarkable end-to-end symmetry – this simple, well-defined and symmetrical morphology is exceptional in this sample. 3C 15 and 3C 424 are of questionable FR II classification. They are notably small ($\lesssim$100 kpc), display no obvious hot spots and house powerful jets which contribute to more than 1/4 of their luminosity.

iv. *Lobe structure and jet-sidedness.* The 1-sided jets show no statistical preferences to be directed to a longer or more luminous lobe – either of which may have been indicative of *intrinsically* asymmetric jets. Indeed, many of these sources with 1-sided jets have lobes which are very symmetric in size: 6 have lobes that differ in length by less than 10%. Of the 4 sources with detected 2-sided jets, 3 are very symmetric, with lobes that differ by less than 15% (the exception, 3C 424, is a rather strange source as mentioned in iii.).

Properties of the hot spots.

v. *No standard hot spot structure prevails.* Even comparing different ends of the same source, hot spots are dissimilar, irregular and highly asymmetric though often embedded in radio lobes of very similar flux, size and shape. The latter property limits the variations in the ambient medium – largescale, side-to-side asymmetries must be small and any smallscale (kpc) inhomogeneities must have a well matched net effect on each side. The variety of hot spot structures therefore suggests that they are *not* steady patterns of shocks and flows and should not be interpreted as such, instead they are transient, evolving features (contrast with the more general analysis of Valtaoja (1984) and Lonsdale and Barthel (1986)).

vi. *The 'primary and secondary' hot spot classification does not fit.* In particular, the choice of *a* primary and *a* secondary in each lobe is meaningless; increasing the resolution, the 'brightest and most compact' primary often changes or becomes ambiguous, with *several possible* hot spot candidates. These hot spots are in marked contrast to those analysed by Laing (1989) in a sample of more distant (z>0.3) and more powerful radio galaxies: from these he drew his morphological definitions of *the* primary and *a possible* secondary hot spot – this choice being confirmed by observations at higher resolution (Laing 1991, private communication.).

vii. *Hot spot structure and jet-sidedness.* Unlike the findings of Laing, *no statistical preference* is found for the jet to be directed to a smaller, brighter or even simpler hot spot (when comparing those in opposite ends of the same source) – as one may naively expect for a Doppler-beamed relativistic jet and associated (non-axisymmetric) relativistic post-shock flow in the hot spots.

It is not yet clear whether any or all of these trends are restricted to the lower-power FR II sources or whether they are more generally applicable to FR II sources of all luminosities.

Figures 1a&b summarise the observational basis for the above comments. The maps are of differing absolute resolution and are taken from Black *et al.* (1992), except for the following to whose observers I am grateful for use of their data and permission to produce diagrams: 3C 33 — Larry Rudnick and Martha Anderson; 3C 277.3, 3C 285 and 3C 327 — Wil van Breugel; 3C 388 — K. Roettiger, J. Burns, D. Clarke and W. Christiansen; 3C 430 — S.R. Spangler.

Acknowledgments I acknowledge support from the SERC which has allowed me to attend this conference. The VLA is part of the National Radio Astronomy Observatory and is operated by Associated Universities, Inc., under co-operative agreement with the National Science Foundation.

References

Black, A.R.S., Baum, S.A., Leahy, J.P., Perley, R.A., Riley, J.M., Scheuer, P.A.G.: 1992, *Mon. Not. R. astr. Soc.* **256**, 186.

Bridle, A.H., Perley, R.A.: 1984, *Ann. Rev. Astron. Astrophys.* **22**, 319.

Carilli, C.L., Dreher, J.W., Perley, R.A., 1989, in *Hot spots in Extragalactic Radio Sources*, eds. K. Meisenheimer, and H.-J. Röser, Heidelberg: Springer-Verlag, p. 51.

Fanaroff, B.L., Riley, J.M.: 1974, *Mon. Not. R. astr. Soc.* **167**, 31.

Muxlow, T.W.B., Garrington, S.T., 1991, in *Beams and Jets in Astrophysics,* ed. P.A. Hughes, Cambridge: Cambridge University Press, p. 69.

Laing, R.A., 1989, in *Hot spots in Extragalactic Radio Sources,* eds. K. Meisenheimer and H.-J. Röser, Heidelberg: Springer-Verlag, p. 27.

Valtaoja, E.: 1984, *Astron. Astrophys.* **140**, 148.

Lonsdale, C.J., Barthel, P.D.: 1986, *Astron. J.* **92**, 12.

Properties of Low Luminosity Radio Galaxies: Polarization and Jets

H. R. De Ruiter [1,2], S. Capetti [3], R. Morganti [1],

P. Parma [1], L. Lazzari [4], R. Fanti [1,5]

[1]Istituto di Radioastronomia del CNR, Bologna, Italy
[2]Osservatorio Astronomico, I-40126 Bologna, Italy
[3]Dipartimento di Fisica Generale, I-10025 Torino, Italy
[4]Dipartimento di Astronomia, I-40126 Bologna, Italy
[5]Dipartimento di Fisica, I-40126 Bologna, Italy

Abstract: A progress report is given on further radio studies of the B2 sample of low luminosity radio galaxies. We present results on the polarization at 20 cm and compare our results with the Laing-Garrington effect. We also discuss new results on the so called *naked jets*.

1 Introduction

Over the years the B2 sample of low luminosity radio galaxies has been studied in detail, first with the Westerbork radio telescope (WSRT) and more recently with the Very Large Array (VLA) (Ekers *et al.*, 1981; de Ruiter *et al.* 1990 and references therein). A discussion of the general properties of the sample is presented in de Ruiter *et al.* (1990).

The dominant radio morphology for these low luminosity radio sources is the edge-darkened FR I type. A high percentage of the sample galaxies show jets which are more symmetrical than in high power FR II galaxies (Parma *et al.*, 1987). Comparison with semi-empirical models (Bicknell *et al.*, 1990) suggests that the jets are turbulent, have a low Mach number, are originally lighter than the external medium, but become heavy by entraining external material, while decelerating rapidly.

For a total of 69 radio sources we have conducted a systematic study of the polarization of the sample at 20 cm, using the same data from which we derived the total intensity maps. These results are presented in Section 2, and discussed in connection with the Laing-Garrington (L&G) effect.

In Section 3 we present new results on the so called *naked jets*, a population of sources which is fairly common at the fainter luminosity end.

2 Polarization

Radio polarization has been used by several authors (see *e.g.* Laing 1988; Garrington *et al.*, 1991) as a diagnostic for the presence of beaming effects in radio sources. In double lobed radio sources with one-sided jets it is generally found that the jet-side is less depolarized than the counter-jet side. The explanation is a combination of orientation effects, beaming and depolarization by Faraday effects in an external halo.

We here present results of a study of polarization for 69 sources of the B2 sample at 20 cm. The complete set of data, the data reduction and the maps are presented in Capetti *et al.* (in prep.). We do not have, for the majority of them, polarization data at a second frequency (*e.g.* 6 cm.), but we argue, from the data of Garrington *et al.* (1991), that the asymmetry at 20 cm in the lobe polarization is an excellent indicator of depolarization asymmetry. In other words, the 6 cm polarization is rather symmetric, so that the 20 cm polarization asymmetry is due to asymmetry in depolarization.

The median fractional polarization, m', at 20 cm in the whole sample is m' = 10.5 %, and ranges between ~ 12 % for the highest polarized side to ~ 9 % for the less polarized one. Sources without evidence of jets are the less polarized ones (~ 6 %), while the sources with one-sided jets show the highest degree of polarization. Both in sources with one-sided and two-sided jets, the main-jet side is generally the more polarized one. Figure 1 shows our polarization results for the sources with jets.

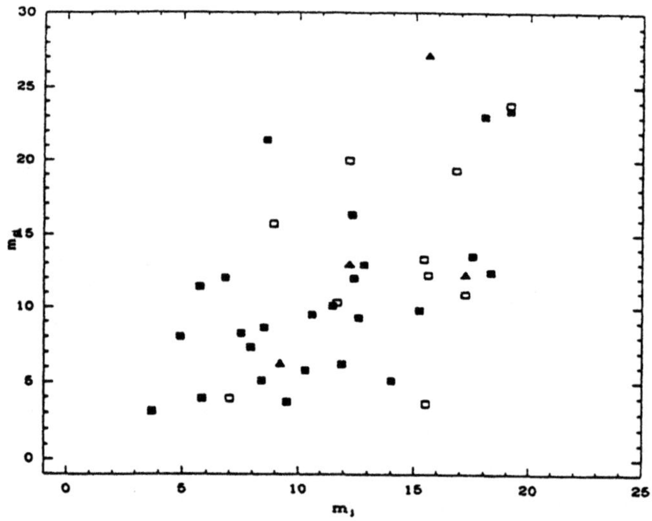

Fig. 1. Fractional polarization m' at 20 cm for jet and counter-jet components. (■ two-sided jet; □ one-sided jet; △ naked jet)

The asymmetry in polarization is not very strong, but definitely clear and in the same sense as the L&G effect. If we compare our results with those of Garrington *et al.*

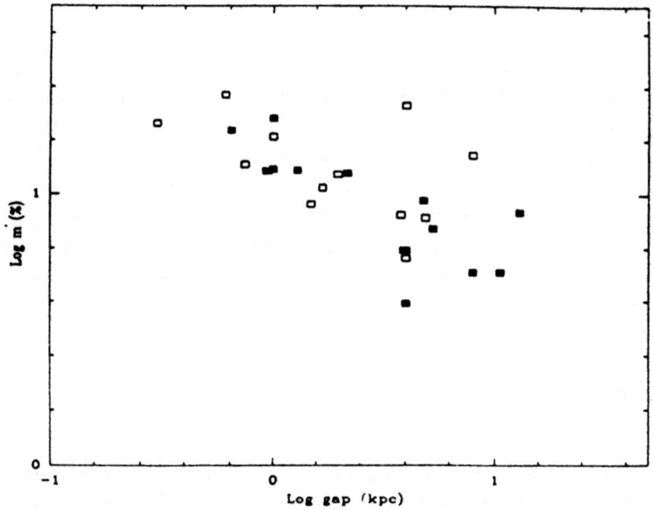

Fig. 2. Fractional polarization at 20 cm, versus size of gap in brightness, for jets and counter-jets (■ refer to counter-jets).

for sources with $0.2 < z < 1.0$ (sample 1), we find very similar results. On the contrary, the sources of Garrington *et al.* with $z > 1$ (sample 2) are definitely more asymmetric in the 20 cm polarization of the two lobes (see Tab. 1). The parameter rm' (m'_j/m'_{cj}) has been compared with other quantities characteristic of the jet. No correlation is found between rm' and the counter-jet to jet arm ratio. This relation was found to be quite strong for high luminosity radio galaxies without a visible jet (Pedelty *et al.*, 1989).

Table 1.

Sample	m'_j	m'_{cj}	<rm'>	rm'> 0.67
one sided jets	15.4	12.8	0.87	3/10
two sided jets	11.0	9.9	0.90	6/29
G sample 1				
galaxies	7.9	7.0	0.80	1/6
quasars	9.1	8.8	0.93	6/17
G sample 2	8.9	1.2	0.15	23/24

On the contrary, a clear correlation has been found (Fig. 2) between the fractional polarization and the size of the brightness gap that is often present in the jets in low luminosity radio galaxies (Parma *et al.*, 1987). In this correlation no distinction appears between the jet side and the counter-jet side.

The existence in the B2 sample of the L&G effect is somewhat intriguing in view of the explanation currently given for the L&G effect in terms of orientation and beaming. The jets in the low luminosity radio galaxies are believed, from jet models (Bicknell *et al.*, 1990), to have low velocities (up to 10^4 km/sec at most) and, for isotropic orientation of the sources, this is not enough to produce significant asymmetry. One has to either reject the above mentioned models, or look for different explanations, perhaps based on intrinsic rather than on orientation effects. The correlation we have found between fractional polarization and size of the brightness gap in the jets may speak in favour of an intrinsic explanation.

3 Naked Jets

The B2 sample of radio galaxies contains five sources whose radio structure consists of a core plus a twin jet and no (or very faint) lobes (Parma *et al.*, 1987). These sources are among the fainter, in absolute total radio power, of the B2 sample ($P_t < 10^{23.5}$ W/Hz). The jets are symmetric in brightness in four out of the five sources. The spreading rates of the jets ($\sim 0.2 - 0.4$) and the core to jet luminosity ratio ($\sim 0.1 - 0.7$) are typical of low luminosity radio galaxies. What makes them peculiar is the very low level or absence of lobe emission. In a sense, this might recall the situation of NAT sources, where, however, the lobe absence is interpreted as being due to the galaxy motion, which prevents the relativistic plasmas carried by the jets from accumulating in a fixed volume. This explanation is clearly not applicable to the *naked jets*, where we do not see any effect of galaxy motion. What kind of sources are they? What are the causes which prevent the formation of radio lobes? Are they young sources, where lobes are not yet formed? Is there anything special about the energy transport in the jets so that the energy is largely dissipated on the way out?

We have carried out a sensitive observing program at the VLA at 20 cm with the C configuration to search for weak lobes undetected in our previous higher resolution observations. Our new observations have a sensitivity to extended brightness emission which is a factor ~ 30 higher. The new maps we have obtained are shown in Fig. 3.

The results are the following: two sources (1122+39 and 2116+26) still do not show evidence of lobes. Two others (1108+27 and 1553+24) do show some lobe emission, but on one side only. The last (1752+32) has weak lobes around the jets. In all cases the overall radio luminosity remains largely due to the jets and core.

We believe it unlikely that the absence of lobes is due to *youth* of the source, since the five sources have sizes which are typical of the low luminosity radio galaxies (tens of kpc). The application of the Bicknell model to three of them showed that their jets are low Mach number and low velocity (few thousand km/sec), so that their observed sizes require an age of $\sim> 10^7$ years, which is contrary to the above explanation.

The application of the Bicknell model also shows that these jets are relatively heavy. This may be explained as being due to entrainment of external material. This implies

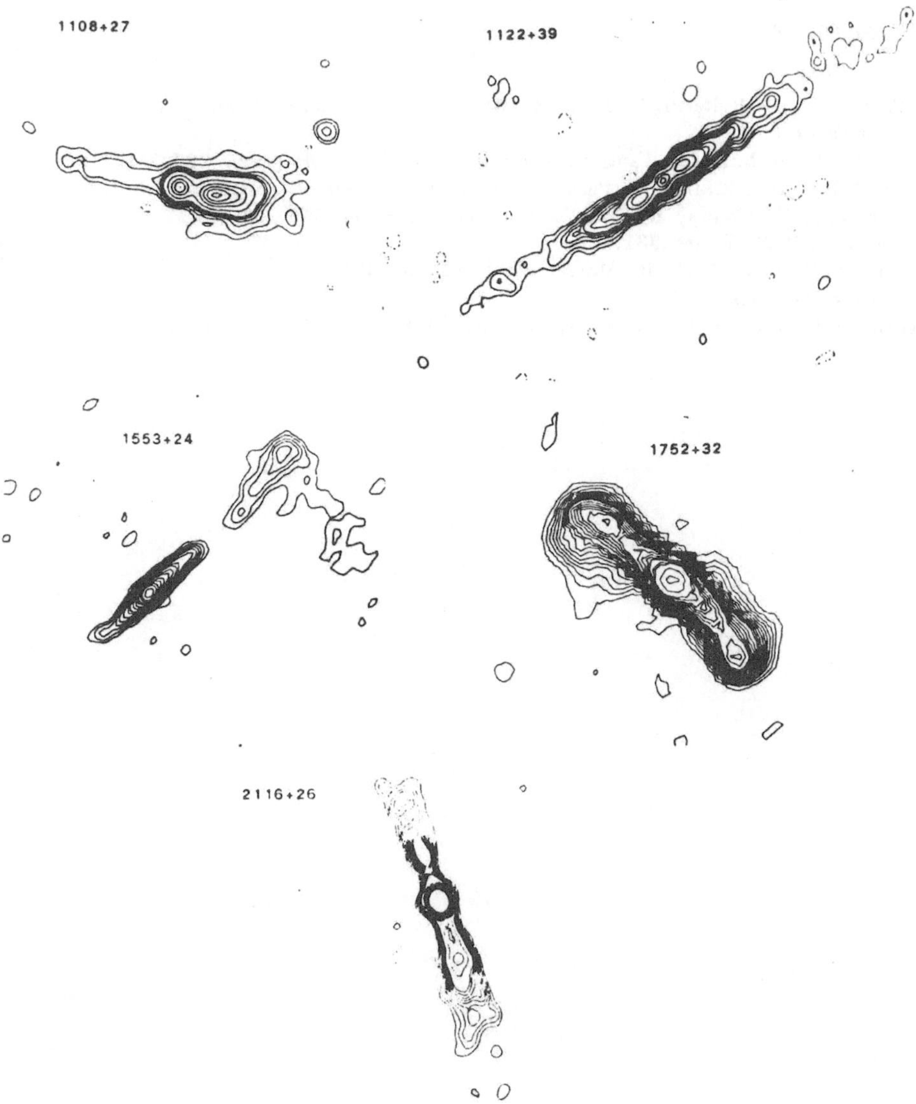

Fig. 3. Maps of *naked jets.*

a considerable dissipation of the energy carried by the jet and this is likely to be the explanation for the absence of lobes (or for their very weak emission).

References

Bicknell, G.V., de Ruiter, H.R., Fanti, R., Morganti, R., Parma, P.: 1990, *Astrophys. J.* **354**, 98.
de Ruiter, H.R., Parma, P., Fanti, C., Fanti, R.: 1990, *Astron. Astrophys.* **227**, 351.
Ekers, R.D., Fanti, R., Lari, C., Parma, P.: 1981, *Astron. Astrophys.* **101**, 194.
Garrington, S.T., Conway, R.G.: 1991, *Mon. Not. R. astr. Soc.* **250**, 198.
Laing, R.A.: 1988, *Nature* **331**, 149.
Parma, P., Fanti, C., Fanti, R., Morganti, R., de Ruiter, H.R.: 1987, *Astron. Astrophys.* **181**, 244.
Pedelty, J.A., Rudnick, L., McCarthy, P.J., Spinrad, H.: 1989, *Astron. J.* **97**, 647.

Compact Steep-Spectrum Radio Sources: A Progress Report ([4])

D. Dallacasa [1,2], C. Fanti [2,3], R. Fanti [2,3]

[1]Dipartimento di Astronomia, via Zamboni 33, I-40126 Bologna, Italy
[2]Istituto di Radioastronomia, via Irnerio 46, I-40126 Bologna, Italy
[3]Dipartimento di Fisica, via Irnerio 46, I-40126 Bologna, Italy

Abstract: The properties of Compact Steep-Spectrum radio sources are reviewed. Particular analysis is placed on radio distortions, which are considered as indicators of strong interaction between jets and ambient medium.

1 Introduction

A large fraction of the sources in flux limited radio samples consists of Compact Steep-spectrum Sources (CSSs). These objects have their dominant radio emission on kiloparsec scales and are found in a variety of active galaxies, including Seyferts, radio galaxies and quasars. Among nearby radio sources they seem to be confined at rather low radio powers ($\lesssim 10^{24}\,\mathrm{W\,Hz^{-1}}$), but appear very common at high powers ($\geq 10^{26}\,\mathrm{W\,Hz^{-1}}$) at $z > 0.2$ (Fanti *et al.*, 1990; Kapahi and Kulkarni, 1990).

Observations of nearby CSSs have shown good correlations between their complex radio structure and optical emission line gas, suggesting strong interaction between the jets feeding the radio source and the gas in the ambient medium (van Breugel *et al.*, 1984). It has been suggested that similar interactions may also occur in distant and more powerful CSSs.

In this paper we review the radio properties of powerful CSSs on the basis of data obtained with VLA, Merlin and VLBI observations.

[4] Paper presented by R. Fanti

2 The Samples

The best studied sample is the one derived from the 3CR catalogue, with the so called *Peacock and Wall* (PW) extension.

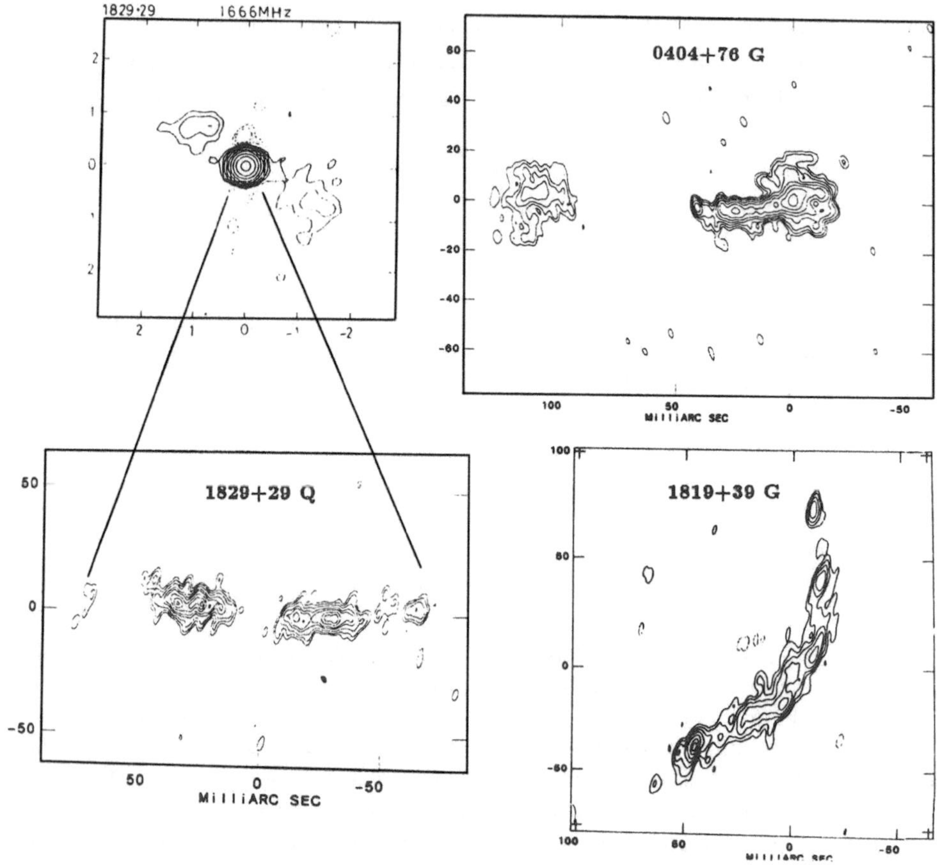

Fig. 1. Examples from the *PW* extension

The 3CR sample contains those sources from the 166 of Jenkins et al., (1977) with projected linear size $\leq$ 15 kpc ($H_0 = 100\,\mathrm{km\,s^{-1}\,Mpc^{-1}}$, $q_0 = 1$), and steep ($\alpha > 0.5$) high frequency spectrum. In order to compensate for spectral biases, we added those sources from the 2.7 GHz catalogue by Peacock and Wall (1982) (the *PW* extension), which according to their high frequency spectrum, would be in the 3CR, but are not, due to a spectral turnover at low frequencies, typically at ~0.4÷1 GHz. We and others have studied the overall sample at several radio wavelengths with the VLA, MERLIN and VLBI. A full set of references on the available data can be found in Fanti *et al.* (1990).

The combined information from all the data available at various frequencies and resolutions allows a very simple morphological classification for the 3CR CSSs (see Fanti *et al.*, 1990 and, earlier, Wilkinson *et al.*, 1984): *galaxies* appear double, with a weak core and weak jets (if any); *quasars* appear dominated by jet emission and their cores are significantly stronger than in radio galaxies (Fanti *et al.*, 1990).

The core fractional luminosities in both radio galaxies and quasars are, within the error margin, the same as in large size radio sources of similar power (~3% for quasars and ~0.3 % for radio galaxies).

Jets, seen essentially in quasars, are asymmetric with a median jet/counter-jet brightness ratio (R_j) greater than ~5 and have small opening angles ($\lesssim 10°$). They can be either straight or sharply bent and their brightness distribution along the axis is generally dominated by knots. The VLA polarization data show the magnetic field to run mostly parallel to the jet. According to these characteristics, CSS jets are very similar to those in larger size quasars, except that in CSSs they are much brighter with respect to the lobe luminosity (in some CSS quasars we see only the jets).

We have recently observed the PW extension, which includes 16 objects, at 1.6 GHz with a global VLBI array (Dallacasa *et al.*, in prep.). These sources are smaller in size than 3CR CSSs, typically less than 1 kpc. Our results do not yet allow a proper morphological classification since, with one frequency only, we are unable to unambiguously identify the source core, which is of major importance for the classification. However there seem to be remarkable exceptions to the morphological scheme drawn for the 3CR CSSs. A number of PW CSS radio galaxies show bright jets, and some quasars appear two-sided. Examples are shown in Fig. 1.

Before drawing any conclusion, however, one needs data at higher frequencies. We have already obtained global VLBI observations at 6 cm and the data will soon be available for analysis.

3 Source Orientation in CSS Samples

It has been argued that CSSs cannot be larger size sources which appear smaller for projection effects, and hence the majority of them have to be physically small in size (*e.g.* Fanti *et al.*, 1990). How are these sources actually oriented with respect to the line of sight? Can there be any kind of orientation bias in the sample?

If we assume that in all radio sources, to some extent, relativistic motions occur close to the *central engine*, then parameters to be checked for orientation biases are: a) the ratio of core to extended luminosity (R_c); b) the ratio of the luminosity of the main jet to that of the counter-jet (R_j) (assuming intrinsic symmetric jet ejection).

The parameter R_c in both CSS radio galaxies and quasars is very similar to that of the larger size ($\gg$ 15 kpc) sources and this strongly suggests that CSSs are oriented in the sky in a similar manner to their larger size cousins.

As far as R_j is concerned, we actually have too many upper limits to draw firm conclusions. On the basis of the number of CSSs with $R_j \lesssim 4$, Spencer and Akujor (1992) conclude that some orientation bias, with respect to random orientation, can be present in the 3CR CSSs. However, this result, apart from being based on limited statistics, refers essentially to quasars. Were the quasars distributed within a limited range of angles with respect to the line of sight, as in the Barthel (1989) model, the predicted values (and upper limits) of R_j for CSS quasars would be consistent with the observed ones.

4 Distortions and Jet Asymmetry

We here consider only the 3CR sample, for which only a morphological classification is possible. We examine two measures of the source distortion.

The first, related to the *large scale* structure, is the angle θ_1 between the two lines connecting each of the outermost emission regions, generally a lobe, to the radio core ($\theta_1 = 0$ for a linear source). This angle is defined for any source with a detected core, which means virtually in all cases for quasars, but for only 5 out of 11 radio galaxies. The distribution of θ_1 shows a significant difference between radio galaxies and quasars: for galaxies $\theta_1 \leq 15°$, while 40% of quasars have $\theta_1 > 35°$.

The other, θ_2, is the angle between the initial direction of the jet (close to the core) and the jet orientation at $\sim$1 kpc from the core. This can be done only for sources with jets, which means essentially only for quasars.

A comparison of θ_1 and θ_2 within each source clearly indicates that most of the distortion occurs within about the first kpc from the core.

We have argued, on the basis of the R_c distributions, that CSSs are likely to be randomly oriented. This implies that the distortions we see must be largely intrinsic and that orientation effects presumably play a minor role.

How do these large distortions arise?

A possible scenario is to assume that interactions with a clumpy ambient medium are important, as recently postulated to explain the structure of 3C 48 (Wilkinson *et al.*, 1991) and perhaps of 3C 343 (Schilizzi *et al.*, in prep). A collimated jet hits a dense interstellar cloud, brightens considerably, may be partially decollimated and bent from its initial direction.

The most impressive cases of bent structure are: 3C 43, 3C 119, 3C 190, 3C 216, 3C 287, 3C 309.1, 3C 454 (Spencer *et al.*, 1991 and references therein) where rather abrupt changes in orientation, up to 90°, are seen. In these sources the median value of R_c is $\lesssim 5\%$, consistent with the median value of the whole 3CR CSS sample, indicating that there is not an anomalous orientation distribution.

It has to be noted, however, that in a number of these cases the source distortion, which consists of multiple bends, is not well justified by random strikes against inhomogeneities in the medium, which should produce a sort of random walk. The overall distortion shows a regularity (*e.g.*: 3C 43, 3C 119, 3C 287, 3C 309.1) which requires systematic effects on the propagating jets such as pressure gradients (either in a static atmosphere or as streaming motions or gas rotation acting via ram pressure). Also, precession models may qualitatively account for the observed systematicity in the distortions, although in such models we could, naively, expect the bends to be more regular and gentle than observed.

We also note that if the bending is due to the jet striking a gas cloud, we would expect the jet to brighten there. This is well observed in 3C 43 but it is not so, for instance, in 3C 119 and 3C 454. One may wonder, on the other hand, whether after a strike the jet momentum flux is reduced and the jet is decollimated, so that at some subsequent stage external pressure gradients, inefficient in a well collimated jet, may start to take over.

For a discussion on jet bending in CSSs see Spencer and Sanghera (1990).

Finally, the majority of jets are one-sided (3C 138 and 3C 286 being possible exceptions). The general interpretation is that asymmetry is due to Doppler favoritism in an intrinsically symmetric jet. Extreme orientations are not required for this. However, if the scenario holds where CSS jets propagate through a dense inhomogeneous medium, collide with inhomogeneities, flare and decelerate, then the jets must be intrinsically one-sided (van Breugel *et al.*, 1988), otherwise it would be difficult to understand why the interaction occurs on one side only (see also Wilkinson *et al.*, 1991, for the case of 3C 48). One-sidedness could therefore be related to an intrinsic mechanism.

5 The Link between CSS and Large Size (LS) Sources

CSSs are, by definition, objects of small linear size. However, we occasionally see a very bright central kpc sized emission region in LS sources which mimics the properties of a CSS. Well known examples are 3C 236 (Schilizzi *et al.*, 1988) and 3C 293 (Bridle *et al.*, 1981). These Steep Spectrum Cores (SSC) are quite rare, nevertheless one should understand their nature and possible connections to CSSs.

Different scenarios are currently being considered in terms of:

YOUTH: CSSs are small because they are still young ($<10^6$ y) and will only later develop into LS sources; in this scenario there is no connection with SSCs. A variant of this possibility is *RECYCLING*: a relic source has its engine restarted and can appear as a CSS or SSC.

FRUSTRATION: the jets are trapped by a dense gas, where they dissipate most of the energy they carry, as in Seyferts; some leakage may occur which builds up some extended emission (SSC source). A variant is *SMOTHERING*: the energy flow to the outer lobes through the jets is stopped by recent acquisition of gas from outside; the outer lobes decay and a small size source is built up; on the basis of the lobe decay time the source may appear as a CSS or SSC.

The distinction between these different possibilities is largely based on the information available on the ambient gas. Information which, at the moment, is rather poor.

References

Barthel, P.D.: 1989, *Astrophys. J.* **336**, 606.

Bridle, A.H., Fomalont, E.B., Cornwell, T.J.: 1981, *Astron. J.* **86**, 1294.

Fanti, R., Fanti, C., Schilizzi, R.T., Spencer, R.E., Nan Rendong, Parma, P., van Breugel, W.J.M., Venturi, T.: 1990, *Astron. Astrophys.* **231**, 333.

Kapahi, V.K., Kulkarni, V.K.: 1990, *Astron. J.* **99**, 1397.

Jenkins, C.J., Pooley, G.G., Riley, J.M.: 1977, *Mem. R. astr. Soc.* **84**, 61.
Peacock, J.A., Wall, J.V.: 1982, *Mon. Not. R. astr. Soc.* **198**, 843.
Schilizzi, R.T., Skillman, E.D., Miley, G.K., Barthel, P.D., Benson, J.M., Muxlow, T.W.B.: 1988, IAU Symp. 129 *The Impact of VLBI on Astrophysics and Cosmology*, eds. M.J. Reid and J.M. Moran, Dordrecht: Kluwer, p. 127.
Spencer, R.E., Sanghera, H.S.: 1990, in *Compact Steep-spectrum and GHz-peaked spectrum radio sources*, ed. C. Fanti, R. Fanti, C.P. O'Dea, R.T. Schilizzi, C.N.R., p. 200.
Spencer, R.E., Schilizzi, R.T., Fanti, C., Fanti, R., Parma, P., van Breugel, W.J.M., Venturi, T., Muxlow, T.W.B., Nan Rendong: 1991, *Mon. Not. R. astr. Soc.* **250**, 225.
Spencer, R.E., Akujor, C.E.: 1992, in Proc. of 7th IAP Conference, eds. G. Pelletier, J. Roland, and H. Sol, p.119.
van Breugel, W.J.M., Miley, G., Heckman, T.: 1984, *Astron. J.* **89**, 5.
van Breugel, W.J.M., Nan Rendong, Schilizzi, R.T., Fanti, C., Fanti, R.: 1988, IAU Symp. 129, *The Impact of VLBI on Astrophysics and Cosmology*, eds M.J. Reid and J.M. Moran. Dordrecht: Kluwer, p. 115.
Wilkinson, P.N., Spencer, R.E., Readhead, A.C.S., Pearson, T.J., Simon, R.S.: 1984, in IAU Symp. 110 *VLBI and Compact Radio Sources*, eds R. Fanti, K.I. Kellerman, G. Setti, Dordrecht: Reidel, p.25.
Wilkinson, P.N., Tzioumis, A.K., Benson, J.M., Walker, R.C., Simon, R.S., Kahn, F.D.: 1991, *Nature* **352**, 315.

Modelling Jets in Class I and Class II Radio Sources

G.V. Bicknell

Mt. Stromlo and Siding Spring Observatories
Institute of Advanced Studies
Australian National University
Private Bag
2606 Woden, ACT
Australia

1 Introduction

The morphological differences between extragalactic radio sources of low and high power, noted by Fanaroff and Riley (1974), are so striking that the conclusion is inescapable that the mode of energy transport in the two types of source is different. It is now accepted that physical differences between the two classes of source are related to two fundamentally different types of jet: in class I sources the jets appear to be turbulent and to entrain the interstellar/intergalactic medium through which they propagate (at least from ~ 1 kpc outwards). In class II jets much of the interesting physics relates to phenomena which are internal to the jet (*e.g.* internal shocks) or to characteristics of the hot spot or backflow. The fact that the physics of the two classes of jet is quite different is reflected in two fundamentally different approaches to the modelling of jets in these sources.

In this review I shall briefly summarize some of the fundamental aspects of class II jets as elucidated in some of the early work and I shall then concentrate on some of the more recent work. This includes the role of passive and active magnetic fields, a summary of approaches to modelling turbulent flow, and some recent results on modelling turbulent jets in class I sources.

2 Fundamental Aspects of Class II Jets

2.1 Scheuer (1974), Blandford and Rees (1974) Beam Theories

Both Scheuer (1974) and Blandford and Rees (1974) recognized the attractiveness of "beam theories" in which the lobes of powerful class II sources are supplied with relativistic particles and field in a quasi-continuous fashion by a "jet" or a "beam". These papers obviously played a seminal role in the study of extragalactic jets and it is apposite to note that they in fact *predicted* their existence. Norman and colleagues realized that the study of super/hypersonic jets was a potentially rewarding one for computational science and this led to a number of important papers (see Norman, Winkler and Smarr 1984 and references therein) containing the results of simulations of jets of varying Mach number ($\mathcal{M}$) and density ratio (η). For a number of reasons, these two parameters are often taken as the two fundamental parameters of a jet. Many of the characteristics of class II jets (*e.g.* the existence of a hot spot at the end of the jet) clearly depend upon their supersonic Mach numbers. Moreover, it is well known that class II jets are required to be light ($\eta \ll 1$) in order to inflate a lobe/cocoon that is large compared to the jet.

2.2 The Dentist's Drill

Although light jets inflate relatively large lobes, Scheuer (1982) suggested that wandering of the jet inside its cocoon would give rise to a larger cocoon more closely related to the lobes observed in class II radio sources together with multiple hot spots (as observed in some sources). In this context two types of hot spots have been identified - splatter spots (see Williams and Gull 1985) and disconnected hot spots. One of the consequences of multiple hot spots is advance of the lobe proportional to $t^{1/2}$ rather than t as pointed out by Cox, Gull and Scheuer (1992). The slower lobe advance, at least qualitatively explains the relatively large backflows in some class II sources.

2.3 The Kelvin-Helmholtz Instability

Much of the differences between class I and class II jets can at least be partially understood in terms of the Kelvin-Helmholtz instability. Payne and Cohn (1985) delineated two types of pinching mode: the *ordinary mode* and the *reflecting mode,* the transition between the two occurring at a critical velocity $v_{\rm crit} = c_{s,ext} + c_{s,int} = \left(1 + \eta^{1/2}\right) c_{s,int}$. The way in which these instabilities saturate is important. The ordinary mode saturates in disruptive transverse shocks; the reflecting mode saturates in travelling "criss-cross" shocks which are not disruptive (Norman *et al.* 1984). Thus, a light, supersonic jet can avoid disruption due to the pinching instability. Nevertheless, other modes of the Kelvin-Helmholtz instability are important and Hardee and Norman (these proceedings) present a simulation of a light supersonic jet which becomes turbulent through the action of non-axisymmetric modes.

3 Passive Magnetic Fields in Class II Jets

Clearly the magnetic field in jets has got something to do with what we see. The emissivity and polarization depend upon the magnitude and direction of the magnetic field, and in the case of a passive magnetic field, this is determined by the dynamics of the underlying flow. A number of workers have included passive magnetic fields in jet simulations with interesting results.

3.1 Hamburger Hot Spots

Matthews and Scheuer (1990a,b) and Clarke, Norman and Burns (1989) have conducted axisymmetric simulations with passive magnetic field and "observed" the resulting surface brightness distributions. Matthews and Scheuer incorporate a tangled magnetic field which is initially uniform in magnitude but random in direction, whereas Clarke *et al.* incorporate a passive helical field. All groups observe "hamburger hot spots" (which may be relevant to the appearance of the hot spots in Cygnus A) as a result of their simulations, but the way in which the Hamburger is served up is different. With Matthews and Scheuer, the ring shaped morphology occurs at the region of maximum compression of the fluid which is a few jet radii from the terminal jet shock. With Clarke *et al.*, the ring shape originates from the compression of the toroidal field which, in their constant current density configuration, is proportional to the distance from the axis. Neither group incorporates particle acceleration in a self-consistent fashion. Matthews and Scheuer transport a passive relativistic fluid; Clarke *et al.* calculate the electron energy density from the pressure of the fluid. As a result of their simulations Matthews and Scheuer speculate that diffusive particle acceleration may not be as important as originally presumed, and it will be interesting to see the effects of particle acceleration included in future simulations.

3.2 Polarization at the Edges of Lobes

An obvious effect from both sets of simulations is that deflection of the jet into the backflow sets up a region of very strong shear producing E-vectors which are perpendicular to the lobe boundary, as observed in many class II radio sources. Matthews and Scheuer introduced synchrotron cooling in order to reduce the emission from highly distorted volume elements, commenting that in more realistic models magnetic reconnection should be included. Such reconnection may also be a source of particle acceleration as has often been speculated.

3.3 Interpretation of the Observational Features of 3C 219

In a related paper Clarke *et al.* (1992) proposed an interesting idea for the morphology of the (predominantly) one-sided jet in 3C 219. This jet appears close to the core with a parallel magnetic field and then disappears after approximately 7 kpc. There is a much smaller region of emission on the other side of the source which is plausibly associated with the counter jet. Eschewing a relativistic jet model, Clarke *et al.* consider a model wherein the main jet produces a series of criss-cross shocks as it comes into equilibrium with the galaxy atmosphere. These enhance the predominantly poloidal magnetic field. As the jet comes into equilibrium with the surrounding medium, the criss-cross shocks disappear, the toroidal component of the field starts to dominate and the jet loses contrast with the lobe due to the r^{-1} dependence of the toroidal field in the lobe. The drop in surface brightness does not happen as quickly in the simulation as is observed. Nevertheless, the proposed mechanism is of interest. It would be worthwhile to see the simulation repeated with the Matthews-Scheuer field configuration.

The simulations do not explain the jet to counterjet surface brightness ratio. This may be related to differing ratios of toroidal to poloidal field, but this, of course, then leads to the question of why the ratios should be different. The most important question to answer in this context is why there should be any parallel field at all? Two possibilities seem to be shearing of an originally perpendicular magnetic field, or the continual processing of a magnetic field through oblique shocks. On the other hand, why should class II jets be so strongly sheared when it is generally accepted that class I jets are the more turbulent and turbulence requires shear to drive it?

4 Active Magnetic Fields

4.1 Helical Magnetic Fields

Since jets clearly contain magnetic fields the question naturally arises as to whether the field is dynamically important or not.

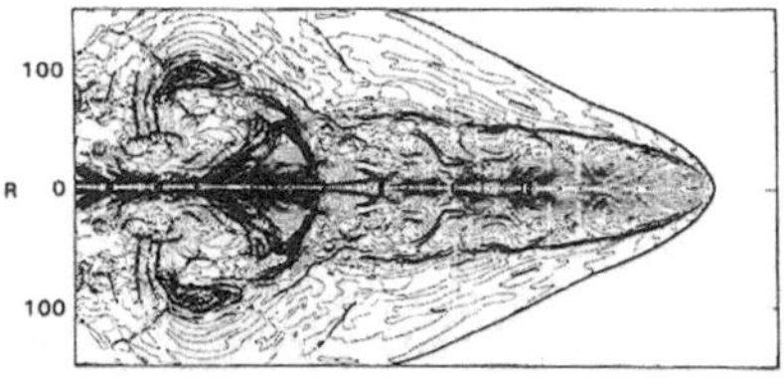

Fig. 1. Pressure contours in a magnetically dominated jet. Reproduced from Lind *et al.* (1989).

Confinement by a toroidal field has often been invoked as a possible explanation for apparently overpressured jets, a problem which was highlighted by Potash and Wardle (1980). However, the inclusion of a dynamically important magnetic field in jet simulations (*e.g.* Clarke, Norman and Burns (1986), Clarke and Burns (1987), Clarke (1988), Kössl (1989), Lind, Payne, Meier and Blandford (1989)) has some strange consequences. Figure 1 shows a contour map of the pressure in the Lind *et al.* simulation. The most pronounced effect is a long "nose cone" to the right of the terminal jet shock. Since this is an axisymmetric simulation the stability of this configuration is unknown. It is sometimes stated that the jet in 3C 273 is a candidate for this sort of jet; however detailed comparisons do not seem to hold up (see Conway, these proceedings).

4.2 Force-Free Magnetic Fields

Königl and Choudhuri (1985a,b) proposed a qualitatively different model for jets when they suggested that some jets may evolve into a magnetic force-free configuration. Subject to certain approximations, they showed that the structure of such a jet would be dominated by the $m = 0$ (parallel field) and $m = 1$ (double helical) modes. The $m = 1$ mode is interesting in that it predicts surface brightness enhancements at spacings ~ 5 jet radii and an asymmetric polarization structure also with this spacing. This behaviour is consistent with the observations of the jet in NGC 6251. Jackson *et al.* (1990) also suggested that the knot structure and polarization in the jet belonging to the quasar 0800+608 may be indicative of a jet with a field structure similar to that of the Königl-Choudhuri $m = 1$ mode. They argued that the observed jet wiggles are not due to a Kelvin-Helmholtz mode.

An alternative view (Clarke, Norman and Burns 1989) is that reflection mode dominated jets usually produce a pattern of criss-cross shocks that are spaced approximately 5 jet radii apart. They suggest that the non-axisymmetric polarization structure could result from superposition of a dominant axisymmetric mode and a weak helical mode. As Clarke *et al.* state: *for every MHD explanation there may lurk an equally valid hydrodynamic, plasma or even ballistic explanation.* I concur with this point of view. In this connection it should be noted that the dominant Kelvin-Helmholtz mode need not be the most rapidly growing. One only has to observe the growth of large surface waves as a result of the Kelvin-Helmholtz instability acting at a plane interface to appreciate this point (see § 5).

One way out of the confinement dilemma (Begelman and Cioffi (1989), Loken *et al.* 1992) is for the jet to be confined by the pressure of its own cocoon, which in turn is overpressured with respect to the surrounding medium by virtue of the advancing bow shock. For 3C 208, Loken *et al.* show that a jet Mach number ~ 40 and a velocity $\sim 0.35\,c$ are consistent with this idea.

5 Modelling FR I Jets

5.1 Some Physical Characteristics of Class I Sources and Jets

The minimum pressures of class I jets and the limited number of detailed comparisons between pressures inferred from X-ray data and jet minimum pressures (*e.g.* Killeen, Bicknell and Carter 1986; Killeen, Bicknell and Ekers 1988) suggest very strongly that class I jets are confined by the pressure of the surrounding medium. In the case of IC 4296, the minimum jet pressure is so close to the X-ray pressure that one may be forced to consider comparable magnetic and non-thermal pressures in dynamical models. The spreading rates of class I jets are generally considerably greater than class II jets (where these have been measured; Bridle 1984) and typical of turbulent laboratory jets. When sufficiently resolved class I jets frequently show an emission gap close to the core and sometimes a narrow well-collimated jet is observed close to the core followed by a more rapidly expanding regime (*e.g.* IC 4296; Killeen, Bicknell and Ekers (1986)).

These facts taken together suggest that class I jets are turbulent (Bicknell 1984), and that the onset of emission is related to the beginning of interaction with the interstellar medium. The latter point is consistent with the extrapolation of core minimum pressures

(Linfield 1982) to kpc scales. Since class I jets are probably initially supersonic, the initial interaction will involve shocks which are capable of reaccelerating relativistic particles, which have cooled on the journey from the core.

5.2 Fundamental Characteristics of Turbulent Flow

Most of the energy in a turbulent flow is contained in the large scales; the turbulent large-scale motions produce a cascade of energy to smaller and smaller scales establishing a power-law spectrum of turbulent energy. Eventually, at some scale, this energy is dissipated. In a normal fluid, viscosity provides the dissipative mechanism; in the astrophysical plasmas we are considering, it is generally envisaged that the turbulence is dissipated by particle acceleration (*e.g.* second order Fermi acceleration or shocks, Bicknell and Melrose 1982). It is worth bearing in mind that there are possibly important differences between incompressible and compressible turbulent flows. In incompressible turbulence, momentum can only be transported via eddies; in compressible flows MHD waves may be important. This may have some bearing on closure schemes.

5.3 Categories of Turbulent Model

5.3.1 Mean Flow Models

One of the most common forms of turbulent modelling (since the work of Prandtl and Taylor) involves statistically-averaged or time-averaged variables: *e.g.*

$$\rho = \bar{\rho} + \rho' \qquad < \rho' >= 0$$
$$v_z = \tilde{v}_z + v_z' \qquad < \rho v_z' >= 0$$

which give a set of averaged equations for the mean flow. For example, the jet z-momentum:

$$\frac{\partial}{\partial z}(\bar{\rho}\tilde{v}_z^2) + \frac{1}{r}\frac{\partial}{\partial r}(r\,\bar{\rho}\,\tilde{v}_r\,\tilde{v}_z) = -\frac{1}{r}\frac{\partial}{\partial r}(r\langle\,\rho\, v_r'\, v_z'\rangle) - \frac{\partial\bar{P}}{\partial z} - \bar{\rho}\, g$$

(see Bicknell (1986b) for this and other related equations). One of the major obstacles to this approach is the problem of closure, that is, what is a good expression for the Reynolds stress term $< \rho v_r' v_z' >$ which appears in this equation and other related terms which appear in the other dynamical equations? Komissarov (these proceedings) reports some interesting work along these lines using a Mach number dependent turbulent viscosity. Nevertheless, it should be remarked that the use of a turbulent viscosity essentially constitutes a *one-zone* model which largely ignores the structure transverse to the jet. Moreover, one would also like to see a turbulent closure method which also takes into account the dependence of turbulence on the pressure gradient.

5.3.2 Self-Similar Turbulent Flow

Self-similarity is a standard technique for incompressible (subsonic) turbulence. The major difficulty in applying self-similar analyses to astrophysical jets is that the collimation properties of the flow change, reflecting a changing non-self-similar level of turbulence. However, a self-similar analysis may be useful for the outer parts of some radio sources such as 3C 31 (and to the backflows from some class-II sources) which exhibit a plume-like structure.

5.3.3 Semi-Empirical Approach

This relies on the fact that a number of characteristics of turbulent flow, *e.g.* spreading rate, deceleration, dissipation, are all related and one can infer one property from another (Bicknell 1984, 86b). The approach is limited by the fact that not all characteristics are calculated from first principles. Its merits are that it does not rely upon poorly understood closure relations, is computationally undemanding and yields useful physical insights.

5.3.4 Direct Simulations

The goal of these can be modestly stated. It is simply to model all of the structure of a turbulent flow from the energy containing to the dissipative scale. For normal fluids, the computational requirements are immense since the dissipative scale can be many orders of magnitude smaller than the energy scale. This may not be such a problem in astrophysical jets because of the likely dissipation of energy at fairly low wave numbers ($k_{\rm diss}/k_0 \sim 100$; Bicknell and Melrose 1982, Biermann, these proceedings)

5.3.5 Large Eddy Simulations

This technique relies upon the fact that most of the energy and momentum in turbulent flow resides in the largest scales. The resolution of the calculation is such that the highest numerical wave number is in the inertial range. Since the entire inertial and dissipative range is not modelled, a prescription for the sub-scale dissipation of energy is required.

So far there have not been any entirely successful direct or large eddy simulations of turbulent, entraining jet flow. The computational demands of such calculations are underlined by Fig. 2 which is a sequence of images from a simulation of the Kelvin-Helmholtz instability in 2D. The large engulfing eddies which are responsible for mixing are similar to those which should be reproduced at the jet boundary by a 3D simulation of an entraining jet. Notwithstanding the computational problems, the relentless advance of computing power into the teraflop regime will no doubt see the publication of realistic calculations of turbulent jet flow in the not too distant future.

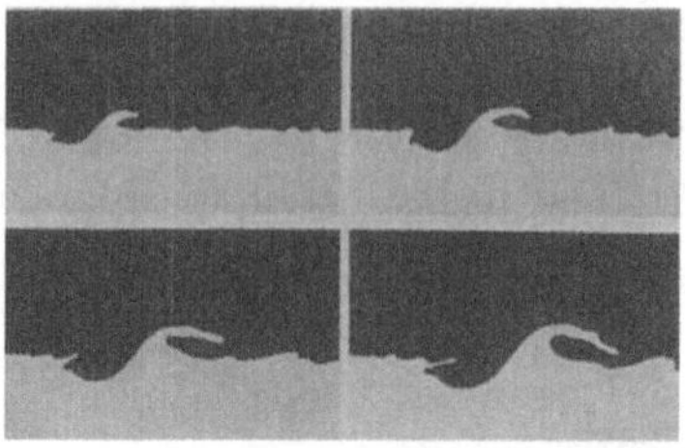

Fig. 2. SPH simulation of the Kelvin-Helmholtz instability for an $\mathcal{M} = 1$, $\eta = 0.1$ stream moving relative to a stationary stream. (The dark stream is moving; the light stream is stationary). Note the appearance of large scale structure which would have to be resolved at the surface of a turbulent jet.

6 Application of the Semi-Empirical Model to a Sample of Class-I Jets

6.1 Fits to the Bologna Sample

The semi-empirical model is based upon conservation laws, obtained by integrating across the jet and using not very restrictive assumptions for the transverse profiles of dynamical variables (see Bicknell (1986b)). One important assumption (which can be justified for low Mach number flows) is that the evolution of the surface brightness is determined by adiabatic effects (the compression of the particle distribution and the perpendicular component of magnetic field). This assumption needs to be relaxed for higher Mach numbers, and the dissipative effects of turbulence are yet to be included in the model. Nevertheless, the model has been applied by Bicknell *et al.* (1990) to a sample of jets observed by the Bologna group (Fanti *et al.* 1987) with some interesting results.

There are a number of parameters in the model: β, the ratio of the galaxy virial temperature to the ISM temperature, η, the ratio of jet to ISM density, $\mathcal{M}$, the jet Mach number, r_c, the core radius of the King potential (with dimensionless potential $W_0 = 10$) and $I_{\nu,0}$ the normalizing surface brightness. Despite the number of parameters, the model is mainly sensitive to β and to some extent η. Acceptable fits are obtained for a wide range of values of the other parameters, except that generally the best fit parameters fall within a reasonable range. (For example, the range of Mach numbers is generally between 1 and 3.)

Perhaps one of the best indicators of the success of this model fitting is the distribution of the parameter β (Fig. 3). The distribution is very similar to that obtained for an independent sample of X-ray ellipticals by Thomas *et al.* (1986).

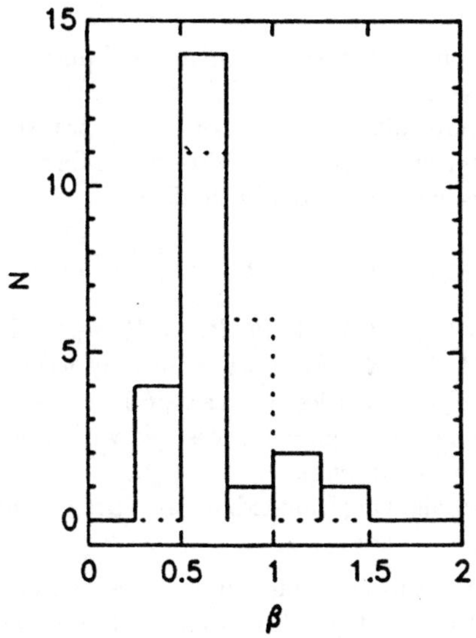

Fig. 3. The distribution of the parameter β from the fit of the semi-empirical turbulent jet model to the Bologna data (solid line), and the distribution of this parameter from the fits of Thomas *et al.* (1985) to X-ray data from an independent sample of ellipticals (dashed line).

6.2 The Energy Budget

It has been shown (Bicknell 1986a) that the radiative output from the lobe tends to equilibrium with the relativistic energy flux in the jet (moderated to some extent by expansion losses) on a radiative time scale. The energy flux calculated from the best fit parameters is more or less uniformly distributed over the range $0 - 1$ indicating a broad consistency between the model and the energy requirements of the source.

6.3 High Mach Number Subset

One interesting result from applying this model is the identification of a subset of 8 jets which can be equally well fit with a range of Mach numbers $1 \leq \mathcal{M} \leq 10$. These are more well collimated than the rest of the sample and have a lower density ratio. They seem to be a transition sample of class I jets with some of the characteristics of class II jets.

7 Summary

It is evident from this review that the mathematical and physical models which are applied to class I and class II jets are quite different. As stated in the introduction, this reflects very much the quite different mode of energy transport in the two classes of source. Another parameter which also highlights the differences between the two types of jet is the velocity. This can be written in the form:

$$v_{\rm jet} = 430\,{\rm km\,s^{-1}}\,\mathcal{M}\,\eta^{-1/2}\,T_{\rm is,7}^{1/2}\,(P/P_{\rm ism})^{1/2}$$

For class I jets with $\mathcal{M} \sim 1$, $\eta \sim 0.1 - 1$ and $P_{\rm jet} \approx P_{\rm ism}$ jet velocities are in the vicinity of $1,000 - 10,000\,{\rm km\,s^{-1}}$. For class II jets it is usually assumed that $\eta \lesssim 0.01$. With $\mathcal{M} \sim 10$, say, and $P_{\rm jet} \gtrsim P_{\rm ism}$ velocities in excess of 0.1 c are not difficult to achieve (all the more so if jets are hypersonic and overpressured with respect to the ISM, confined by the pressures of their own cocoons).

Let me finally give the following (non-exhaustive) list of some key directions and key questions for jet physics:

- Higher resolution 3D simulations, including large-eddy or direct simulations.
- Three dimensional studies of jet disruption and the transition to turbulence in jets: Norman, Burns and Sulkanen (1988) have shown that the passage of a 2D jet through a shock can disrupt it. However, it may be more appealing to make use of the larger number of unstable modes in 3D to drive a jet into instability, hence turbulence and/or disruption.
- The inclusion of particle acceleration in conjuction with MHD.
- The inclusion of more general **B** in both axisymmetric and 3D calculations.
- The origin of parallel field in class II sources (and in class I sources close to the base of the jet).
- If BL-Lacs are beamed class I, what causes the deceleration of the initially relativistic jets from the pc to the kpc scale and where exactly does it occur? Can this occur without intense dissipation?
- Is the transition from two-sided behaviour to one-sided jets across the Fanaroff-Riley break a consequence of relativistic beaming or some intrinsic effect?

References

Begelman, M.C., Cioffi, D.F.: 1989, *Astrophys. J. (Lett.)* **345**, L21.
Bicknell, G.V.: 1984, *Astrophys. J.* **286**, 68.
——: 1986a, *Astrophys. J.* **300**, 591.
——: 1986b, *Astrophys. J.* **305**, 109.
Bicknell, G.V., deRuiter, H., Fanti, R., Morganti, R., Parma, P.: 1990, *Astrophys. J.* **354**, 98.
Bicknell, G.V., Melrose, D.B.: 1982, *Astrophys. J.* **262**, 511.
Blandford, R.D., Rees, M.J.: 1974, *Mon. Not. R. astr. Soc.* **169**, 395.
Bridle, A.H.: 1984, *Astron. J.* **89**, 979.
Clarke D., Norman, M.L., Burns, J.O.: 1986, *Astrophys. J. (Lett.)* **311**, L63.
——: 1989, *Astrophys. J.* **342**, 700.

Clarke, D.A., Bridle, A.H., Burns, J.O., Perley, R.A., Norman, M.L.: 1992, *Astrophys. J.* **385**, 173.
Cox, C.I., Gull, S.F., Scheuer, P.A.G.: 1991, *Mon. Not. R. astr. Soc.* **252**, 558.
Davis, R.J., Muxlow, T.W.B., Conway, R.G.: 1985, *Nature* **318**, 343.
Fanaroff, B.L., Riley, J.M.: 1974, *Mon. Not. R. astr. Soc.* **167**, 31P.
Fanti, C., Fanti, R., de Ruiter, H.R., Parma, P.: 1987, *Astron. Astrophys. Suppl. Ser.* **69**, 57.
Jackson, N., Browne, I.W.A., Shone, D.L., Lind, K.R.: 1990, *Mon. Not. R. astr. Soc.* **244**, 750.
Killeen, N.E.B., Bicknell, G.V., Carter, D.: 1986, *Astrophys. J.* **309**, 45.
Killeen, N.E.B., Bicknell, G.V., Ekers, R.D.: 1986, *Astrophys. J.* **302**, 306.
———: 1988, *Astrophys. J.* **325**, 180.
Königl, A., Choudhuri, A.R.: 1985a, *Astrophys. J.* **289**, 173.
———: 1985b, *Astrophys. J.* **289**, 188.
Kössl, D., 1989, in *Hot Spots in Extragalactic Radio Sources*, ed. K. Meisenheimer and H.-J. Röser, Heidelberg: Springer-Verlag, p. 225.
Linfield, R.: 1982, *Astrophys. J.* **244**, 436.
Lind, K.R., Payne, D.G., Meier, D.L., Blandford, R.D.: 1989, *Astrophys. J.* **344**, 89.
Loken, C., Burns, J.O., Clarke, D.A., Norman M.L.: 1992, *Astrophys. J.* **392**, 54.
Matthews, A.P., Scheuer, P.A.G.: 1990a, *Mon. Not. R. astr. Soc.* **242**, 616.
———: 1990a, *Mon. Not. R. astr. Soc.* **242**, 623.
Norman, M.L., Burns, J.O., Sulkanen, M.: 1988, *Nature* **335**, 146.
Norman, M.L., Winkler, K-H.A., Smarr, L.L.: 1984, in *Physics of Energy Transport in Extragalactic Radio Sources*, ed. A.H. Bridle and J.A. Eilek, N.R.A.O, Greenbank,p. 150.
Payne, D.G., Cohn, H.: 1985, *Astrophys. J.* **291**, 655.
Scheuer, P.A.G.: 1974, *Mon. Not. R. astr. Soc.* **166**, 513.
———: 1982, in IAU-Symp. 97 *Extragalactic Radio Sources*, ed. D.S. Heeschen and C.M. Wade, Dordrecht: D. Reidel, p. 163.
Thomas, P.A., Fabian, A.C., Arnaud, K.A., Forman, W., Jones, C.: 1986, *Mon. Not. R. astr. Soc.* **222**, 655.
Williams A., Gull, S.F.: 1984, *Nature* **310**, 33.

Supersonic Extragalactic Jets with Turbulent Boundary Layers: Preliminary Consideration

Sergey S. Komissarov

Astro-Space Center
Lebedev Physical Institute
Leninsky prospect 53, 117924, Moscow B-333, Russia

Abstract: Numerical calculations are used to study the properties of stationary pressure-matched supersonic jets with boundary layers, propagating through interstellar and intergalactic mediums. It is shown that for low initial Mach numbers, $M = 1 - 2$, the jets become slow, fully turbulent, wide, and subsonic on a scale typical of FR I radio sources. These slow, distant parts of the jets could be identified with the edge-darkened lobes of FR Is. For slightly higher M the jets remain quite narrow and supersonic up to a distance of 150 kpc and more but still become fully turbulent. It is possible that the jets produce radio sources having a transitional morphology between classical FR Is and FR IIs. Edge-brightened lobes with extended 'warm spots' and jets that are more prominent and wider than those in FR IIs should be distinguishable features of these sources. A rise of brightness near the jet boundary seems to be possible only in the case of low initial contributions of relativistic electrons to the full jet pressure and effective particle acceleration in the turbulent layer.

1 Introduction

The well known global division of extragalactic radio sources into the classes FR I and FR II could be a result of the different dynamics of their jets (a fluid dynamical description is appropriate in each case). The main properties of FR II radio sources are well described by the laminar supersonic/hypersonic jet model (Blandford and Rees, 1974; Norman *et al.*, 1982; Loken *et al.*, 1991), and the peculiarities of FR I radio sources seem to be consequences of the dynamical importance of turbulent viscosity (Bicknell, 1984, 1986; Komissarov, 1985, 1990a, 1990b). Numerous investigations have been made of the dynamics of laminar extragalactic jets, and important progress has been achieved. In contrast, only the first steps in developing a turbulent jet model have been taken to date.

Bicknell (1984, 1986) developed the so-called integral approach to reconstruct a turbulent flow pattern using the observed radio brightness distributions of extragalactic jets. The main advantages of this approach are its simplicity (the final equations are algebraic) and the absence of a need to specify a model for the turbulent viscosity. Unfortunately, only self-similar flows with negligible dissipation can be described in this manner. Moreover, the reconstructed jet pattern depends strongly on the physical assumptions concerning the evolution of the radiating components — relativistic particles and magnetic fields. Further work on this problem is clearly needed. Komissarov (1988, 1990ab) used the Reynolds equations closed by a "modified" Prandtl expression for the turbulent viscosity. The main disadvantage of this and similar approaches is the use of a phenomenological model for the turbulent viscosity. In fact, the accuracy of the model should in principle be tested for every new kind of turbulent flow. However, because the limitations of the integral and differential approaches are quite different, the two methods may be considered to complement each other.

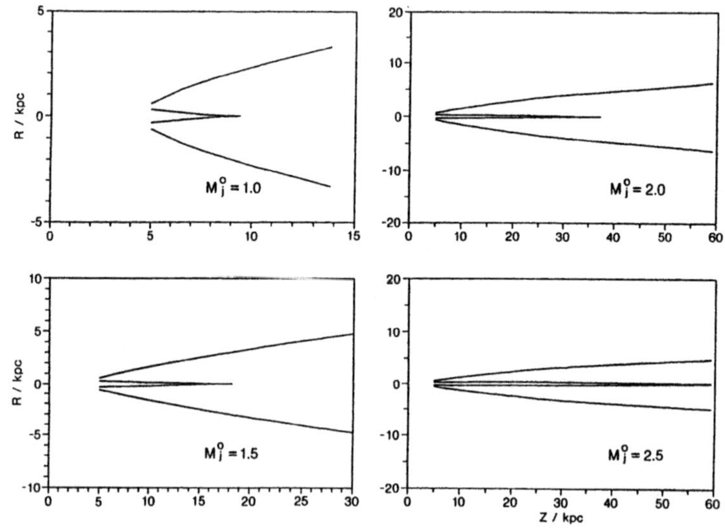

Fig. 1. Spreading of turbulent boundary layer for various values of the initial Mach number M_j^0.

Both authors considered only fully developed turbulent jets. However, jets with turbulent boundary layers may be more realistic, since significant limb-brightening has been observed in several carefully investigated jets (*e.g.* Reid *et al.*, 1989; Owen *et al.*, 1989; Clarke *et al.*, 1986). In this report the results of numerical calculations of stationary supersonic jets with turbulent boundary layers, propagating through intragalactic and intergalactic mediums, are described. The same approach and hydrodynamical code are used as in Komissarov (1990ab). One of the important approximations of the model is the assumption of pressure balance between the jet and the external gas: $P_j = P_{ext}$. This is a reasonable approximation for the case of jets which enter a uniform external medium and are initially pressure matched. However, if the external pressure decreases

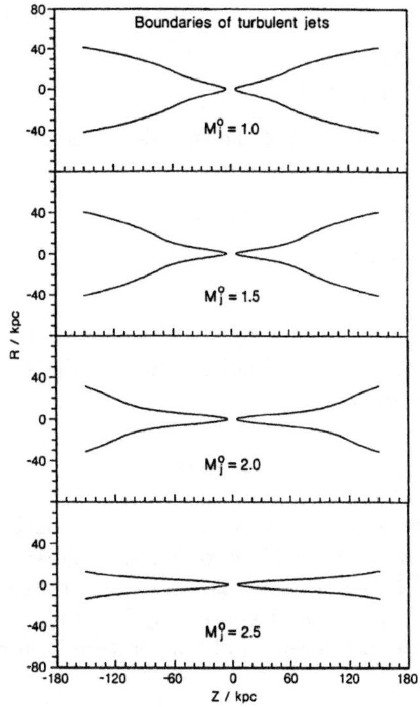

Fig. 2. Boundaries of steady turbulent jets at a scale larger than in Fig. 1

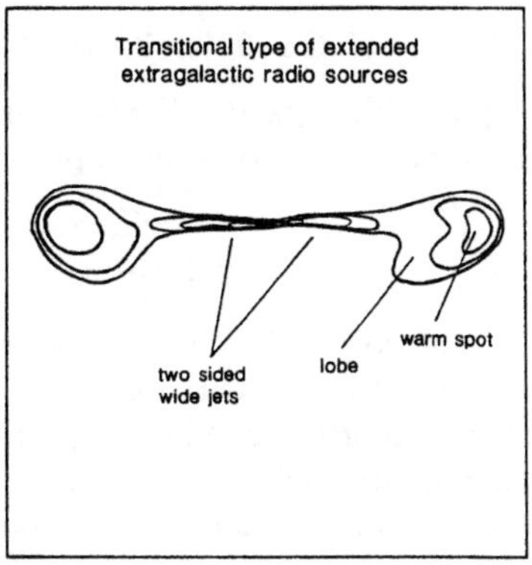

Fig. 2a. The expected morphology of a radio source supported by a supersonic turbulent jet.

in the direction of the jet, the balance may be destroyed. Wilson and Falle (1985) have examined this problem for laminar jets and found a pattern of internal stationary shocks produced by jet reconfinement. Similar behaviour might also be expected for jets with boundary layers. Indeed, bright knots are often observed in FR I jets. Therefore, the assumption of pressure balance is rough and the results described below must be considered as preliminary.

2 Parameters of Numerical Models

2a Turbulent Viscosity Model

The standard mixing length model by Prandtl was successfully developed for incompressible flows. To describe supersonic turbulent jets the model may be modified by introducing a dependence of the mixing length, l, on the Mach number, M:

$$E_t = l^2 |\partial U_z / \partial r|; \qquad l = a(M^c) \cdot R,$$

where E_t is the turbulent viscosity, U_z is the z-component of the mean velocity and R is the width of a turbulent layer (Abramowich, 1984). Departing from the model I employed in previous papers, I use here the following expression for the function $a(M)$:

$$a(M) = \begin{cases} 0.06 \text{ for } M < 1, \\ 0.06/M \text{ for } M > 1. \end{cases}$$

For supersonic jets this dependence leads to a saturation of the turbulent velocity at subsonic levels. Indeed,

$$v_t \approx l(\partial U_z/\partial r) \approx l \cdot U_z^c/R \approx a(M^c) \cdot M^c \cdot V_s = const \cdot V_s,$$

where v_t is the turbulent velocity, V_s is the sound speed.

2b External Gas Distribution

For gas inside a galaxy, the following static isothermal distribution satisfying the data from X-ray observations of elliptical galaxies is used:

$$\left.\begin{array}{l} \rho(z) = \rho^o(1 + (z/z^o)^2)^{-3/4} \\ T = 10^7\text{K} \end{array}\right. \qquad \text{for } z < z^g = 25\text{kpc}$$

where the core radius $z^o = 1$ kpc, and the central density $n^o = 10^{-1}\text{cm}^{-3}$. For larger distances, one should think of the external gas as intergalactic rather than galactic. Since the distribution of such gas has a typical length scale which is much larger than a galactic atmosphere, the uniform model, ρ = const, T = const, may be used as a first approximation for $z > z^g$. Continuity of the gas density and temperature is then assumed at the transition point z^g.

2c Initial Parameters of Jets

The parameters were chosen mainly to satisfy the observational constrains on the physical parameters of FR I radio sources. Therefore, most of them are listed below in dimensional form:

Input distance	$z = 5$kpc
Jet radius	$R = 0.6$kpc
Core radius	$R_c = 0.3$kpc
Velocity ($r = 0$) ([5])	$U_z^o = 10^{10}$cm/s ([6])
Mach number ($r = 0$)	$M^o = 1 - 2.5$
Density contrast ($n = \rho_j/\rho_{ext}$)	$n = (1 - 6) \cdot 10^{-4}$
Kinetic energy flux	$T^o = 2 \cdot 10^{41} - 4 \cdot 10^{42}$erg/s
Nonthermal energy flux	$Q^o = 5 \cdot 10^{40} - 5 \cdot 10^{41}$erg/s
Cosmic ray pressure	$P_c^o = (.1 - .9) \cdot P_{tot}^o$
Efficiency of cosmic ray gas heating	$\lambda = 0.0 - 0.9$

[5] 'r' = radial cylindrical coordinate.

[6] The upper indices 'o' and 'c' correspond to the values of a variable on the symmetry axis at the initial section and downstream respectively.

3 Results

3a Dependence upon Mach Number

Figure 1 shows the calculated evolution of a turbulent boundary layer for models with different initial Mach numbers, M^o. It is evident that the jets become narrower, with a longer potential core when M^o increases. This property is obviously connected with the above mentioned dependence of the mixing length on the Mach number. An additional factor is the increase of M along the jet axis for $z < z^g$, due to the pressure gradient in the atmosphere of the galaxy (Fig. 3). The corresponding drop in the rate of turbulent spreading produces visible "pseudo- collimation" reflecting more the curvature of the turbulent zone boundary than curved flow lines. This is a possible interpretation of the frequently observed "collimation" of extragalactic jets at kiloparsec scale (see also Bicknell, 1986). For $z > z^g$ and $z > z^c$, where z^c is the length of the potential core, the behaviour is reversed. The decrease of the jet's Mach number caused by turbulent dissipation is followed by a gain in the rate of spreading: at larger distances the latter approaches the value for incompressible jets (Fig. 2).

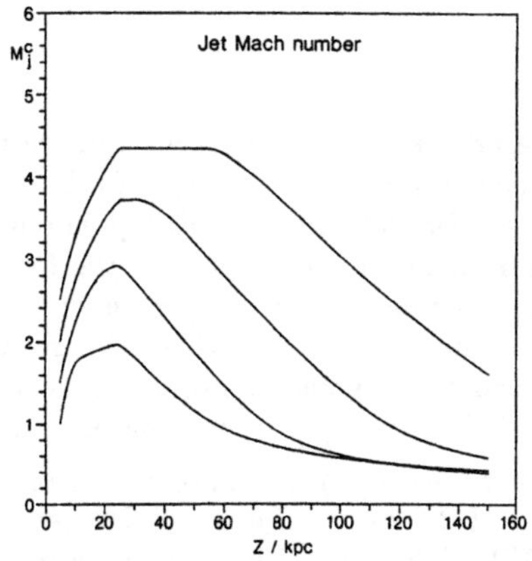

Fig. 3. Variation of Mach number along the axis of symmetry for various values of M_j^0.

In more detail, this process looks like decollimation with subsequent recollimation. Indeed, there is a maximum in the spreading rate near the point at which $M^c = 1$. Perhaps this is connected with the unusual profile of the momentum density which forms during propagation through the galactic atmosphere and gradually takes on a normal shape after entering the IGM (Fig. 4). It seems natural to identify the wide subsonic final parts of the jets considered here with the edge darkened lobes of some FR I radio sources.

It is evident that the length of the supersonic region increases with the initial Mach number. For $M^o = 2.5$ the calculated stationary jet remains quite narrow and supersonic up to $z = 150\,\mathrm{kpc}$ — the typical size of FR I lobes. Therefore, the interpretation of FR-I lobes as external subsonic parts of jet-like flows demands low values of M^o. For larger

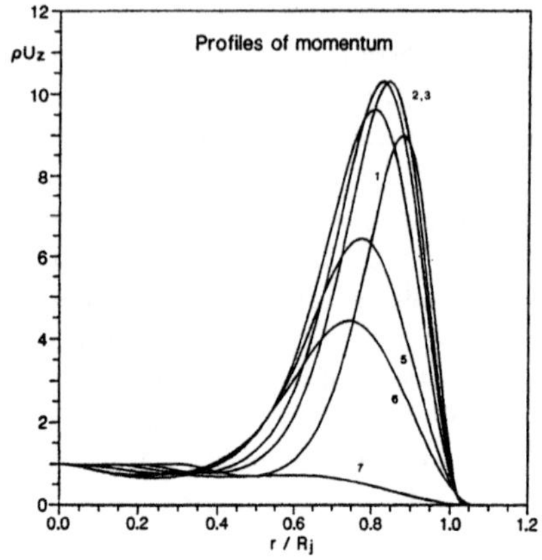

Fig. 4. Variation of the distribution of momentum in cross section of jet with distance from center of galaxy. The curves are shown for distances z_i exponentially spaced from $z_1 = 5\,\mathrm{kpc}$ to $z_7 = 150\,\mathrm{kpc}$.

M^o, time-dependent models describing the interaction between a supersonic turbulent jet and the IGM seem to be appropriate. Formation of a bow-shock and a cocoon may be expected for this case (see Fig. 2a). However, the spreading of fluxes over the much bigger cross-section of a turbulent jet should prevent the formation of a compact hot spot such as is seen in laminar jet models. Instead, one may expect much more extended features having a lower brightness ('warm spot'). For hypersonic jets — $M \gg 1$ — the dynamical importance of turbulence should be negligible and only thin turbulent layers are expected. This case probably corresponds to classical FR II sources.

One is thus led to propose the existence of a population of radio sources corresponding to the transition between FR I and FR II classes. Members of this population should have lobes reminiscent of FR II lobes but with extended 'warm spots', and jets reminiscent of FR I jets but narrower and less conspicuous. 1528+29, 1005+28, and 0908+37 (Parma *et al.*, 1987) look like such hypothetical sources. The linear sizes of these sources may depend upon the jet power and the age of the source.

Figures 5 and 6 show the evolution of the calculated jet velocity and density. It should be noted that for models with $M^o = 1 - 1.5$, the final velocity of the jets is less than $1 - 2 \cdot 10^3\,\mathrm{km/s}$ — a value typical of cluster velocity dispersions. The central density of the region is about one order of magnitude less than the external density. Therefore, strong distortions of the flow caused by the buoyancy force and the motion of the parent galaxy through the IGM should be expected.

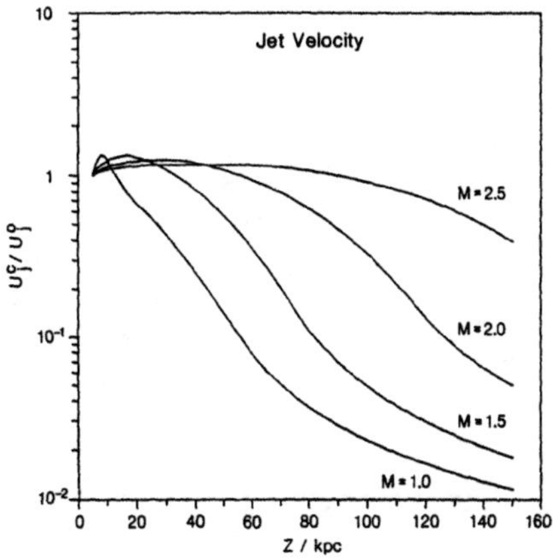

Fig. 5. Variation of jet velocity along the symmetry axis for models with various values of M_j^0.

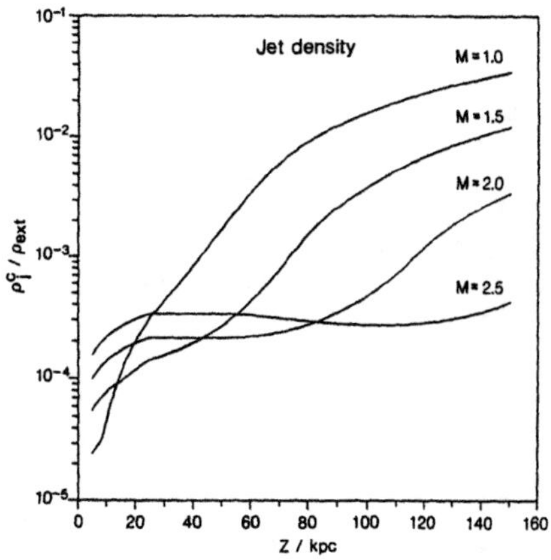

Fig. 6. Variation of density ratio along the symmetry axis for models with various values of M_j^0.

3b Dependence upon Efficiency of Particle Acceleration (λ)

The parameter λ introduced by Komissarov (1990a) gives the fraction of the kinetic energy of bulk motion which is dissipated as internal energy of the relativistic gas. The energy density and pressure of relativistic particles are the variables most sensitive to this parameter. Figure 7 shows the evolution of the relativistic gas pressure (P_c) for models with different λ and $\chi^o = P_c^o/P_{tot}$ — the initial ratio of relativistic to total gas pressure. Inside the potential core there is the simple adiabatic behavior:

$$\chi = P_c/P_{tot} = [1 + (P_c^o/P_g^o) \cdot (V/V^o)^\kappa]^{-1},$$

where $\kappa = (\gamma_c - \gamma_g)/(\gamma_c \gamma_g)$, P_g is the pressure of the thermal component and V is the volume. Outside the core region the evolution strongly depends on λ, and both flows (lobes) dominated by thermal pressure for low $\lambda(\lambda = 0.1, M^o = 2$ in Fig. 7) and flows dominated by relativistic pressure for high $\lambda(\lambda = 0.9, M^o = 2$ in Fig. 7) can be obtained. The two following factors are important to understand this behaviour. Firstly, the jet is accelerated by pressure forces inside the galaxy. Therefore, the energy contained in the random motion of both relativistic and thermal particles is converted into the kinetic energy of bulk motion. Secondly, because of turbulent dissipation, the energy of bulk motion is continuously converted back into internal energy, that part which is converted to relativistic particles being determined by λ. Therefore, even in the case of constant total flux of kinetic energy in bulk motion, a significant variation of χ is possible inside the galaxy.

The shape of the radial profile of χ is also sensitive to λ. The evolution of the profile for a model with $\lambda = 0.5$, $\chi^o = 0.5$, and $M^o = 2$ is presented in Fig. 8a (increase of a curve index corresponds to increase of distance from the input section). One can see that everywhere the profile has an almost "top-hat" shape with a little bump in the

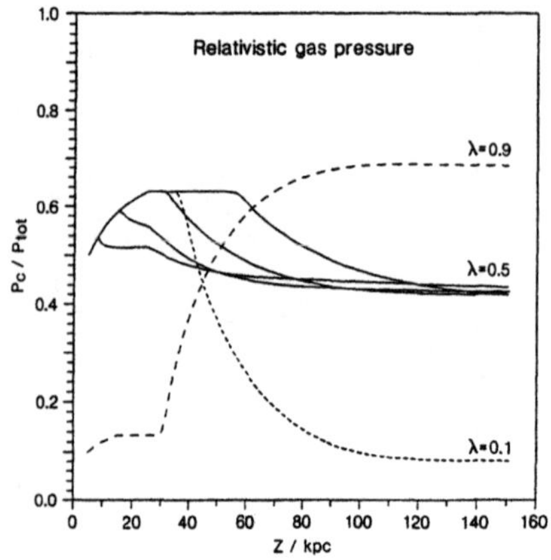

Fig. 7. Variation of pressure ratio (cosmic ray pressure / total jet pressure) along the symmetry axis for models with various values of λ.

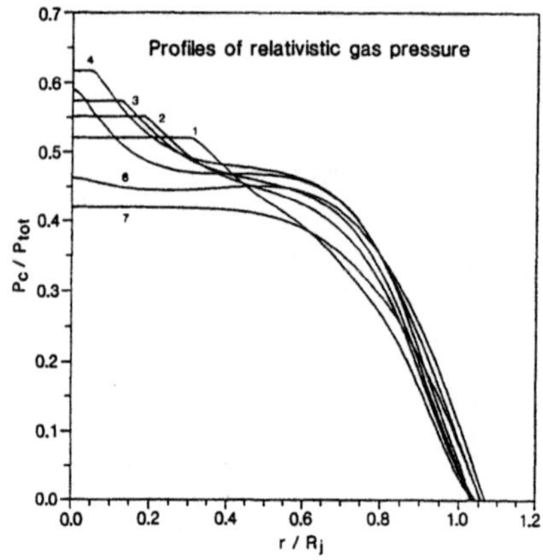

Fig. 8a. The variation of distribution of relativistic gas pressure in cross section of jet for a model with $\chi = 0.5$, $\lambda = 0.5$. The curves are shown for the same distances z_i as in Fig. 4.

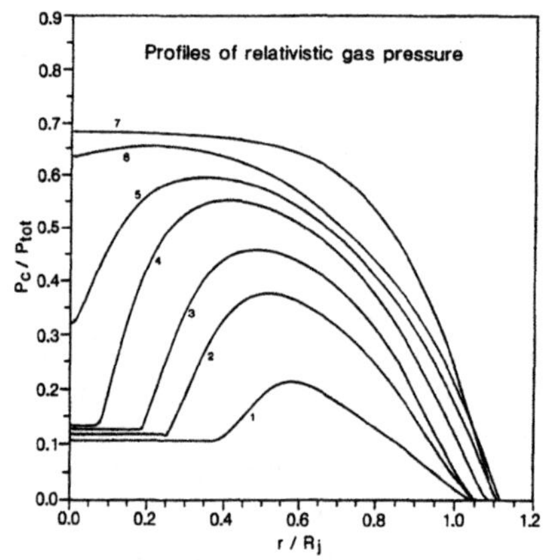

Fig. 8b. As in Fig. 8a but for a model with $\chi = 0.1$, $\lambda = 0.9$.

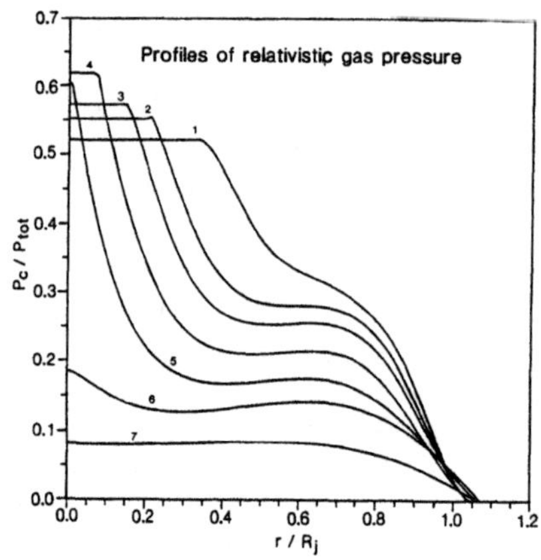

Fig. 8c. As in Fig. 8a but for a model with $\chi = 0.9$, $\lambda = 0.0$.

potential core region. A significant rise of relativistic particles energy density in the boundary layer can arise only for models with $\chi^o \ll \lambda$. The evolution of the profile for a model with $\lambda = 0.9, M^o = 2$, and $\chi^o = 0.1$ is presented in Fig. 8b. In contrast, in the opposite case, $\lambda \ll \chi^o$, a drop of P_c in the boundary layer is observed (Fig. 8c).

Thus the presence of a turbulent boundary layer does not guarantee that the distribution of relativistic particle energy will be suitable to explain the observed limb-brightened extragalactic jets. This fact could be used to obtain constrains on parameters of the jets. In principle, limb-brightening may also be caused by an exotic spatial distribution across the jet of the magnetic field or of the spectral index of relativistic electrons. However, these explanations look more artifical. In the case of relativistic jets, on the other hand, limb brightening is a natural consequence of relativistic beaming (Komissarov, 1990c).

References

Abramowich, G.N.: 1984, "Theory of turbulent jets", Mechanical engineering, Moscow.
Bicknell, G.V.: 1984, *Astrophys. J.* **286**, 68.
Bicknell, G.V.: 1986, *Astrophys. J.* **305**, 109.
Blandford, R.D., and Rees, M.J.: 1974, *Mon. Not. R. astr. Soc.* **169**, 395.
Clarke, D.A., Burns, J.O., and Feigelson, E.D.: 1986, *Astrophys. J.* **300**, 141.
Komissarov, S.S.: 1985, preprint FIAN,N 263.
——: 1988, *Astrofizika* **28(2)**, 261.
——: 1990a, *Astrophys. Space Sci.* **165**, 313.
——: 1990b, *Astrophys. Space Sci.* **165**, 325.
——: 1990c, *Sov. Astron. Lett.* **16(4)**, 284.
Loken, C., Burns, J.O., Clarke, D.A., and Norman, M.L.: this volume.
Norman, M.L., Winkler, K.-H.A., and Smith, M.D.: 1982, *Astron. Astrophys.* **113**, 285.
Owen, F.N., Hardee, P.E., and Cornwell, T.J.: 1989, *Astrophys. J.* **340**, 698.
Parma, P., Fanti, C., Fanti, R., Morganti, R., and de Ruiter, H.: 1987, *Astron. Astrophys.* **181**, 244.
Reid, M.J., Biretta, J.A., Junior, W. *et al.*: 1989, *Astrophys. J.* **336**, 112.
Wilson, M.J., and Falle, S.A.E.G.: 1985, *Mon. Not. R. astr. Soc.* **216**, 971.

VLBI Observations of Jets

J. A. Zensus

National Radio Astronomy Observatory
Socorro, NM 87801, USA

Abstract: VLBI images of compact radio sources show parsec-scale jets that extend continuously from the central radio core to the jets and extended emission found on arcsecond scales. Detailed imaging and monitoring show that these jets are curved and complex, and that superluminal motion is prominent in core-dominated objects.

1 Introduction

Many compact radio sources exhibit parsec-scale radio structure that can be studied with Very Long Baseline Interferometry (VLBI). Images from such observations often show features that can be identified as "jet-like", although sometimes the definition of a jet must be applied rather loosely (cf. the definition adopted by Patrick Leahy at this meeting). This is because "standard" VLBI images tend to have limited dynamic range (seldom better than 100:1), and tend to show only relatively little structural detail. However, there is now a small but growing number of sources of which high-dynamic range images have been made that show remarkably complex features in their parsec-scale jets. The highest fidelity in imaging of VLBI jets has been achieved at the intermediate wavelengths of 18 and 6 cm. At 18 cm, observations with a "World VLBI Array" have shown dramatically that pushed to its limits VLBI is able to produce image quality that reaches that of the VLA, but with three orders of magnitude higher resolution (*e.g.*, Reid *et al.*, 1989; Benson *et al.*, 1988). Apparent superluminal motion is still the most startling property of compact VLBI sources, providing direct evidence for relativistic motion, and substantial effort in current research is spent focusing on the relationship between this effect and the physics of the underlying continuous jets.

In this paper, I will not give a comprehensive review, but merely highlight some of the properties of VLBI jets and discuss results for several objects that are thought to be representative examples of parsec-scale jets. For further reading, see the reviews by Pearson (1990), Cawthorne (1991), and Witzel (1992), and the contributions of Krichbaum, Pearson, Fanti, Biretta, and Wardle at this meeting. For extensive treatment of the subjects of superluminal motion and parsec-scale radio jets, see Zensus and Pearson (ed., 1987 and 1990).

2 VLBI Monitoring of Compact Radio Jets

So far, still only a small number of sources have been extensively monitored with VLBI. However, systematic VLBI studies of samples of compact sources are underway that include different morphological classes, and in particular, BL Lac objects and OVVs: Witzel and colleagues (Witzel *et al.* 1988) are investigating a sample of 13 flat-spectrum sources from the MPIfR 5-GHz survey with $\delta > 70°$, $b_{\rm II} > 10°$, and $S_{5\,\rm GHz} > 1$ Jy. Pearson and colleagues (Pearson and Readhead 1988, and references therein) are studying a sample of 65 objects with $\delta > 35°$, $b_{\rm II} > 10°$, and $S_{5\,\rm GHz} > 1.3$ Jy. Cohen, Wehrle and colleagues are monitoring a sample of strong sources selected at 5 GHz (Wehrle and Cohen, 1989; Wehrle, Cohen, and Unwin 1990). Gabuzda, Wardle, Roberts and colleagues have been monitoring the polarization structure of a large sample of compact sources (Gabuzda and Cawthorne 1992; Roberts *et al.* 1990). Bååth and colleagues have been investigating a sample of BL Lac objects (Bååth, 1992).

The parsec-scale structures found in compact sources can be sorted into a small number of morphological classes (*e.g.*, Pearson and Readhead 1988): (a) the "core-jet" sources, that are characterized by a compact flat-spectrum core, and a typically one-sided steep-spectrum jet; (b) the "compact steep-spectrum" sources that have a steep spectrum below 5 GHz and a (relatively) small overall size $\lesssim 15$ pc; (c) the "compact double" sources containing two separated regions of emission of comparable flux density and spectrum. In addition to this double morphology, this group is characterized by gigahertz-peak component spectra and faint galaxy identification. A new distinct class may be the "compact triples" (see Pearson, this volume) that may be similar to the compact doubles, but with a prominent core.

3 Superluminal Motion

The systematic studies show that apparent superluminal motion is a common phenomenon in quasars and BL Lac objects (*e.g.*, Zensus 1989; Witzel *et al.* 1988; Pearson and Readhead 1988). Among the morphological groups, it occurs most frequently in core-jet sources. It is also seen in compact steep-spectrum sources, but never in compact-double objects. The motions always appear as expansion, *i.e.*, outward, never as contraction. They typically occur aligned with or even along an underlying smooth jet. Different velocities have been measured for individual components in some sources, as well as accelerations and decelerations. The observed tracks can be curved, and in some cases appear to differ within a source. The observed speed may of course reflect a "phase velocity" of the brightness peak within a jet, not a physical speed.

A comment on the difference between quasars and BL Lac objects: there appears to be a trend towards slower motion in BL Lac objects as compared to quasars and OVVs, but this still needs to be confirmed (see Mutel 1990). On average, BL Lacs and OVVs have velocities of $\beta_{\rm app}h = v/c \sim 3$–5, whereas for quasars, the average is $\beta_{\rm app}h \sim 6$ (*e.g.*, Zensus 1989). The work of Gabuzda and colleagues (Gabuzda and Cawthorne 1992) especially will be important in this respect. Such a difference in speeds has been explained by a particularly close orientation along the line-of-sight (*e.g.*, Impey 1992; cf. Mutel 1990). The evidence for this hypothesis is still inconclusive. However, note that

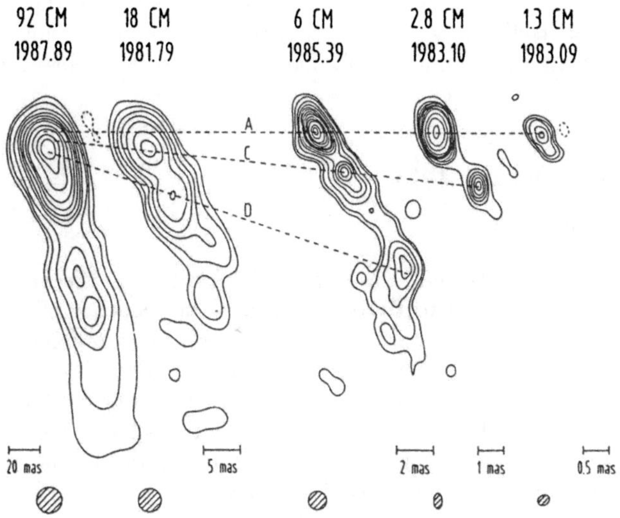

Fig. 1. VLBI images of the jet in the quasar 0836+71 at different wavelengths and epochs (Hummel and Krichbaum 1990). The lines connect the moving features as they were registered by the authors. Note the increase in angular scale.

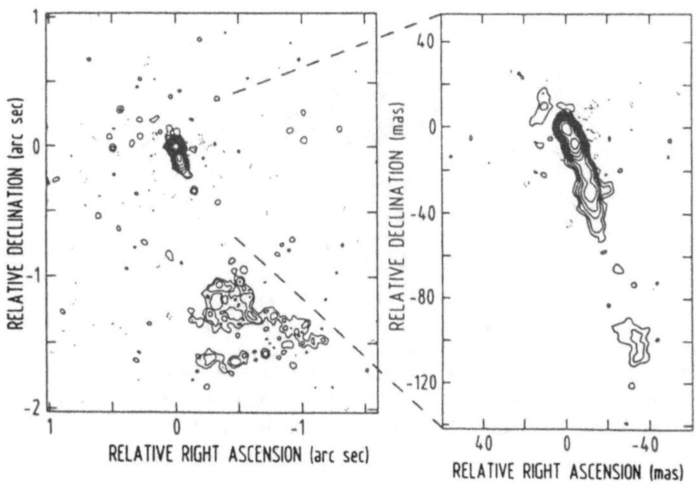

Fig. 2. Combined MERLIN-VLBI images of 0836+71, observed at 18 cm in 1988.7 (Hummel, in preparation).

(a) the redshift distribution of BL Lacs appears to differ from that of other superluminal sources (but see the results of Impey on HPQs/LPQs (1992); (b) the angle to the line of sight for BL Lac itself is large; (c) less core-dominated sources should be faster; but 2007+77 is the least core-dominated BL Lac object and lies below the average with $\beta_{\rm app} h = 2.3$.

4 Examples of VLBI Jets

a The Quasar 0836+71

A good example for the typical range of detail obtained at the various frequencies at which VLBI imaging is currently being done, is the sequence of images of the jet in the high-redshift ($z = 2.16$) quasar 0836+71 by Hummel and coworkers (Figure 1; Hummel and Krichbaum 1990). The source exhibits a continuous jet that is traced out to 160 mas from the bright core. Combining the data from a true "World VLBI Array" with data from the MERLIN interferometer, the authors have recently obtained the image shown in Figure 2 (Hummel *et al.*, in preparation), tracing the jet structure over a scale extending from a few milliarcsec to 200 arcsec!

b The Quasar 3C 345

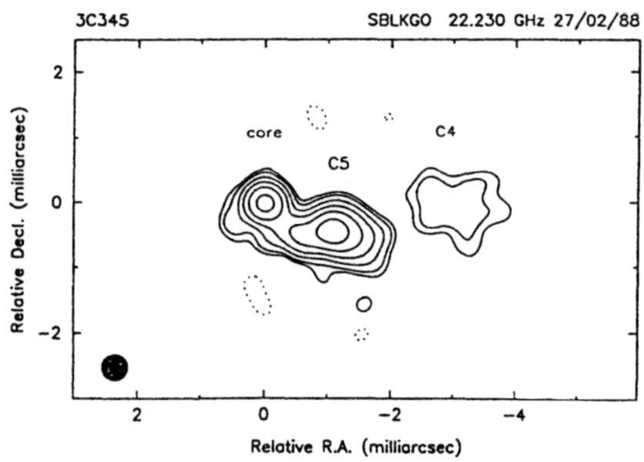

Fig. 3. VLBI image at 22 GHz of the superluminal quasar 3C 345, in February 1988 (Zensus, Unwin and Cohen, in preparation). The image shows the compact (stationary) core, and the two superluminal features C4 and C5. The resolution is about 0.4 mas (FWHM of restoring beam).

The quasar 3C 345 ($m_b = 16$, $z = 0.595$) is a prototype of core-dominated superluminal radio sources and an excellent example of a parsec-scale jet (Zensus, Unwin and Cohen, in preparation). The milliarcsecond structure of this source contains a core that is unresolved at the highest frequencies and stationary. Multiple features are moving (at apparently superluminal speeds of $\beta_{\rm app}h \sim 1.4$–$9.5$, for $H_0 = 100h$ km s^{-1} Mpc^{-1}, $q_0 = 0.5$) from the core in the direction of the outer structure, and the best images have confirmed the existence of an underlying complex jet (Unwin and Wehrle, 1992). Extensive monitoring at cm wavelengths has provided a detailed record of the structural evolution. Observations at mm-wavelengths (Krichbaum at this meeting; see also Bååth *et al.*, 1992) provide the relatively new possibility of studying the evolution of the superluminal components in the immediate vicinity of the core.

At least 4 superluminal components have been followed in 3C 345 so far, and the mm-observations indicate that new components continue to emerge from the compact core. Component sizes and core-separations are frequency dependent. Accelerations are seen and the component sizes increase with core distance. Whereas the outer components appear to roughly follow the same path towards the outer jet, distinctly different trajectories are found for the inner components C4 and C5. These trajectories appear to be non-monotonically curved.

From the high-frequency (10 and 22 GHz) monitoring we have established that the jet in 3C 345 is not only curved, but that different superluminal features in the close vicinity of the stationary, unresolved core appear to be moving on different curved trajectories (Zensus, Cohen and Unwin, in preparation). At larger distances from the core, the older superluminal features appear resolved and fade away, but they seem to continue to move in the same general direction on a curved path towards the outer arcsecond-structure. It is tempting to analyze the trajectories of the superluminal features in terms of 3-dimensional, helical kinematic models (*e.g.*, Qian et al., in preparation), and to apply shock-in-jet models to the record of structural evolution.

An interesting recent result from an analysis of the monitoring of the superluminal motion in 3C 345 is that the speed of the superluminal motion appears to continuously increase with separation from the core (Zensus, Cohen and Unwin, in preparation). This could reflect a continuous change in orientation of the source with respect to the line-of-sight or instead true acceleration.

Recent observations of 3C 345 at 6 cm have led to the production of two images with exceptional dynamic range (Unwin and Wehrle, 1992; see Figure 4). These not only show the previously known superluminal features, consistent with their motion as determined at higher frequencies, but also reveal an extended component at a radius of about 18 mas with rather unusual morphology: they show an apparently flat surface facing the core where the visible jet flares, possibly as the result of a shock, as apparent in the jet of M 87. The older superluminal features clearly appear to form a continuous jet, and the identification of "components" is becoming increasingly difficult with greater distance from the core.

c The Quasar 3C 273

The quasar 3C 273 ($z = 0.158$) is another classic example of sources showing apparent superluminal motion (Zensus *et al.* 1990). VLBI observations have shown components in the narrow jet to separate from the core region with apparent speeds of $v/c = 5$–$7h^{-1}$ (for $H_0 = 100h$ km s^{-1} Mpc^{-1}, $q_0 = 0.5$). This superluminal motion can be traced to at least $46h^{-1}$ pc from the core at 6 and 18 cm (Zensus *et al.*, 1988, Unwin 1990). The jet is non-monotonically curved, extends continuously to more than 150 mas ($275h^{-1}$ pc)

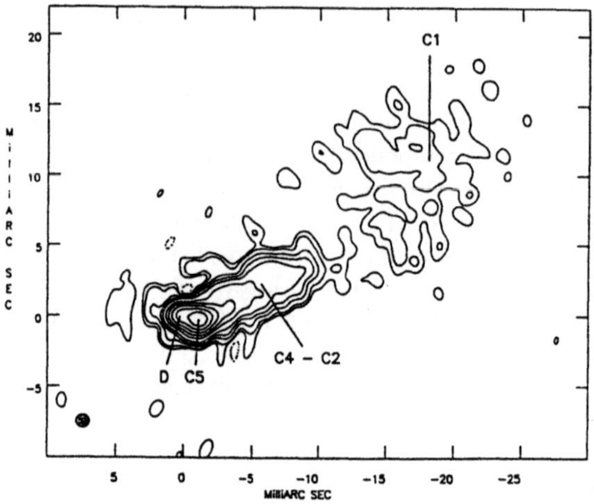

Fig. 4. 6 cm VLBI image of the jet in quasar 3C 345, at epoch 1989.28 (Unwin and Wehrle 1992). The dynamic range in this image is greater than 1000:1, and the angular resolution is 1 mas (4 pc).

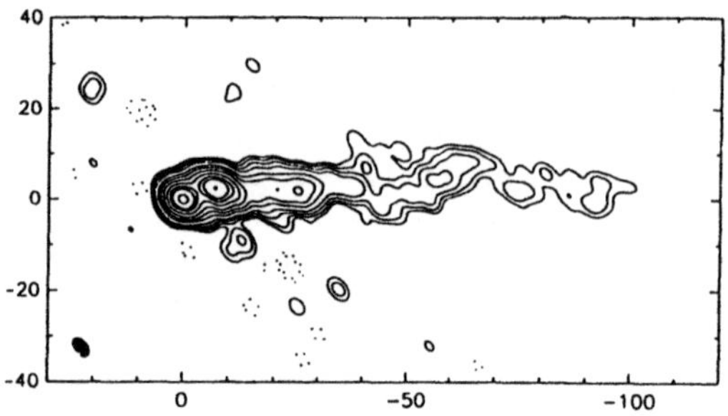

Fig. 5. Hybrid image of 3C 273 at 18 cm (1988.88), achieving an angular resolution of 3 mas, and with dynamic range better than 1000:1 (Unwin 1990).

from the core (Unwin 1990), and is closely aligned with the position angle of the 23″ jet.

Once again, the superluminal components appear to emerge from the core region coincident with major flux density increases at mm-wavelengths indicating a causal connection. The trajectories of components appear to be determined by the ridge line of the jet which undergoes repeated bending, probably amplified by projection and beaming effects. There is no significant evidence against one single path of the motion.

The compact structures seen in the VLBI images—especially the superluminal features—appear to be related to the regions of the jet that cause the variability in total flux density. Shock models, which have been successful in explaining total flux variability in other sources (*e.g.*, BL Lac) also offer an attractive interpretation of the structure of 3C 273. These models would interpret the moving components as disturbances in an underlying relativistic flow, rather than, *e.g.*, relativistically moving clouds. The observations so far are consistent with some of the predictions of such a scenario, although alternative models have not been ruled out.

d The Radio Galaxy Cygnus A

Carilli, Bartel, and Linfield (1992) have recently revisited the VLBI jet in the archetype of powerful radio galaxies, Cygnus A. They find that the jet is partially resolved and "knotty", extending to about 25 mas from the core, strongly collimated, and aligned with the kiloparsec-scale jet. There is no counterjet visible in their image, at the level of 1/15 times the flux density in the jet. They derive an upper bound of the expansion velocity of about $2c$ (Hubble constant of $60\,\mathrm{km\,s^{-1}\,Mpc^{-1}}$). Their observations do not contradict the notion that this source is oriented very closely to the sky plane.

5 Interpretation of the Source Structure

It is now widely accepted (although a prime subject of discussion at this conference) that the apparent superluminal motion and other properties of compact radio sources are a consequence of the presence of bulk relativistic motion in a plasma jet originating in the nucleus of the source (*e.g.*, Blandford and Königl 1979; Blandford 1987). In this picture, the "core" in VLBI images represents the optically thick base of the jet (located near the apex of the jet cone), and the superluminal "components" are regions of enhanced emission moving down the jet. The latter can be modeled as clouds of plasma in the jet which are by some mechanism accelerated to relativistic speeds. For example, this approach has been applied to 3C 345 and reasonable physical parameters consistent with observed properties can be derived (Zensus, Cohen, and Unwin, in preparation). In particular, the expected frequency dependence of the radial positions of components measured relative to the "core" (due to varying opacity of that region; *e.g.*, Königl 1981) is indeed seen in 3C 345.

When the axisymmetry is broken and the jet is curved, as seen in many core-jet sources, frequency-dependence of positions transverse to the jet can also occur. The sense of the effect, for optically-thin moving components, is that higher frequency images show the component peaks on the *outside* of the bend, relative to lower frequencies. In general, for jets which have greatly increased curvature at small radii (like 3C 273 and 3C 345) opacity effects will cause the jet to appear more curved at higher frequencies. This is separate from any effects of varying resolution.

Recently, the "shock-in-jet" models (which are based on the model of Blandford and Königl) have been remarkably successful in reproducing the radio variability behavior of several sources (cf. Marscher 1990, 1992; Daly and Marscher 1988; Hughes, Aller, and Aller 1989*a*,*b*). These include 3C 273: Marscher and Gear (1985) modeled the flux outburst that occurred in 1983. In these models, the underlying jet flow is assumed to be relativistic and confined to a cone of constant opening angle, and the physical conditions are functions solely of distance from the jet's apex. Variability of total flux and polarization, and moving or stationary features in the jet (but likely not the core itself) correspond to shocks which are caused by disturbances of the underlying steady jet flow. The magnetic field is thought to be random or to some degree parallel to the jet flow originally, but perpendicular in the regions of the shocks, owing to compression of the field structure, which has been confirmed with polarization VLBI in some sources (*e.g.*, Roberts *et al.* 1990).

A consequence of the shock models is that the Lorentz factor of the shocks can be different from that of the underlying flow, and it can vary between different shocks according to the strength of the disturbances (*e.g.*, Marscher 1990, 1992). From the measured apparent speeds β_{app}, we can derive only limits for the Lorentz factor γ and the component motion's angle to the line of sight θ (Pearson and Zensus 1987): $\gamma_{\mathrm{min}} = (1+\beta_{\mathrm{app}}^2)^{1/2}$, and $\cot(\theta_{\mathrm{max}}/2) = \beta_{\mathrm{app}}$. According to Marscher and Gear (1985), the lowest Lorentz factors which would be observed for the shocks correspond to those of the undisturbed flow. The actual values are expected to be higher in the early shock phases, although only by about a factor of two for the most extreme disturbances.

The shock model predicts that the superluminal components become resolved at centimeter wavelengths in the adiabatic phase of the shock's evolution (cf. Marscher and Gear). They are then expected to expand isotropically, and their Lorentz factors should decrease with time until they equal that of the flow, *i.e.*, the components should appear to decelerate as they separate from the core. The expansion of components is consistent with our images of superluminal sources, but deceleration is generally not observed; however, depending on the shock properties this may be explained by the fact that they can typically be followed only for 2–3 yr.

The "wiggling" observed in several compact radio sources could be due to precession of a decelerating jet as might be the case for the kpc-scale structure of 3C 273 (Davis *et al.* 1985). For large inclination angles, suitable magnetic field configuration (*e.g.*, Königl and Chouduri 1985) could cause the observed bends as well. Finally, Kelvin-Helmholtz instabilities from interaction with the environment and oblique reflection shocks may be relevant in this context (Hardee and Norman 1989, and the discussion at this meeting).

6 Conclusion

Recent work has shown that VLBI images have begun to approach the quality we have come to expect from the VLA and MERLIN: the advent of the Very Long Baseline Array in early 1993 can be expected to enable us to make routinely high-dynamic range VLBI images of a large number of sources.

Acknowledgements: I am grateful to the hosts of the conference for support. The National Radio Astronomy Observatory is operated by Associated Universities, Inc. under cooperative agreement with the U. S. National Science Foundation.

References

Parsec-Scale Radio Jets, eds. J.A. Zensus and T.J. Pearson, 1990, Cambridge: Cambridge University Press.

Physics of Active Galactic Nuclei, eds. S. Wagner and W.J. Duschl, 1992, Heidelberg: Springer Verlag.

Superluminal Radio Sources, eds. J.A. Zensus and T.J. Pearson, 1987, Cambridge: Cambridge University Press.

Variability of Blazars, eds. E. Valtaoja and M. Valtonen, 1992, Cambridge: Cambridge University Press.

Bååth, L.B.: 1992, in *Variability of Blazars*, op. cit., p.196.

Bååth, L.B., *et al.*: 1992, *Astron. Astrophys.* **257**, 31.

Benson, J.M., Walker, R.C., Unwin, S.C., Muxlow, T.W.B., Wilkinson, P.N., Booth, R.S., Pilbratt, G., Simon, R.S.: 1988, *Astrophys. J.* **334**, 560.

Blandford, R.D.: 1987, in *Superluminal Radio Sources*, op. cit., p.310.

Blandford, R.D., Königl, A.: 1979, *Astrophys. J.* **232**, 34.

Carilli, C.L., Bartel, N., Linfield, R.: 1992, *Astron. J.* **102**, 1691.

Cawthorne, T.V.: 1991, in *Beams and Jets in Astrophysics*, ed. P.A. Hughes, Cambridge: Cambridge University Press, p.187.

Daly, R.A., Marscher, A.P.: 1988, *Astrophys. J.* **334**, 539.

Davis, R.J., Muxlow, T.W.B., Conway, R.G.: 1985, *Nature* **318**, 343.

Gabuzda, D.C., Cawthorne, T.V.: 1992, in *Variability of Blazars*, op. cit., p.238.

Hardee, P.E., Norman, M.L.: 1989, *Astrophys. J.* **342**, 680.

Hughes, P.A., Aller, H.D., Aller, M.F.: 1989a, *Astrophys. J.* **341**, 54.

——: 1989b, *Astrophys. J.* **341**, 68.

Hummel, C.A., Krichbaum, T.K.: 1990, in *Parsec-Scale Radio Jets*, op. cit., p.38.

Impey, C.D.: 1992, in *Variability of Blazars*, op. cit., p.55.

Königl, A.: 1981, *Astrophys. J.* **243**, 700.

Königl, A., Chouduri, A.R.: 1985, *Astrophys. J.* **289**, 188.

Marscher, A.P.: 1990, in *Parsec-Scale Radio Jets*, op. cit., p.236.

——: 1992, in *Physics of Active Galactic Nuclei*, op. cit., p.510.

Marscher, A.P., Gear, W.K.: 1985, *Astrophys. J.* **298**, 114.

Mutel, R.L.: 1990, in *Parsec-Scale Radio Jets*, op. cit., p.98.

Pearson, T.J.: 1990, in *Parsec-Scale Radio Jets*, op. cit., p.1.

Pearson, T.J., Readhead, A.C.S.: 1988, *Astrophys. J.* **328**, 114.

Pearson, T.J., Zensus, J.A.: 1987, in *Superluminal Radio Sources*, op. cit., p.1.

Reid, M.J., Biretta, J.A., Junor, W., Walker, R.C., Unwin, S.C., Muxlow, T.W.B., Spencer, R.E.: 1989, *Astrophys. J.* **336**, 112.

Roberts, D.H., Wardle, J.F.C., Brown, L.F., Gabuzda, D.C., Cawthorne, T.V.: 1990, in *Parsec-Scale Radio Jets*, op. cit., p.110.

Unwin, S.C.: 1990, in *Parsec-Scale Radio Jets*, op. cit., p.13.

Unwin, S.C., Wehrle, A.E.: 1992, *Astrophys. J.*, in press.

Wehrle, A.E., Cohen, M.H.: 1989, *Astrophys. J. (Lett.)* **346**, L69.

Wehrle, A.E., Cohen, M.H. Unwin, S.C.: 1990, in *Parsec-Scale Radio Jets*, op. cit., p.49.

Witzel, A.: 1992, in *Physics of Active Galactic Nuclei*, ed. S. Wagner and W. J. Duschl, Heidelberg: Springer Verlag, p.484.

Witzel, A., Schalinski, C.J., Johnston, K.J., Biermann, P.L., Krichbaum, T.P., Hummel, C.A., Eckart, A.: 1988, *Astron. Astrophys.* **206**, 245.

Zensus, J.A.: 1989, in BL Lac Objects, ed. L. Maraschi, T. Maccacaro, and M.-H. Ulrich, Heidelberg: Springer Verlag, p.1.
Zensus, J.A., Bååth, L.B., Cohen, M.H., Nicolson, G.D.: 1988, *Nature* **334**, 410.
Zensus, J.A., Unwin, S.C., Cohen, M.H., Biretta, J.A.: 1990, *Astron. J.* **100**, 1777.

Parsec Scale Jets in FR I Radio Galaxies

Gabriele Giovannini [1,2], Tiziana Venturi [2],

Luigina Feretti [1,2], Gianni Comoretto [3], Ann E. Wehrle [4]

[1]Dipartimento di Astronomia, via Zamboni 33, I-40126 Bologna, Italy
[2]Istituto di Radioastronomia, via Irnerio 46, I-40126 Bologna, Italy
[3]Osservatorio Astrofisico di Arcetri, Largo E. Fermi 5, I-50125 Firenze, Italy
[4]California Institute of Technology, Pasadena, CA 91125, USA

Abstract: We present new VLBI observations obtained with the European VLBI network (EVN) and US network (USN) of the two FR I radio galaxies NGC 315 and 3C 338. They belong to a complete sample of low-intermediate power radio galaxies selected at 6 cm, which is presently under study on the parsec scale. NGC 315 shows a one-sided parsec scale jet, while 3C 338 is two-sided. We present these new data and briefly summarize the properties of parsec scale jets in FR I radio galaxies.

1 Introduction

High luminosity radio sources (FR II) show a parsec scale structure strictly connected with the kpc scale one. They are, in fact, characterized by a small scale one-sided jet, aligned with the large scale jet, and on the same side of the nucleus. The behaviour of low-intermediate luminosity radio galaxies FR I is, however, still puzzling. In the few cases, which have so far been well studied with the VLBI, the structure is found to be essentially one-sided, or at least strongly asymmetric, while kpc scale jets are mostly two-sided (even if in some cases a jet counter-jet asymmetry is present, see Parma *et al.* 1987) and with velocities of the order of thousands km/sec (Bicknell *et al.*, 1990). From the unified scheme models, if jets are relativistic and intrinsically symmetric, with a uniform distribution of angles to the line-of-sight, we expect to see a few sources with two-sided parsec scale jets. Throughout, we use H_0=100 km s^{-1} Mpc^{-1}.

2 NGC 315

NGC 315 is optically classified as a dusty elliptical (z=0.0167) The major axis of the optical galaxy is perpendicular to the major axis of the radio emission. Its total radio power at 408 MHz is 8.9 $\times 10^{23}$ W/Hz. The arcsec/arcmin scale radio structure of NGC 315 is well known. It is a very large, two-sided radio galaxy ($\sim 1°$ in the sky), with the radio axis aligned at position angle P.A. $\sim -45°$ (Bridle *et al.*, 1979). On the arcmin scale the northwestern jet is stronger and more collimated than the southeastern one (Fig. 1).

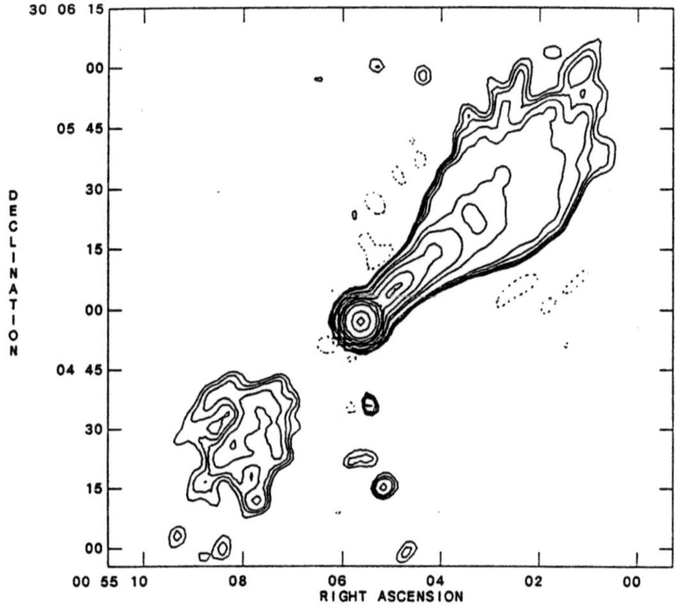

Fig. 1. VLA map at 6 cm of NGC 315. The main jet and some emission from the counter-jet are visible.

We observed this radio galaxy at 1.6 GHz, 5 GHz, 8.4 GHz with a global VLBI array (respectively 12, 10 and 8 telescopes from the EVN and USN) to map the milliarcsecond structure at different frequencies and resolution. The MarkII recording system was used at each frequency at all stations. In addition, simultaneous MarkIIIB recordings were made at 5 GHz in the 5 EVN stations and Owens Valley.

The milliarcsecond morphology of NGC 315 is a core-jet, with the jet in position angle $-45° \pm 1°$, well aligned with the strongest arcsec/arcmin jet. The total extent of the VLBI radio emission, and therefore the length of the mas jet, is strongly dependent on the frequency. At 18 cm it is ~ 40 mas, which corresponds to a projected distance of 10 pc from the core. The jet presents many knots of enhanced radio emission (see Fig. 2). No indication of radio emission on the opposite side of the jet is present at the noise level. At 6 cm, the jet has an extension of $\sim$ 10 mas with a knot at $\sim$ 8 mas (Fig. 3). At 3.6 cm only the core and the very beginning of the jet are visible.

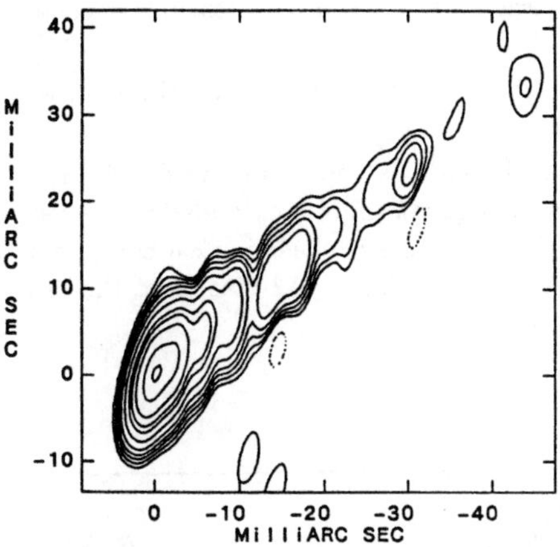

Fig. 2. VLBI map of NGC 315 at 18 cm. The resolution is 8.7×2.4 mas in PA − 11°

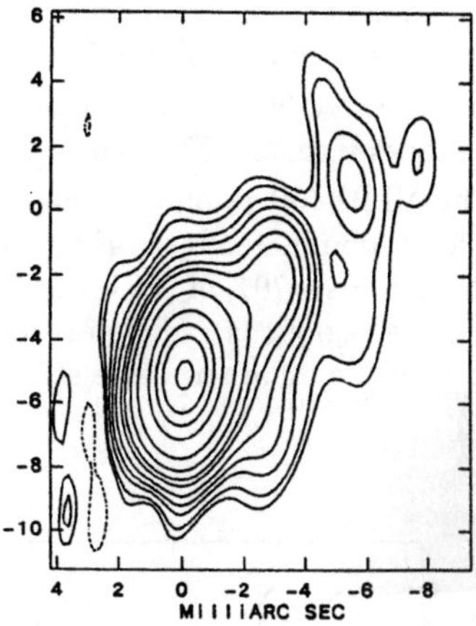

Fig. 3. VLBI map at 6cm of NGC 315. The resolution is 3×0.8 mas in PA − 4°

Comparing our maps at different frequencies we derive a slightly inverted spectrum for the core ($\alpha_{5.0}^{8.4} = 0.19$) and a jet spectral index in the first 10 mas, constant within the errors, with a value of −0.65.

A simple model to explain the asymmetric parsec scale structure is to assume an intrinsically symmetric two-sided jet with a relativistic velocity, oriented at a small angle to the line-of-sight. In this case, only the jet pointing toward us is visible for the Doppler boosting effect. The jet to counter-jet flux density ratio $R = S_j/S_{cj}$, derived at 18 cm, is $R \geq 50$. Assuming a jet spectral index $\alpha = -0.65$, we derived a limit for the product $\beta cos\theta \geq 0.63$, where β is the ratio of the jet flow to the speed of light, and θ is the angle to the line-of-sight. With such a constraint assuming a large value for β (0.8) only, we have an angle $\theta \sim 40°$. These values are very extreme in the contest of unified models and do not explain the jet asymmetry in the kpc scale structure where the jet velocity cannot be so high. Moreover, the angular size of NGC 315 would be more than 1.5° corresponding to 1.4 Mpc.

3 3C 338

This radio galaxy has a total radio luminosity at 408 MHz of 1.78×10^{25} W/Hz, and it is associated with the multiple-nuclei cD galaxy NGC 6166 (z=0.031), located at the center of the cooling flow cluster of galaxies A2199. The large scale structure is a symmetric relaxed double with an unusually steep spectral index. No jet-like feature is present between the core and lobes (Burns, 1983).

We observed 3C 338 at 6 cm on September 25, 1989 for 8 h, in MarkIIIB mode, with 9 telescopes from the EVN and USN.

The VLBI full resolution map obtained with AIPS is shown in Fig. 4. The radio emission is dominated by a bright source (C in the figure), slightly extended in P.A. 90°, which contains ~70% of the total VLBI flux. On both sides of the brightest component, two knotty regions of radio emission are detected, extending ~ 12 mas from the radio peak, with the main structure still along p.a. 90°.

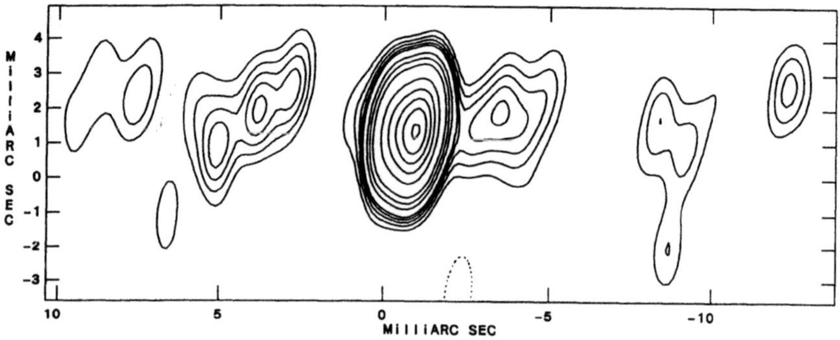

Fig. 4. Full resolution VLBI map of 3C 338 at 6 cm. The HPBW is 2.4 × 0.9 mas ($PA - 10°$). Contour levels are: −1, 1, 1.5, 2, 2.5, 3, 5, 7, 10, 20, 30, 40, 50, 60 mJy/beam

The structure within 5–7 mas from the strong central source is on both sides at the 10σ level, while the outermost blobs are at 5σ. The total linear extent of the structure is ~10 pc. The total flux density in this map is 133 mJy, which corresponds to ~90% of the flux density of the VLA arcsec core, measured at the same epoch (154 mJy, Ge, private communication).

In the low resolution VLBI map obtained by cutting the longest baselines (Fig. 5), the low brightness structure is better visible, thanks to the better signal to noise ratio. The structure is still elongated in p.a. ~90°, and the total extent is about 12 pc.

We altered the reduction procedure in many ways to see if a source with asymmetric structure is still a good representation of the data. In particular we forced the CLEAN on a one-sided box and used this model in the self-calibration routine. No satisfactory result was achieved. Even after many iterations the final map had a high r.m.s. and,

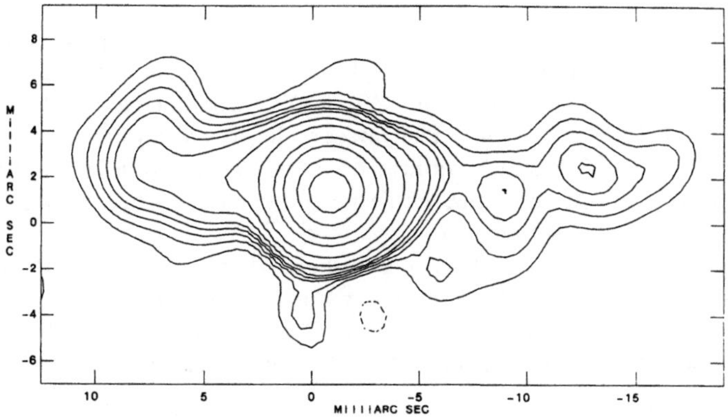

Fig. 5. VLBI contour map at 6cm of 3C 338 smoothed at the resolution of 3.2 mas

with respect to the cleaning box. some flux was still present on the opposite side of the nucleus.

3C 338 is the first case where we have two jets, very similar in strength and length. This implies that Doppler boosting is not important, leading to the conclusion that either the angle with the line of sight is close to 90° or the jets are non-relativistic. At present no information concerning the jet velocity can be obtained, nor can an estimate be made of this. A source orientation in the plane of the sky is supported by the symmetry detected on the large scale. Higher resolution observations are necessary to resolve the central component to separate the real core from the jet beginning.

4 Other FR I Radio Galaxies

Observations of FR I radio galaxies are scarce. Detailed maps are published of Virgo A; we observed NGC 2484 and 1144+35 (Giovannini *et al.*, in preparation). The first one shows an elongated structure with three blobs in the same direction of the two-sided kpc-jet. However, the lack of spectral index information does not allow us to derive unambiguous information on the parsec structure of this galaxy. If the core is the brightest component and if the radio structure is asymmetric for Doppler boosting, we have the following constraint: $\beta cos\theta \geq 0.5$. It implies $\beta \sim 0.7$ with an angle $\theta \sim 45°$. The radio galaxy 1144+35 shows a complex parsec structure with two compact components and an extended lobe. No jet-like feature is visible in our map, so more observations are necessary to understand the parsec structure of this source.

Bartel (private communication) observed 3C 66B and Bååth (private communication) 3C 465. Both galaxies show a one-sided parsec scale jet in the same position angle as the main kpc scale jet.

5 Conclusions

We have presented new VLBI data on 2 FR I radio galaxies. We have found:

a. A one-sided jet structure in NGC 315. It implies severe constraints to reconcile the parsec scale asymmetry with Doppler boosting: β has to be larger than 0.8 for angles to the line-of-sight larger than 40°. Similar asymmetric structures are present in Virgo A, 3C 66B and 3C 465. The parsec scale jet is always on the same side of the main kpc jet. If this result is confirmed by observations of a larger sample of FR I radio galaxies, it will imply that either pc and kpc scale jets are both relativistic or jets are intrinsically asymmetric.

b. A two-sided jet in 3C 338. This source shows that two-sided VLBI jets do exist, though rare. With the better sensitivity and uv-coverage now available for VLBI observations we should be able to find more two-sided sources and to test whether the proportion of sources with counter-jets is consistent with the unified scheme.

These data add new information to the parsec scale structure of faint radio galaxies, however, more observations are necessary to determine the velocity of parsec scale jets in FR I radio galaxies and to test the prevision of unified models.

References

Bicknell, G.V.: 1986, *Astrophys. J.* **305**, 109.

Bicknell, G.V., de Ruiter, H., Fanti, R., Morganti, R., Parma, P.: 1990, *Astrophys. J.* **354**, 98.

Bridle, A.H., Davis, M.M., Fomalont, E.B., Willis, A.G. Strom, R.G.: 1979, *Astrophys. J. (Lett.)* **228**, L9.

Burns, J.O., Schwendeman, E., White, R.A.: 1983, *Astrophys. J.* **358**, 159.

Parma, P., Fanti, C., Fanti, R., Morganti, R., de Ruiter, H.: 1987, *Astron. Astrophys.* **181**, 244.

Probing the Sub-Parsec Scale Structures in Extragalactic Radio Jets: VLBI at 43 GHz

T.P. Krichbaum, A. Witzel, and D.A. Graham

Max-Planck-Institut für Radioastronomie
Auf dem Hügel 69
D-5300 Bonn 1, Germany

Abstract: Global VLBI observations at 7 mm wavelength provide images of extragalactic radio jets with an angular resolution of $\sim 100\mu$as. The objects observed so far show enhanced jet-curvature in the vicinity of their 'VLBI-cores'. Monitoring the structural evolution reveals in many cases strong evidence for motion along curved paths and flux density outburst preceding the ejection of new components.

1 Introduction

VLBI at 43 GHz has been developed continuously since the early eighties, when first fringes at 43 GHz were detected. Meanwhile — despite difficult observing conditions and the small number of telescopes available — it reveals reliable images of radio sources with an angular resolution of $\sim 100\,\mu$as (for more details see *e.g.* Marcaide *et al.*, 1985, Bartel *et al.*, 1988, Dhawan *et al.*, 1990, Krichbaum 1990, Krichbaum *et al.*, 1990, Krichbaum and Witzel, 1992(a,b)). The present 7 mm VLB-array consists of 8 telescopes, some of whose technical properties are summarized in Tab. 1. The table also includes the 30 m-millimeter telescope at Pico Veleta and the VLBA antenna at Pie Town, with which successful 43 GHz VLBI test experiments were recently performed. Presently, the sensitivity limit of the 43 GHz VLBI observations still restricts the number of sources accessible, however, future millimeter-VLBI observations — including new sensitive instruments like Pico Veleta and the VLBA — will substantially lower the detection threshold and improve the *uv*-coverage, allowing the investigation of more and fainter objects in greater detail.

2 Results

Intercontinental 43 GHz VLBI observations aimed at imaging radio sources have been performed annually since 1986. These observations targeted the evolution of the sub-milliarcsecond structures of *e.g.* 3C 84, 3C 273, 3C 345, 1803+78, BL Lac. The first VLBI images at 43 GHz were obtained in 1990/91 for sources like OJ 287, 4C 39.25, 1928+73 and Cygnus A. In the following we will shortly summarize some of the major results:

Table 1. Antenna specifications: U/CS= Un/Cooled Schottky-Diode mixer, M= Maser amplifier, SIS= SIS-tunnel-junction mixer, HEMT= HEMT-amplifier. The illuminated diameter is given for the 100 m-telescope in Effelsberg.

Station	Code	D [m]	T_{sys} [K]	η_A [%]	Receiver
Effelsberg	B	60	400	40	CS
Pico Veleta	V	30	200	51	CS
Onsala	T	20	220	48	SIS
Haystack	K	36	200	19	M
Maryland Point	N	26	740	5	US
Owens Valley	O	40	140	7	M
Pie Town	P	25	100	50	HEMT
Nobeyama	X	45	120	62	HEMT

At 43 GHz the VLBI structure of **3C 84** (NGC 1275, z=0.018) consists of a knotty jet-like structure of $\sim$ 1 mas length, which is embedded in a halo-like region of extended emission (see Krichbaum and Witzel 1992(a,b) and Krichbaum *et al.*, 1992b for further details). In Fig. 1 the structural evolution of the northern mas-jet is shown. Several jet-components (labelled B, C, D and E) separate with subluminal speeds ($v_{app} \sim 0.1 - 0.2c$ (for $H_0 = 100\, kms^{-1} Mpc^{-1}$, $q_0 = 0.5$)) from component A, which, considering its flux density variability, compactness and inverted spectrum, has been identified with the 'core' of 3C 84. VLBI imaging at cm-wavelengths shows a lobe-like structure extending up to 10 mas south of A. This region is at 43 GHz nearly resolved by the interferometer beam, however, Gaussian model fits to the subset of short interferometer baselines strongly indicate the existence of distinct components embedded in the southern lobe. Consistent with 22 GHz VLBI observations (Readhead *et al.*, 1990, Marr *et al.*, 1989), the 43 GHz data suggest subluminal motion of these components with $\beta_{app} \simeq 0.5$. This is considerably faster than the motion in the northern 'mas-jet'. The two regions of different velocity-patterns are separated by a major kink in the source structure of about 60°. A bending of the jet-flow towards the observer's line of sight would provide a simple explanation for the apparent acceleration.

Figure 2 shows the variations of the total intensity and the temporal evolution of the sub-parsec structure in the quasar **3C 273** (z=0.158). 43 GHz VLBI observations shortly after a major optical/radio flux density outburst in early 1988 (Courvoisier *et al.*, 1990) revealed a new jet component, C9, being ejected with $\beta_{app} = 5.5 \pm 0.8$ (Krichbaum *et al.*, 1990). Thus, a direct correlation between large flux density outbursts and the genesis of new jet components is indicated (as in the case of BL Lac (Mutel *et al.*, 1990)). In 3C 273

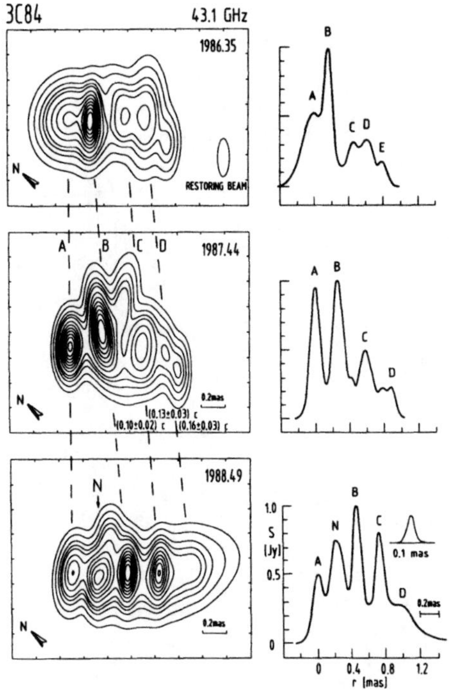

Fig. 1. Left: Gaussian component models of 3C 84 for observing epochs 1986-1988 at 43 GHz. A surrounding extended halo component has been omitted. The three models are rotated counter-clockwise by 50°. Straight lines indicate the component identification, numbers denote apparent component velocities relative to A. Right: normalized profiles of the jet's brightness distributions for the models seen on the left. In this identification scheme a new component N appeared in 1988 (for details see Krichbaum *et al.*, 1992b).

the pairs of jet components C7/C8 and C5/C4 seem to move along a *common* bent path, in contrast to 3C 345 where components move on *different* trajectories. The appearance of C9 at a position located significantly further north than the outer jet components and the curvature of the paths of C4-C8 (see Fig. 2) strongly suggest increasing jet bending in the vicinity of the core of 3C 273 (*e.g.* Krichbaum and Witzel, 1992b).

3C 345 (z=0.595) is the archetype of a source with jet components moving along *different* bent trajectories. In Fig. 3a, 43 GHz maps are shown from observations in 1988 and 1990 (Krichbaum and Witzel, 1992(a,b) and Zensus 1992). It is apparent that a new component C6 emerged from the core and moved to the north-west. The appearance of another component C7 in 1990 may be related to the recently observed increase in total flux density of 3C 345. In Fig. 3b we plot the positions of the components C7 to C4 with respect to the core, together with data taken at 10.7 and 22 GHz (see

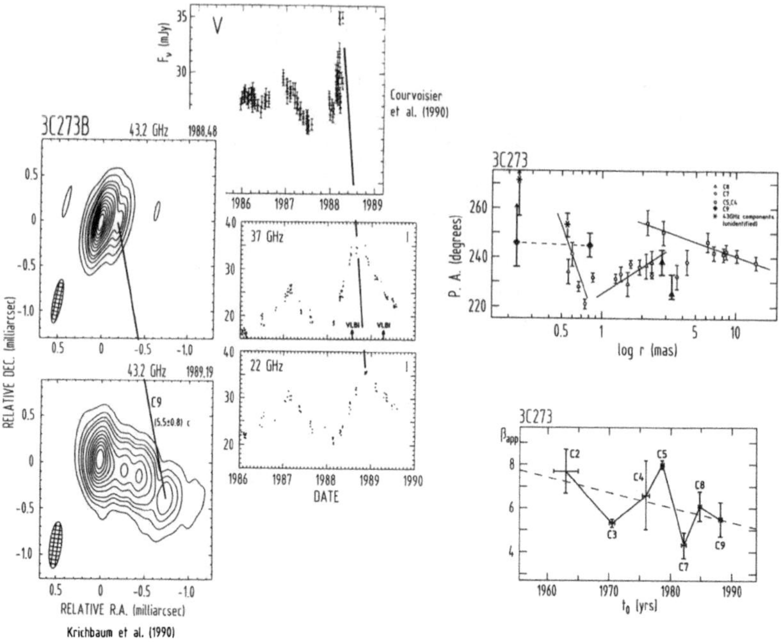

Fig. 2. 43 GHz-maps of 3C 273 at 1988.48 and 1989.9. The straight lines indicate the motion of C9. Optical (V-band) and radio (22 and 37 GHz) lightcurves of the pronounced flux density outburst in 1988 are shown (Courvoisier *et al.*, 1990). A plot of position angle versus core separation is shown for components C9-C4 at the top right. Straight lines indicate a possible jet-ridgeline. On the bottom right β_{app} is plotted versus (linearly extrapolated) time of ejection t_0 for C2-C9. The slope of the dashed line is $-(6.9 \pm 6.1) \cdot 10^{-2} yr^{-1}$ (for details see Krichbaum *et al.*, 1990, 1992a).

Zensus, this conference). The 43 GHz data for components C4 and C5 complement earlier measurements, and confirm that both components move along 'quasi-helically' bent paths. The new data for C6 and C7 indicate that these components move along paths different from those of their precursors, thus suggesting that *each* component may follow its own trajectory.

OJ 287 (z=0.306) is one of the most violently variable AGN. At cm-wavelength the source is nearly pointlike, showing a faint jet-like extension of $\sim$ 4 mas length to the west (Gabuzda *et al.*, 1989). 3 mm-VLBI observations in 1988 and 1989 (Bååth, 1992) showed a very compact, basically unresolved structure on μas-scale. In Fig. 4 we show the first 43 GHz map of OJ 287, obtained in April 1991 with a VLB-array of four telescopes. The source exhibits a core-jet structure of about 1 mas length at an orientation similar to that found at lower frequencies. The map reveals some evidence for jet curvature close to the core. Further observations with better *uv*-coverage are, however, needed to confirm this structure.

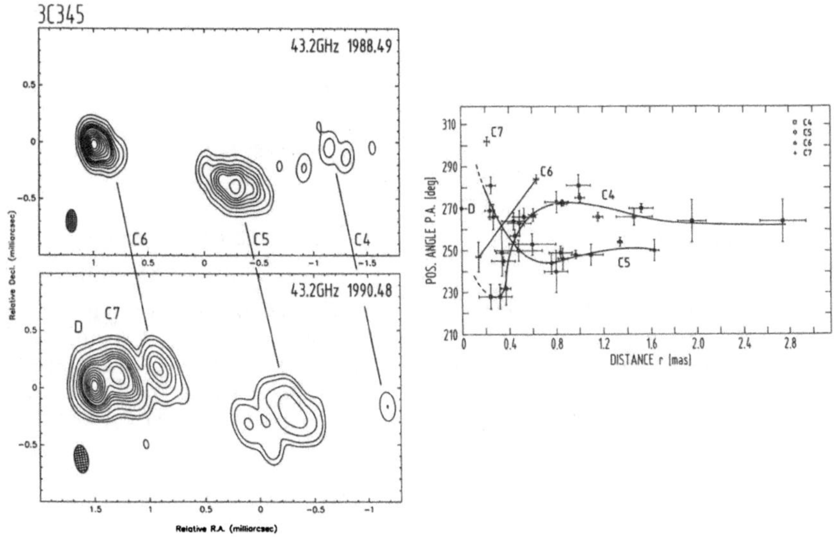

Fig. 3. (a): 43 GHz maps of 3C 345 at 1988.49 and 1990.48. Straight lines indicate the motion of C4-C6 with respect to the stationary assumed core D. (b): Position angle versus core separation plotted for C4-C7. Data points are from 10.7 and 22 GHz and 43 GHz. The lines sketch a possible mean path (for details see Zensus 1990 and 1992, Krichbaum *et al.*, 1992a and Witzel, 1992).

BL Lac (z=0.0695) has been observed with VLBI at 43 GHz in 1990.5 and 1991.3. Due to the small number ($N = 4$) of participating antennas, the *uv*-coverage and signal strength is poorer than for the sources discussed above, so that future observations will have to confirm the preliminary results obtained so far:

In Fig. 6 a possible scenario for the structural evolution in BL Lac is shown. The combination of the 43 GHz models with 10.7 GHz maps (kindly provided by R. Mutel) suggests superluminal motion of two components with $\beta_{app} = 2.6 \pm 0.7$ (at $r \simeq 1.5$ mas) and with $\beta_{app} = 3.6 \pm 1.4$ (at $r \simeq 4$ mas). These velocities are similar to the velocities quoted for the preceding components S1-S3 (see Mutel *et al.*, 1990). Furthermore, due to the higher angular resolution, the 43 GHz data reveal a new component close to the core at $r = 0.28 \pm 0.10$ mas (1990.47). Adopting the above identification scheme, it is very probable that this component could be identified with a component located at $r = 1.37 \pm 0.15$ mas in 1991.26 (see Fig. 6). Its apparent velocity would then be $\beta_{app} = 4.3 \pm 0.7$.

Cygnus A (z=0.0562) is commonly regarded as the 'prototype' of a powerful double-lobed radio-galaxy. Recent VLBI imaging at 5 GHz showed a one-sided core-jet structure extending 25 mas to the north-west (Carilli *et al.*, 1991). From two epochs of observa-

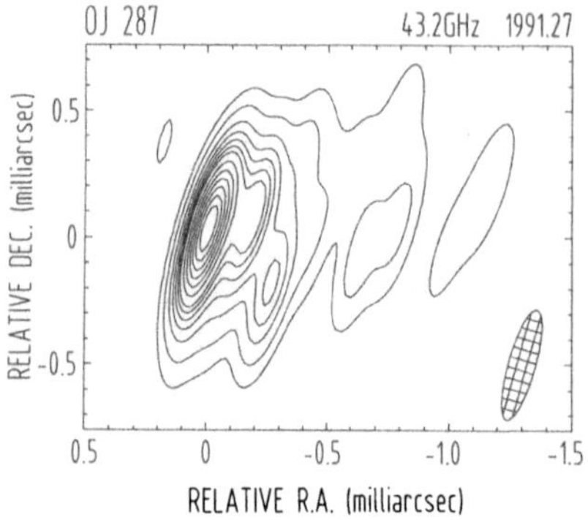

Fig. 4. Clean-map of OJ 287 at 43 GHz (1991.27). Contour levels are 1, 2, 5, 10, 15, 20, 25, 30, 40, 50, 60, 70, 80 and 90 % of the peak brightness of 0.63 Jy/beam. The restoring beam size is 0.45×0.11 mas, $PA = -16°$.

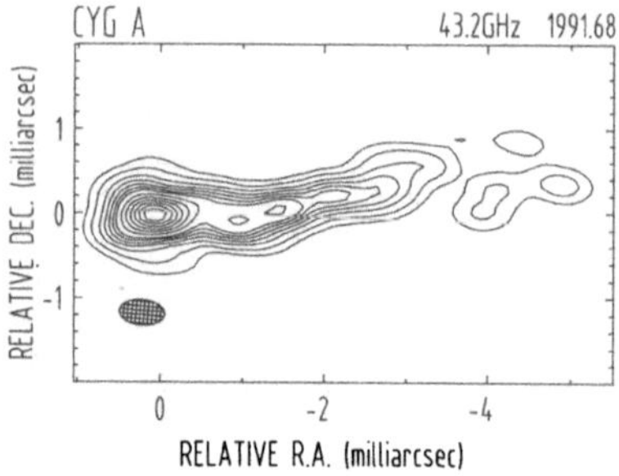

Fig. 5. 43 GHz map of Cygnus A at 1991.68. Contour levels are at 2, 5, 10, 15, 20, 25, 30, 40, 50, 60, 70, 80 and 90 % of the peak brightness of 0.38 Jy/beam. The restoring beam size is 0.55 × 0.30 mas, $PA = 85.4°$.

tions, the authors find an upper limit for possible structural variability of $\beta_{app} \leq 2$. The shape of the total intensity spectrum of Cygnus A (a spectral turnover between 20-60 GHz) and the decreasing contribution of the lobes to the total flux density at mm-wavelength make Cygnus A a promising candidate for 43 GHz VLBI observations.

In September, 1991 a 43 GHz VLBI experiment was carried out between the telescopes at Effelsberg, Onsala and the IRAM 30 m-millimeter telescope at Pico Veleta in order to test the VLBI-performance of the 30 m telescope. The success of the experiment and the resulting excellent data quality enabled us to make reliable maps/models for some of the objects observed: in Fig. 5 we show the (first) 43 GHz map of Cygnus A, obtained after 5 hours of mutual observing. The source consists of a core-jet structure of ~ 4 mas length. The jet is oriented nearly east-west for $r \leq 1.5$ mas, but seems to bend to the north-west at larger core separations, thus meeting the VLBI structure observed at 5 GHz. The visible 'knots' in the jet and the small observing beam size will facilitate future studies of structural variability, which will also have to address the problem posed by the slow apparent motion and the absence of any visible counter-jet. A possible solution to the problem could be — if intrinsically one-sided jet emission is not assumed — a change in the viewing angle of the jet, *e.g.* increasing jet-curvature towards the core, which is already indicated by the 43 GHz map.

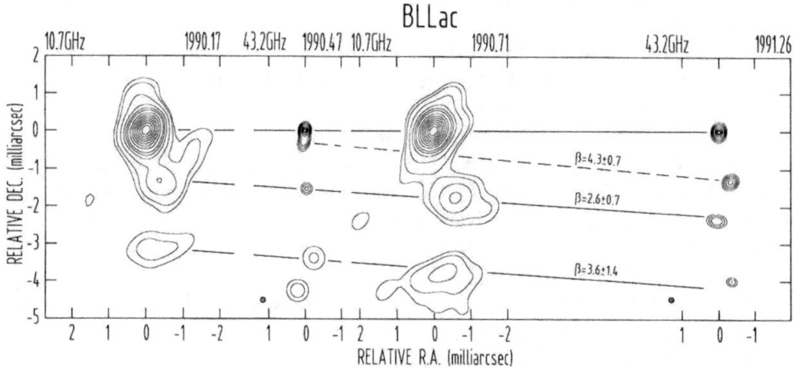

Fig. 6. A sequence of maps of BL Lac at 10.7 and 43 GHz. Straight lines indicate the motion of components, labels denote β_{app}. The 10.7 GHz maps were provided by R. Mutel. The 43 GHz images (epochs 1990.47 and 1991.26) are based on Gaussian model fits to the data.

3. Conclusion

VLBI imaging with sub-parsec scale resolution reveals further evidence for enhanced jet bending in the vicinity of the 'VLBI-cores' of AGN. For the best observed sources (3C 84, 3C 273, 3C 345), jet-components move along bent paths or along paths not intersecting the core. The shape of the paths often resembles projected 'quasi-helical' trajectories. The apparent velocities of components located at different jet positions are often different. The velocity of a single component can change as its path curves. It is therefore likely that in the sub-parsec region of the jets of AGN VLBI components move along *three-dimensionally* bent trajectories. Facilitated by its high angular resolution, millimeter-VLBI observations provide a unique tool for measuring the relative positions of jet components with a high degree of accuracy. However, monitoring the motion will require more frequent observations in the future.

Acknowledgements: We thank all collaborators of the joint 43 GHz VLBI campaigns for their efforts and enthusiasm, and R. Mutel for providing the 10.7 GHz maps of BL Lac prior to publication. We also thank the members of staff of the observatories who helped to make the observations possible. The work of T.K. was supported in part by the BMFT-Verbundforschung.

References

Bååth, L., 1992, in *Frontiers of VLBI*, ed. H. Hirabayashi, M. Inoue and K. Kobayashi, Tokyo: Universal Academy Press, p.285.

Bartel, N.: *et al.*, 1988, *Nature* **334**, 131.

Carilli, C.L., Bartel, N., Linfield, R.P.: 1991, *Astron. J.* **102**, 1691.

Courvoisier,T.J.-L., *et al.*: 1990, Å234 73.

Dhawan, V., *et al.*: 1990, *Astrophys. J. (Lett.)* **360**, L43.

Gabuzda, D.C., Wardle, J.F.C., Roberts, D.H.: 1989, *Astrophys. J. (Lett.)* **336**, L59.

Marcaide, J.M. *et al.*: 1985, in *Proc. IRAM-ESO-Onsala Workshop on (Sub-) Millimeter Astronomy*, ed. P.A. Shaver and K. Kjär, p. 157.

Marr, J.M. *et al.*: 1989, *Astrophys. J.* **337**, 671.

Mutel, R.L. *et al.*: 1990, *Astron. J.* **352**, 81.

Krichbaum, T.P.: 1990, in *Parsec-scale radio jets*, ed. J.A. Zensus and T.J. Pearson, Cambridge: Cambridge University Press, p. 83

Krichbaum, T.P. *et al.*: 1990, *Astron. Astrophys.* **237**, 3.

Krichbaum, T.P., Witzel, A.: 1992a, in *Frontiers of VLBI*, ed. H. Hirabayashi, M. Inoue and K. Kobayashi, Tokyo: Universal Academy Press, p.297.

——: 1992b, in *Variability of Blazars*, ed. E. Valtaoja and M. Valtonen, Cambridge: Cambridge University Press, p.205.

Krichbaum, T.P. *et al.*: 1992a, in *Physics of Active Galactic Nuclei*, ed. W.J. Duschl and S.J. Wagner, Heidelberg: Springer Verlag, p.574,

Krichbaum, T.P. *et al.*: 1992b, *Astron. Astrophys.* **260**, 33.

Readhead, A.C.S. *et al.*: 1990, in *Parsec-scale radio jets*, ed. J.A. Zensus and T.J. Pearson, Cambridge: Cambridge University Press, p. 71

Witzel, A.: 1992, in *Physics of Active Galactic Nuclei*, ed. W.J. Duschl and S.J. Wagner, Heidelberg: Springer Verlag, p.484.

Zensus, J.A.: 1990, in *Parsec-scale radio jets*, ed. J.A. Zensus and T.J. Pearson, Cambridge: Cambridge University Press, p. 28

——: 1992, in *Variability of Blazars*, ed. E. Valtaoja and M. Valtonen, Cambridge: Cambridge University Press, p.187.

Symmetric Structure in Active Galactic Nuclei

T. J. Pearson

California Institute of Technology
Pasadena, CA 91125, U.S.A.

Abstract: We have identified a new class of active galactic nucleus, with symmetric nuclear radio structure (two opposed jets), and little evidence for relativistic effects. These objects are most readily explained as young (or rejuvenated) radio galaxies, less than 10^4 years old. Study of these and similar objects is likely to lead to a better understanding of the formation of double-lobed radio galaxies and the creation and collimation of jets.

1 Introduction

VLBI surveys of the nuclear structure of core-dominated radio galaxies and quasars (*e.g.* Pearson and Readhead 1988; Witzel *et al.* 1988) show that the majority have a one-sided core-jet structure consistent with the idea that they are strongly beamed objects in which we see only the approaching jet. However, in our survey of a flux-density limited sample of 65 objects selected at 5 GHz (Pearson and Readhead 1988), in which we made images of the 45 objects with compact cores stronger than about 0.5 Jy, we have found several objects that show none of the characteristic signs of a Doppler-boosted jet such as high surface brightness, large and rapid flux variability, high polarization, or optical non-thermal continuum. Most of these objects are dominated by two resolved emission regions of roughly equal brightness, separated by 10–100 pc, and are called "compact double" sources. This class was first identified by Phillips and Mutel (1982), and their characteristic properties were summarized by Hodges and Mutel (1987). Both components have similar gigahertz-peaked radio spectra showing synchrotron self-absorption at low frequencies and a steep optically-thin part at high frequencies. They have low integrated radio polarization ($\lesssim 2\%$), low variability ($\lesssim 10\%$), and no kiloparsec-scale extended structure in most cases; they are identified with faint narrow-emission-line galaxies; and they show no superluminal motion. Hodges and Mutel suggested that the compact double sources are young radio galaxies and that the two emission regions are two "lobes" powered by unseen jets from an invisible central engine.

Here I shall present observations of two objects from our survey, the radio sources 0710+439 and 2352+495, that share the characteristic properties of the compact doubles except that they have a symmetrical *triple* structure. I shall argue that these compact triple sources are indeed young radio galaxies, differing from the compact doubles only in the prominence of the central component, which I identify with the central engine.

2 Properties of Compact Triple Sources

As part of our VLBI survey (Pearson and Readhead 1988) we have observed both 0710+439 and 2352+495 at 5.0 GHz (three epochs), 1.7 GHz, and 10.7 GHz. Figure 1 presents high-resolution images of both sources showing their linear triple structure, while Fig. 2 presents a new 1.7 GHz image of 2352+495, sensitive to lower-brightness emission, that shows the symmetric "lobes" clearly for the first time. Of the 45 sources that we have observed only two show the triple structure, and a further two (0108+388 and 1031+567) appear to be true "compact doubles" as defined by Phillips and Mutel. Four other sources (0153+744, 0711+356, 0923+392, 2021+614) have double structure at 5 GHz, but additional observations suggest that they do not belong to the same class.

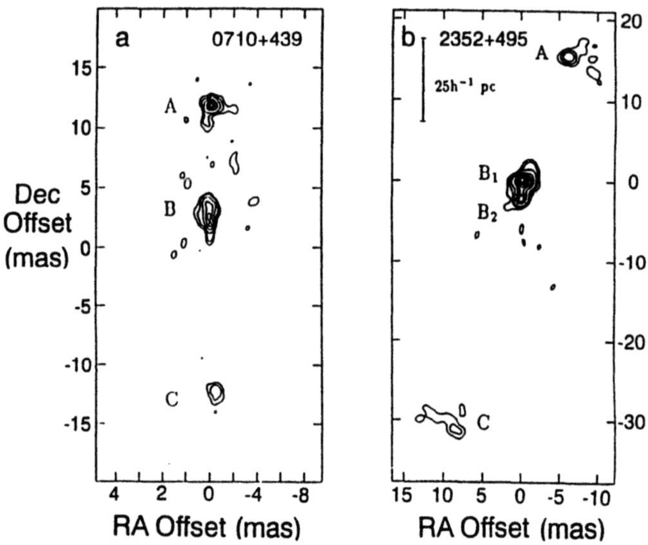

Fig. 1. Images of 0710+439 and 2352+495 on the same linear scale (Conway *et al.* 1992). (a) 0710+439, 10.7 GHz, beam FHWM 0.8 × 0.5 milliarcsec, p.a. −20°, lowest contour 5% of peak. (b) 2352+495, 5.0 GHz, beam FWHM 1 milliarcsec circular, lowest contour 2% of peak.

We can summarize the properties of the compact triple sources as follows:

1. They have a linear triple morphology with angular sizes 24.1 milliarcsec (0710+439) and 45.7 milliarcsec (2352+495), corresponding to linear sizes of 84 and 107 pc at redshifts 0.518 and 0.237 (assuming $H_0 = 100\,\mathrm{km\,s^{-1}\,Mpc^{-1}}$).

2. They have no emission on larger scales visible in VLA and MERLIN images of high dynamic range.

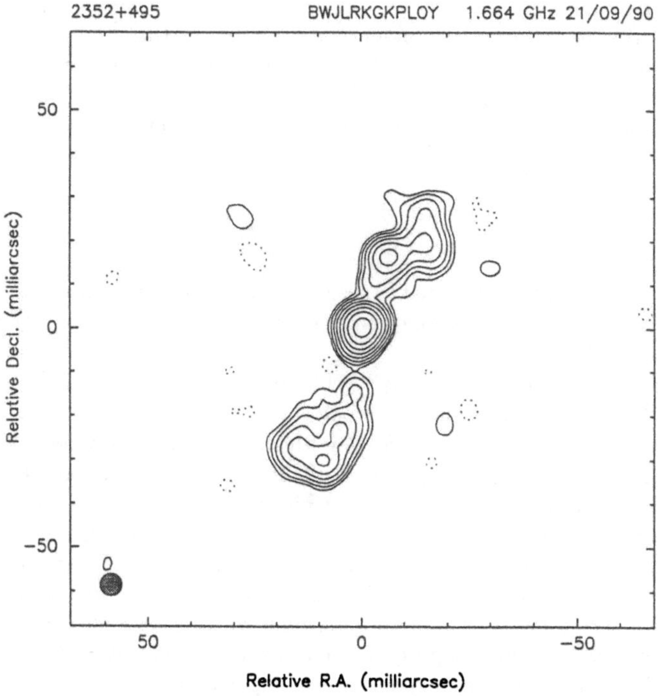

Fig. 2. Image of the radio galaxy 2352+495 at 1.7 GHz made from a "snapshot" VLBI observation using 12 antennas (Polatidis *et al.* 1992). The contour levels are spaced by factors of two in brightness, with the lowest contour at 0.5% of the maximum (0.86 Jy/beam). The gaussian restoring beam has FWHM 0.005 arcsec.

3. They are only weakly polarized (0.1–0.2% integrated over the source).

4. All three components have a gigahertz-peaked spectrum with a steep optically thin part: there is no evidence for a flat-spectrum component at high frequencies, as seen in the superluminal sources.

5. They are identified with $\sim$ 20-mag galaxies with narrow emission lines and stellar absorption features, but no broad emission lines or non-thermal continuum.

6. There is no evidence for relative component motion; we place 3σ upper limits on movement of A relative to B of $v < 0.41c$ for 0710+439 and $v < 1.0c$ for 2352+495.

7. In both sources the central component (B) is brighter than the outer two (A and C), has substructure, and is elongated along the source axis. The relative brightnesses of the subcomponents in the central component of 2352+495 (and possibly of 0710+439) have changed, but there is a possibility that this is due to refractive interstellar scintillation, not an intrinsic effect.

8. Figure 2 appears to show emission between the central component B and the outer components A and C in 2352+495. It is tempting (but premature) to identify this as emission from two jets. We are conducting further observations to try to confirm this identification.

9. From minimum-energy arguments and assuming that the low-frequency spectral cutoff is due to synchrotron self-absorption, we estimate minimum pressures $\sim 10^{-4}$ dyne cm^{-2}, magnetic fields $\sim 10^{-2}$–10^{-1} G, and synchrotron life-times < 3000 yr; the central components have similar properties to the outer ones.

3 Interpretation

In order to understand the nature of these compact triple radio galaxies, we would like to be able to answer the following questions. For further discussion of these points, see Polatidis *et al.* (1992).

1. Where is the center of activity? While it is possible that these are asymmetric objects in which we see three knots in a one-sided, Doppler-boosted jet, it is much more likely that they are symmetric objects in which the center of activity – the primary energy source – lies in the middle component. In both objects the middle component is elongated along the source axis and has the highest brightness, most compact substructure and strongest evidence for flux variability. In 2352+495 the middle component is clearly a center of symmetry. In the following discussion we will assume that the two outer components are "lobes" powered from the central engine via unseen jets.

2. How old are they? We want to determine whether they are small because they are young, or because some property of the central engine or of the environment of the galactic nucleus has suppressed expansion to kiloparsec scales. A lower limit on the age comes from the upper limit on the expansion speed: $v < 0.4c$ implies an age > 1000 yr. At present, we cannot place an upper limit on the age, but it is probable that an analysis of the energetics of the sources may provide such a limit.

3. How are the lobes confined? The minimum pressure in the lobes is too high to be confined by a typical galactic narrow-emission-line region. While it is possible that a much higher-pressure medium in the central few parsecs of the galaxy is confining the lobes, it is more plausible that they are confined by ram pressure; *e.g.*, if they are advancing at $v = 0.1c$ they could be confined by a medium of density $n = 0.1\,\mathrm{cm}^{-3}$. This would be consistent with an age of $\sim 10^4$ yr.

4. Is there any evidence for relativistic motion? The upper limits on component motion are subrelativistic. It is worth noting that the arm-length asymmetry in both objects is not due to relativistic advance of the lobes, which would make the brighter lobe appear further from the nucleus, in contradiction to the observations. There is some evidence for relativistic motion within the central component of 2352+495 ($\gamma \lesssim 2$) from variability, unless the variations are due to interstellar scintillation.

5. What causes the S-shaped symmetry? One of the two sources is clearly S-shaped; the other may be. Possible explanations include precession or rotation of the central engine, or ejection at an angle to the galactic pressure gradient.

6. How are the compact triple sources related to the compact doubles? We need to look harder for a third component in the compact doubles, but it seems likely that we

have to account for similar objects both with and without strong central components. It does not seem likely that the central components in the triples are enhanced by relativistic beaming, but even if they are, there do not appear to be enough compact doubles to constitute the parent unbeamed population.

7. Why do the central components have similar gigahertz-peaked spectra to the lobes? Although there are differences between the component spectra, both the central components and the lobes have steep high-frequency spectra and similar low-frequency cutoffs, in spite of their rather different physical origin. If the low-frequency cutoff is due to synchrotron self-absorption, it is apparently just a coincidence that the central components and lobes have similar sizes and hence similar cutoff frequencies. External thermal absorption by a dense medium might provide an explanation.

Acknowledgments: This work has been carried out in collaboration with my colleagues at Caltech (J.E. Conway, C.R. Lawrence, A.C.S. Readhead, S.C. Unwin, and W.Xu) and at Jodrell Bank (A. Polatidis and P.N. Wilkinson), and I am grateful for their permission to present these unpublished results at this meeting. A fuller account will be published in papers by Conway *et al.* (1992) and Polatidis *et al.* (1992). I am grateful to the Netherlands NWO for financial support and to the Kapteyn Laboratory, Groningen, for their hospitality.

References

Conway, J.E., Pearson, T.J., Readhead, A.C.S., Unwin, S.C., Xu, W., Mutel, R.L.: 1992, *Astrophys. J.* **396**, 62.

Hodges, M.W., Mutel, R.L.: 1987, in *Superluminal Radio Sources*, ed. J.A. Zensus and T.J. Pearson, Cambridge: Cambridge University Press, p.168.

Pearson, T.J., Readhead, A.C.S.: 1988, *Astrophys. J.* **328**, 114.

Phillips, R.B., Mutel, R.L.: 1982, *Astron. Astrophys.* **106**, 21.

Polatidis, A., Wilkinson, P.N., Xu, W., Readhead, A.C.S., Pearson, T.J.: 1992, in preparation.

Witzel, A., Schalinski, C.J., Johnston, K.J., Biermann, P.L., Krichbaum, T.P., Hummel, C.A., Eckart, A.: 1988, *Astron. Astrophys.* **206**, 245.

What Does Variability of the Central Sources Tell Us About Jets?

Joel N. Bregman

Astronomy Department
University of Michigan
Ann Arbor, MI 48109
USA

Abstract: For radio-loud active galactic nuclei, and blazars in particular, the nonthermal radio continuum shows slow variation (months) that is successfully modeled by a shock propagating down a partially opaque jet. At frequencies above the millimeter band, the plasma is transparent to synchrotron radiation and the entire broadband continuum responds to brightness fluctuations together, suggesting that this part of the continuum is produced in a fairly uniform region. Optical outbursts precede radio flares by about a year in several objects, indicating that these spectral regions may represent two physically distinct parts of the same jet. However, the statistical properties of variability are different for the two wavebands, which may indicate the influence of shocks, optical depth effects, or radiative cooling.

1. Introduction

Imaging studies of astronomical objects such as radio jets have proven to be among the most powerful approaches to studying this phenomenon. The large-scale structure of radio jets on sizes of kpc have been beautifully studied with synthesis array telescopes, and the structure on parsec scales has been probed with VLBI efforts. Unfortunately, technical limitations to the sizes that can be studied arise well before one is able to study the central engines of these objects. Even for one of the closest AGN, M 87 at a distance of about 15 Mpc, and for the highest VLBI angular resolution that can be achieved from the ground, 0.1 mas, the size of structure expected for the central engine is still at least 10–100 times smaller (*e.g.*, an accretion disk). Until radio astronomers can extend their baselines by one or two orders of magnitude, or until optical/infrared interferometry becomes feasible at low light levels, we shall need to rely upon indirect methods of deducing information about the very important central region. It is this motivation that has led to the considerable effort of understanding the broad-band continuum emission from the central region, and to use the variability of this emission to probe the structure. Here I review the current state of understanding of how variability studies relate to the jet phenomenon and discuss where future work might lead.

2. Radio-Quiet Active Galactic Nuclei

First I would like to begin with a brief overview of the basic understanding of continuum emission from the core of various types of active galactic nuclei (for a detailed review, see Bregman 1990). If this talk had been given a decade ago, it would have been claimed that the continuum from all AGN was essentially all nonthermal in origin. However, this point of view has changed dramatically, and there now is a strong current of thought, justified by observation, that there are two types of AGN, those that emit nearly all thermal emission, and those where the nonthermal emission is dominant. As an example of an AGN with thermal emission, we consider the radio-quiet objects, which have three distinct broad-band features that suggest various thermal mechanisms.

2.1 Dust Emission in the Infrared

When plotting the power per logarithmic bandwidth versus frequency, one sees that the typical radio-quiet AGN has two broad maxima, one in the infrared near 25–6μm and the other in the ultraviolet region (*e.g.*, Sanders *et al.* 1989). It is essential to note that the local minima between the two regions occurs at about 1um in the rest frame of the object. The long wavelength falloff of the far infrared emission occurs consistently in the 100–350μm region (also for rest frame wavelengths), and the slope of the 1mm–100μm continuum can be steeper than 2 ($F_\nu \propto \nu^2$), the maximum permitted by a blackbody (*e.g.*, Chini *et al.* 1987, 1988, 1989; Barvainis and Antonucci 1989). These types of features are precisely those expected from thermal emission by dust grains. Dust cannot survive at temperatures above about 1600 K, with a peak wavelength of about 2 μm, which is consistent with the falloff and minimum in the power emitted near 1 μm. At long wavelengths ($\lambda > 100\mu$m), the dust rapidly becomes transparent with increasing wavelength, which would lead to a sharp falloff in the continuum, also consistent with the observations. The final ingredient required is that there be a wide range of dust grain temperatures, and although this is an inferred requirement, the distribution of dust mass with radius might be naturally realized with a warped disk of dust and gas (Barvainis 1987; Phinney 1990).

The expected distribution of radiating dust with radius can be tested directly with time variability studies over the infrared region. For a covering factor near unity, the radius of the dust emitting near 1 μm is about 1 (LIR/10^{45} erg/sec)^1/2 (λ/1μm)^2 lt mo, and variations near 1 μm are commonly seen. This model also predicts that the timescale for variation becomes increasingly longer with wavelength, and observations show that flux density variations of radio quiet objects is quite small at 2.2 μm and nearly unmeasurable at 10 μm (Cutri *et al.* 1985). In the 60 μm and 100 μm IRAS bands, no variation would be expected during the 9 month lifetime of IRAS, and no such variations were observed (Edelson and Malkan 1987; Clement *et al.* 1988). So the time variability data provide very strong support for the distributed dust emission model for the infrared continuum.

2.2 The Continuum of an Accretion Disk?

The second broad peak in the optical to soft X-ray region contains a comparable and often larger amount of the total power than from the infrared region. In the rest frame, the shape of this continuum in the optical-ultraviolet region is remarkably similar from object to object, and this has led to the suggestion that we are viewing a single well-defined emission process (*e.g.*, Malkan 1989). The most popular model posits the existence of an opaque accretion disk spinning around a black hole of mass 10^8–10^9 $M_\odot$ (Rees 1984; Czerny and Elvis 1987; Wandel and Petrosian 1988; Sun and Malkan 1989). Due to viscous friction and differential rotation, where the rotation rate increases toward the black hole, the resulting temperature also increases toward smaller radii. A maximum temperature is anticipated approaching the last stable orbital radius of the black hole, which might explain why the peak in the power occurs at roughly the same frequency. However, there are some problems with the simple opaque accretion disk model in that the maximum temperature is lower than expected, no continuum discontinuity is seen at 912 Å, and the percentage polarization is lower than initially anticipated (Kolykhalov and Sunyaev 1984; Laor and Netzer 1989; Antonucci, Kinney and Ford 1989). None of these problems are particularly disastrous for the theory and can easily be accommodated with reasonable modifications to the model (Laor, Netzer and Piran 1990). However, central to the accretion disk concept is that most of the emission at a particular wavelength can be identified with a particular radius, so if a continuum outburst were to occur at one wavelength and evolve to a longer wavelength, the rate of the evolution would be a critical test of the model.

The effort to search for time delays in the variability at different wavelengths in the optical and ultraviolet region has only just begun, but the results are already intriguing. For NGC 5548, the optical and ultraviolet variations are tightly correlated and with a time delay of less than 212 days (Clavel, Reichert and Alloin 1991). The same result is obtained for NGC 4151 (Clavel, Boksenberg and Bromage 1990), while for Fairall 9, the time delay is $\leq (6 \pm 86)$ days (Clavel, Wamsteker and Glass 1989; Courvoisier and Clavel 1991; Malkan 1992). The upper limits to the time delays for NGC 5548 and NGC 4151 are shorter than predicted with the simple disk model, and should additional observations support this result, a serious modification to the disk model may be required, such as the introduction of a thick disk component.

Not all of the continuum from a radio-quiet AGN is thermal in origin. The radio emission, which accounts for only a small fraction of the total power, is probably synchrotron in origin, and although the radio luminosity is low, it is still greater than the radio luminosity of starburst galaxies (Wilson and Heckman 1985; Wilson 1988). Finally, the X-ray origin of the X-ray emission is poorly understood, although the rapid variability points to a small region, which would occur in a model where Compton scattering takes place in an atmosphere around the inner part of an accretion disk (McHardy and Czerny 1987; Pounds 1992).

3. Radio-Loud Active Galactic Nuclei

The other types of AGN are those featured at this conference, the radio-loud objects, which, in their purest form, are dominated by nonthermal emission. The extreme objects in this category are the BL Lac objects, which have rapidly variable continua that are highly polarized, weak or absent broad emission lines, and compact radio emission, often with a flat spectrum (Miller 1989). It is worth reconsidering the certainty with which this emission is nonthermal, since the interpretation of the continua in radio-quiet AGN changed from a nonthermal to a thermal origin during the past decade. Unlike the radio-quiet AGN, the BL Lac objects do not possess a local minimum near 1 μm, a rapid falloff in the 100–300 μm region, or a peak in the power in the ultraviolet (*e.g.*, Landau *et al.* 1986; Brown *et al.* 1989a). There are few distinctive features in the spectra of BL Lac objects, and none that appear over and over again at the same rest wavelength. Also, the continuum can be highly polarized (typically 10-20%), as one might expect from synchrotron emission, while the polarization is quite low for radio-quiet AGN ($< 3\%$; Stockman *et al.* 1984; Smith *et al.* 1986, 1987). However, the most compelling evidence for nonthermal emission comes from radio VLBI imaging and from temporal variability studies at a range of wavelengths.

The radio VLBI studies reveal brightness temperatures in the vicinity of 10^{12} K (Zensus, this meeting), which is lower than that inferred from variability sizescales (Altschuler 1989), although the two become equivalent if modest relativistic motion is introduced as well. Continuum flux variations were seen in some BL Lac objects with the IRAS telescope, and the inferred brightness temperature in the 60–100 μm waveband is $> 10^6$ K (Edelson and Malkan 1987), more than four orders of magnitude greater than the temperature of dust emitting at these wavelengths, once again demonstrating a deep difference with the radio-quiet AGN. The infrared continuum progresses smoothly from the infrared into the optical-ultraviolet continuum (*e.g.*, Brown *et al.* 1989a), with the added feature that the timescale of variation becomes shorter (for the same amplitude change; *e.g.*, Bregman *et al.* 1988; Neugebauer *et al.* 1989; Brown *et al.* 1989b; Edelson *et al.* 1991; Tagliaferri *et al.* 1991). This reaches an extreme situation for the X-ray region, where for a few objects the timescale of variation is less than the light crossing time for a supermassive black hole (Doxsey *et al.* 1983; Feigelson *et al.* 1986). Again, this can be explained by a combination of relativistic motion and nonthermal emission (*e.g.*, Ghisellini *et al.* 1985; Cellotti, Maraschi and Treves 1991). Finally, it is worth noting that the peak in the power can occur anywhere between the far infrared to the X-ray region, and with the recent GRO observation of 3C 279, the peak can also occur in the gamma-ray region (Hartman *et al.* 1991).

If the radio-quiet AGNs are largely thermal emitters, and the BL Lac objects are non-thermal emitters, then the radio-loud quasars are intermediate objects. For such objects the two distinct emitting structures are both visible, and this is the basis for some of the most recent attempts to create unified schemes (*e.g.*, Barthel 1989). In the remainder of this contribution, I will forego further discussion of the thermal component and instead concentrate on the nonthermal emission (the blazar), with an eye toward the jet mechanism.

3.1 A Host of Components

One of the purposes of a single multifrequency spectrum of a blazar is that one can hope to fit a continuum model to the data to deduce the basic plasma parameters. To do so, one hopes to identify a single important spectral feature, the frequency at which the source first becomes completely transparent. The expectation is that a low frequency falloff should occur where optical depth unity is first reached. Studies of the GHz radio region of blazars showed that the spectra were not perfectly smooth (flat), but had fluctuations that could be traced to separate components, each of which became transparent at a different frequency (*e.g.*, Cotton *et al.* 1980; Mutel *et al.* 1990). These studies, some conducted about a decade ago, extended to about 15 GHz, with no end in sight for the separate components. When fitting continuum models to the data, one usually made an educated guess as to the maximum frequency where opacity effects were significant. Fortunately, with the technological advances made in the millimeter and submillimeter region and with an extensive effort by Robson and collaborators, we now have important new observations concerning this issue (Brown *et al.* 1989a,b; Robson 1991; Robson, this meeting).

Some of the blazars show nearly perfectly smooth continua from the GHz region into the ultraviolet region, but several do not, and perhaps the finest example of this to date is 0736+017 (Fig. 1). In this object, the spectrum declines from the submillimeter to the millimeter region, and a sharp change in the spectrum is required if it is to connect to the higher frequency infrared and optical data. This spectral feature can be understood as being due to opacity effects resulting from synchrotron-self-absorption and implies a sizeable magnetic field ($\approx 0.1\,\mathrm{G}$) associated with the synchrotron emission at that frequency.

3.2 Flares Along a Jet

The entire broadband continuum from the radio through at least the ultraviolet region cannot be modelled with a homogenous synchrotron emission model and one often resorts to an inhomogeneous model where the plasma is distributed within a jet-like structure (Blandford and Königl 1979; Ghisellini *et al.* 1985; Marscher, Gear and Travis 1991). Plasma parameters are assumed to vary with distance along the jet in a power-law fashion. Depending upon the jet geometry and the power-law dependencies of the density, magnetic field, and bulk motion, the high frequency emission can originate in the inner or outer region of the jet. One's intuition suggests that the high frequency emission comes from the more compact inner part of the jet, and the smaller variability timescales at higher frequencies (for the same fractional amplitude variation) support this point of view. Although there are too many parameters to constrain the model uniquely from just a single multifrequency spectrum, variability studies offer the opportunity to make powerful quantitative tests of the jet paradigm.

If the higher frequency regions originate from the inner part of the jet, then, as a disturbance such as a shock propagates outward, one would expect the higher frequency region to brighten first. Formally, if a sufficiently long data train is obtained over a range of adjacent frequencies, it should be possible to determine the time delays between the various frequencies and thereby map out the jet. While this is already the focus of several efforts in the near future, there is little data at hand, so this is an easy topic to

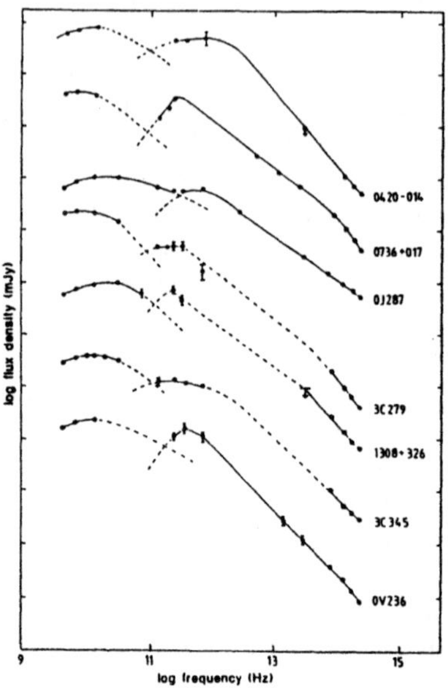

Fig. 1. The radio through optical continuum of seven blazars (from Robson 1991) shows the lack of continuity between the mm and cm regions, especially for 0736+017.

review, the three most interesting objects being 1156+295, AO 0235+164, BL Lac, and 3C 273.

During 1981, the blazar 1156+295 underwent a major flare, which was studied by an army of observers from about the period of peak flux through a dimming phase (Glassgold *et al.* 1983; Wills *et al.* 1983). Observations on a timescale of hours to days in the optical and ultraviolet region showed a spectrum that changed brightness at constant slope. Over a period of weeks to months, there was some spectral steepening, but this may be telling us about the synchrotron cooling time (106 sec, if this is the correct interpretation) rather than the connection between different regions along the jet.

One of the most intensely studied outbursts was that of AO 0235+164 during 1975–76, when it was well observed in both the optical and infrared region (0.36 μm to 10 μm; Rieke *et al.* 1976). These data permit a comparison between the fluctuations near 1 μm with that at 10 μm, and at face value, it appears that the 10 μm variations follow those at 0.44 μm by 2–3 days, but this result depends heavily on one or two 10 μm datum, at a time when it was still challenging to obtain accurate 10 μm measurements. Tantalizing as this result is, it must be viewed with caution.

One of the only blazars for which there is both long-term optical and infrared monitoring data is BL Lac, which makes it possible to search for a time delay by using a cross-correlation function that employs hundreds of separate observations. No measurable time delay is seen, with a formal shift of $\Delta t = 1.5 \pm 2.0$ days (Bregman *et al.*

1990), which indicates that the spectrum changes intensity, not shape, and that any propagation of a disturbance from the optical to infrared regions occurs in less than a few days.

Perhaps the most extensively studied object is also the brightest quasar, and one of the most luminous, 3C 273 (Clegg *et al.* 1983; Robson *et al.* 1983, 1986; Courvoisier *et al.* 1987, 1988, 1990). Although 3C 273 often does not show dramatic variations, it was particularly cooperative during 1988, when there was a series of three flares from February to April, with two of the flares separated by only two days. Frequent observations at several wavelengths show that the infrared emission is simultaneous with the optical flare to within 12 hours. Also, observations at 800 μm and 1000 μm can vary on a timescale of hours to days, and with no perceptible time delay.

The observations thus far collected indicate that the ultraviolet through infrared continuum (and sometimes to the millimeter region) respond together with approximately the same timescale and with no measurable time delay (less than days). This suggests that optical depth effects are not important in these spectral ranges, and that this broad-band continuum is dominated by a single region of plasma (thus making estimates of the plasma properties more valid). These rapid variations, especially at the longer wavelengths in 3C 273, might be explained if there is a relativistic jet bending into and out of our line-of-sight, rather than by radiative synchrotron cooling. Also, if one is to contemplate using time variability to map out the jet region, it will be necessary to push to shorter timescales, such as hours.

3.3 The Shock-in-Jet Model

The variability situation changes in the radio region, where optical depth effects become critical in creating time delays between outbursts at different spectral regions (*e.g.*, Aller *et al.* 1985, Altschuler 1989). It is in this spectral region that the most detailed physical model has been applied, the shock-in-jet model. In general terms, a shock propagating down a jet will compress the magnetic field and the plasma in its path as well as increasing the bulk velocity of the material (review by Marscher, Gear and Travis 1991). This shock may, through the first-order Fermi process, accelerate particles in the vicinity of the shock, thus increasing the radiative intensity. Energy losses can occur radiatively through the synchrotron and Compton processes, and by adiabatic expansion. For the models that have been successful in the GHz radio region (*e.g.*, Hughes, Aller and Aller 1989a), the energy losses are dominated by adiabatic expansion and particle acceleration is not considered important. In their model, independent parcels of plasma are shocked, causing them to be compressed and accelerated, which increases their apparent brightness, causing a radio outburst.

To constrain the parameters of this model requires that one observe a well-defined outburst event in both total flux density and polarized fraction, and at several frequencies that cover the continuum from the opaque to the transparent regions. Through a considerable flux density monitoring effort at 5–15 GHz, several such events have been observed in a few objects, most notably BL Lac (Hughes, Aller and Aller 1989a,b). The remarkably successful fit of this simple model to the data (Fig. 2) provides considerable support for the shock-in-jet model and suggests a Lorentz factor for the first shock transition of 2.5. Additional observations show that the shocked regions identified with outbursts correspond to superluminal VLBI components (Mutel and Phillips

1987). With this observed structural information, the picture of the physical state becomes fully constrained so that one may determine the distance to BL Lac by forcing the angular motion of the superluminal components to correspond to the velocity determined by the shock model. This leads to a Hubble constant of 100 km/sec/Mpc or greater.

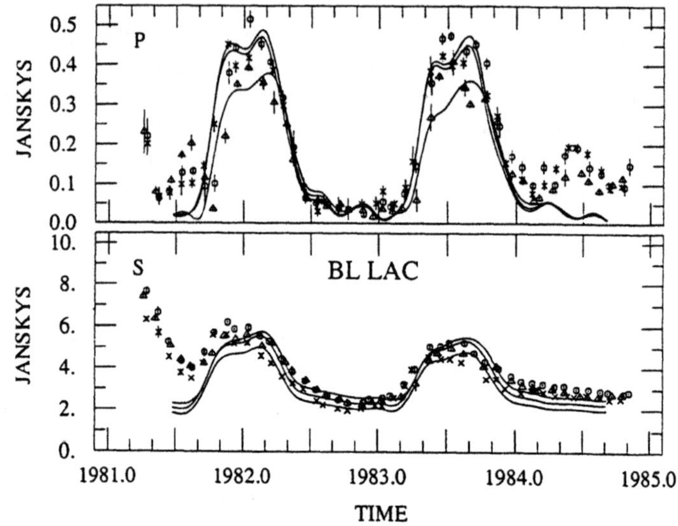

Fig. 2. Curves from the shock-in-jet model are superimposed upon the monthly averaged polarized flux (top panel) and total flux (lower panel) for BL Lac (Hughes *et al.* 1989b). Three frequencies are shown: 4.8 GHz as open triangles and short-dashed curves; 8.0 GHz as crosses and long-dashed curves, and 14.5 GHz as crosses and long-dashed curves.

The shock-in-jet model seems to be quite a successful model for the radio emission, and it can also accommodate a distributed plasma that produces the infrared through ultraviolet continuum, but first we consider whether there is a compelling reason to believe that the radio and IR-UV regions are closely connected. One of the important pieces of evidence is the polarization data, which supports such a connection. Although the optical polarization of blazars is quite variable in both percentage polarization and polarization angle, there is a most common or "preferred" angle of polarization. This preferred angle of polarization is generally aligned with the VLBI structural axis, indicating that the magnetic field knows about the jet and is perpendicular to it (Rusk and Seaquist 1985; Smith *et al.* 1986; Impey 1987).

3.4 Flare Propagation from the Optical to the Radio Region

The connection between the IR-UV region and the radio jet can be investigated further by examining whether flares in the IR-UV region are eventually seen in the radio region after some time delay. Hufnagel and Bregman (1991) have recently investigated this issue by using the existing monitoring data in the optical and radio regions for the five best studied objects, several of which have two decades of data: 3C 345, 3C 446, OJ 287, BL Lacertae, and 0735+178. We investigated both time delays between spectral regions as well as the nature of the variability itself. In discussing the type of variability, we make use of the structure function, SF(Δt), which is a calculation of the mean variance between all pairs of observations separated by Δt. For timescales longer than ≈ 1000 days, the structure function becomes independent of time, indicating that the longest important variability timescale is resolved. However, for shorter timescales, the structure function is roughly described as a power-law in time, SF$\propto (\Delta t)^a$. The familiar flicker noise has a slope of $a = 0$ while shot noise has a slope of $a = 1$. The optical fluctuations have a slope of about $a = 0.3$–0.5 while the radio variations are significantly different, with a mean slope for the 8 GHz data of $a = 0.7$–1.5. This indicates that the very nature of the variations is different in the two spectral domains, which may simply reflect the effect of opacity or possibly some other physical mechanism, such as shock reprocessing or the reduced importance of synchrotron cooling in the radio region. For whatever the reason, this generic difference in the variability between the IR-UV and radio domains will adversely affect the ability to determine accurate time delays between the two regions.

Time delays between the optical and radio regions were calculated using the discrete Fourier transform technique (Krolik and Edelson 1988) and one other method (Hufnagel and Bregman 1992). Weak correlations were usually found, with the optical data preceding the radio variations by about a year. The normalized amplitude of the peak of the cross-correlation function was typically 0.6–0.7 (from a maximum of 1.0), which is greater than expected by random chance, but is not the signature of a strong correlation (for comparison, the cross-correlation peak between 15 GHz and 8 GHz observations was always near 1.0). Also, the peak of the cross-correlation function is often broad, indicating that the time delay is not narrowly confined to a single reproducible value, but may take on a range of values that is comparable to the shift itself. The amplitude and shape of the cross-correlation function may be a consequence of the difference in optical depth effects between the two spectral regions (or reprocessing, etc. as mentioned above). This result can be improved in the future by more adequately sampling the optical variations (the radio variations are already fully resolved), but we also need a more sophisticated model. The implicit model used above is that the radio flare is similar to the optical flare, but shifted in time. A more realistic model that includes optical depth and evolutionary effects of a flare needs to be developed to determine the time delay in the mapping of an optical to a radio flare.

Finally, the most exciting development in the optical-radio connection is the short time scale variations that are seen by Quirrenbach *et al.* (1989, 1991) and discussed in this meeting by Stefan Wagner (also, Witzel 1991). Intraday radio variability has been known for several years in many core-dominated radio-bright AGN (Witzel *et al.* 1986; Heeschen *et al.* 1987), and for most of these objects, interstellar scintillation seemed like a reasonable explanation. However, there are some objects whose variation properties are different from that expected from a simple scintillation model (Quirrenbach *et al.* 1989), although a more complicated scintillation model might be viable. A breakthrough occurred in these studies when, during a one-month period, the BL Lac object 0716+714 was observed at optical and radio wavelengths simultaneously (Quirrenbach *et al.* 1991). It was found that the nature of the variability in both wavebands changed at approximately the same time, and that the small amplitude variations were correlated (although not strongly) with a time delay of 3–4 days. These observations indicate that either the emitting region is extremely small and with a brightness temperature exceeding 10^{17} K, or that other effects are at work, such as a rippled jet where the beaming factor changes as the emitting plasma moves down this imperfect jet. In this very exciting area, we look forward to the additional observations that are being obtained and the detailed modeling of these results that will help determine the most likely explanation for this phenomenon.

Acknowledgements: JNB would like to thank the meeting organizers for their efforts in providing all the necessary ingredients for such an enjoyable and productive workshop and to the support provided by the Max-Planck Gesellschaft and by NASA through grant NAGW-2135.

References

Aller, H.D., Aller, M.F., Latimer, G.F., Hodge, P.E.: 1985, *Astrophys. J. Suppl.* **59**, 513.
Altschuler, D.R.: 1989, *Fund. of Cosmic Physics* **14**, 37.
Antonucci, R.R.J., Kinney, A.L., Ford, H.C.: 1989, *Astrophys. J.* **342**, 64.
Barthel, P.D.: 1989, *Astrophys. J.* **336**, 606.
Barvainis, R.: 1987,*Astrophys. J.* **320**, 537.
Barvainis, R., Antonucci, R.: 1989, *Astrophys. J. Suppl.* **70**, 173.
Blandford, R.D., Königl, A.: 1979, *Astrophys. J.* **232**, 34.
Bregman, J.N.: 1990, *Astron. Astrophys. Rev.* **2**, 125.
Bregman, J.N. *et al.*: 1988, *Astrophys. J.* **331**, 746.
Bregman, J.N. *et al.*: 1990, *Astrophys. J.* **352**, 574.
Brown, L.M.J. *et al.*: 1989a, *Astrophys. J.* **340**, 129.
Brown, L.M.J., Robson, E.I., Gear, W.K., Smith, M.G.: 1989b, *Astrophys. J.* **340**, 162.
Celotti, A., Maraschi, L., Treves, A.: 1991, *Astrophys. J.* **377**, 403.
Chini, R., Kreysa, E., Salter, C.J.: 1987, *Astron. Astrophys. (Lett.)* **182**, L63.
Chini, R., Steppe, H., Kreysa, E., Krichbaum, T., Quirrenbach, A., Schalinski, C., Witzel, A.: 1988, *Astron. Astrophys. (Lett.)* **192**, L1.
Chini, R., Kreysa, E., Biermann, P.L.: 1989, *Astron. Astrophys.* **219**, 87.
Clavel, J., Boksenberg, A., Bromage, G.E.: 1990, *Mon. Not. R. astr. Soc.* **246**, 668.
Clavel, J., Reichert, G.A., Alloin, D.: 1991, *Astrophys. J.* **366**, 64.
Clavel, J., Wamsteker, W., Glass, I.C.: 1989, *Astrophys. J.* **337**, 236.
Clegg, P.E. *et al.*: 1983, *Astrophys. J.* **340**, 129.
Clement, R., Sembay, S., Hanson, C.G., Coe, M.J.: 1988, *Mon. Not. R. astr. Soc.* **230**, 117.
Cotton, W.D. *et al.*: 1980, *Astrophys. J. (Lett.)* **238**, L123.
Courvoisier, T.J-L. *et al.*: 1987, *Astron. Astrophys.* **176**, 197.
Courvoisier, T.J-L. *et al.*: 1988, *Nature* **335**, 330.
Courvoisier, T.J-L. *et al.*: 1990, *Astron. Astrophys.* **234**, 73.

Courvoisier, T.J-L., Clavel, J.: 1991, *Astron. Astrophys.* **248**, 389.
Cutri, R.M., Wisniewski, W.Z., Rieke, G.H., Lebofsky, M.J.: 1985, *Astrophys. J.* **296**, 423.
Czerny, B., Elvis, M.: 1987, *Astrophys. J.* **321**, 305.
Doxsey, R. *et al.*: 1983, *Astrophys. J. (Lett.)* **264**, L43.
Edelson, R.A., Malkan, M.A.: 1987, *Astrophys. J.* **323**, 516.
Edelson, R.A., Saken, J., Pike, G., Urry, C.M., George, I.M., Warwick, R.S., Miller, H.R., Carini, M.T., Webb, J.R.: 1991, *Astrophys. J. (Lett.)* **372**, L9.
Feigelson, E.D. *et al.*: 1986, *Astrophys. J.* **302**, 337.
Ghisellini, G., Maraschi, L., Tanzi, E., Treves, A.: 1985, *Astron. Astrophys.* **146**, 204.
Glassgold, A.E. *et al.*: 1983, *Astrophys. J.* **274**, 101.
Hartman, R.C. *et al.*: 1991, *B.A.A.S.* **23**, 1344.
Heeschen, D.S., Krichbaum, T., Schalinski, C.J., Witzel, A.: 1987, *Astron. J.* **94**, 1493.
Hufnagel, B.R., Bregman, J.N.: 1992, *ApJ*, in press.
Hughes, P.A., Aller, H.D., Aller, M.F.: 1989a, *Astrophys. J.* **341**, 54.
———: 1989b, *Astrophys. J.* **341**, 68.
Impey, C.D.: 1987, in *Superluminal Radio Sources*, ed. J.A. Zensus, T.J. Pearson, Cambridge: Cambridge University Press, p. 233.
Kolykhalov, P.I., Sunyaev, R.A.: 1984, *Adv. Space Res.* **3**, 249.
Krolik, J.H., Edelson, R.A.: 1988, *Astrophys. J.* **333**, 646.
Landau, R. *et al.*: 1986, *Astrophys. J.* **308**, 78.
Laor, A., Netzer, H.: 1989, *Mon. Not. R. astr. Soc.* **238**, 897.
Laor, A., Netzer, H., Piran, T.: 1990, *Mon. Not. R. astr. Soc.* **242**, 560.
Malkan, M.: 1989, in *Dynamics of Astrophysical Disks*, ed. J. Sellwood, Cambridge: University Press, p. 89.
Malkan, M.: 1992, in *Physics of Active Galactic Nuclei*, ed. S.J. Wagner and W.J. Duschl, Hdeidelberg: Springer Verlag, in press.
Marscher, A.P.: 1990, in *Parsec-Scale Radio Jets*, ed. J.A. Zensus and T.J. Pearson, Cambridge: Cambridge University Press, p. 236.
Marscher, A.P., Gear, W.K., Travis, J.P.: 1991, in *Variability of Blazars*, ed. Valtaoja, Cambridge: Cambridge University Press, in press.
McHardy, I., Czerny, B.: 1987, *Nature* **325**, 696.
Miller, J.S.: 1989, in *BL Lac Objects*, ed. L. Maraschi, T. Maccacaro, and M.-H. Ulrich, Heidelberg: Springer-Velag, p. 395.
Mutel, R.L., Phillips, R.B.: 1987, in *Superluminal Radio Sources*, eds. J.A. Zensus and T.J. Pearson, Cambridge: Cambridge Univerity Press), p. 60.
Mutel, R.L., Phillips, R.B., Su, B., Bucciferro, R.R.: 1990, *Astrophys. J.* **352**, 81.
Neugebauer, G., Soifer, B.T., Matthews, K., Elias, J.H.: 1989, *Astron. J.* **97**, 957.
Phinney, E.S.: 1990, in *Theory of Accretion Disks*, eds. W. Duschl, F. Meyer, and J. Frank, Dordrecht: Kluwer, in press.
Pounds, K.A.: 1992, in *Physics of Active Galactic Nuclei*, ed. S.J. Wagner and W.J. Duschl, Heidelberg: Springer-Verlag, in press.
Quirrenbach, A., Witzel, A., Qian, S.J., Krichbaum, T., Hummel, C.A., Alberdi, A.: 1989, *Astron. Astrophys. (Lett.)* **226**, L1.
Quirrenbach, A., Witzel, A., Wagner, S., Sanchez-Pons, F., Krichbaum, T.P., Wegner, R., Anton, K., Erkens, U., Haehnelt, M., Zensus, J.A., Johnston, K.J.: 1991, *Astrophys. J. (Lett.)* **372**, L71.
Rieke, G.H., Grasdalen, T.L., Kinman, T.D., Hintzen, P., Wills, B.J., Wills, D.: 1976, *Nature* **260**, 754.
Robson, E.I. *et al.*: 1991, in *Variability of Blazars*, ed. Valtaoja, Cambridge: Cambridge University Press, in press.

Robson, E.I. *et al.*: 1983, *Nature* **305**, 194.
Robson, E.I. *et al.*: 1986, *Nature* **323**, 134.
Rusk, R., Seaquist, E.R. 1985: *Astron. J.* **90**, 30.
Sanders, D.B., Phinney, E.S., Neugebauer, G., Soifer, B.T., Matthews, K.: 1989, *Astrophys. J.* **347**, 29.
Smith, P.S., Balonek, T.J., Heckert, P.A., Elston, R.: 1986, *Astrophys. J.* **305**, 484.
——: 1987, *Astrophys. J. Suppl.* **64**, 459.
Stockman, H.S., Moore, R.L., Angel, J.R.P.: 1984, *Astrophys. J.* **279**, 485.
Sun, W.-H., Malkan, M.A.: 1989, *Astrophys. J.* **346**, 68.
Tagliaferri, G., Stella, L., Maraschi, L., Treves, A., Celotti, A.: 1991, *Astrophys. J.* **380**, 78.
Wagner, S.J.: 1991, in *Physics of Active Galactic Nuclei*, ed. S.J. Wagner, S.J. and W.J. Duschl, Heidelberg: Springer-Verlag, p.598.
Wandel, A., Petrosian, V.: 1988, *Astrophys. J. (Lett.)* **329**, L11.
Wills *et al.*: 1983, *Astrophys. J.* **274**, 62.
Wilson, A.S.: 1988, *Astron. Astrophys.* **206**, 41.
Wilson, A.S., Heckman, T.M.: 1985, in *Astrophysics of Active Galaxies and Quasi-Stellar Objects*, ed. J.S. Miller, Mill Valley: University Science Books, p. 39.
Witzel, A.: 1991, in *Physics of Active Galactic Nuclei*, ed. S.J. Wagner and W.J. Duschl, Heidelberg: Springer-Verlag, p.484.
Witzel, A., Heeschen, D.S., Schalinski, C., Krichbaum, Th.: 1986, *Mitt. Astr. Ges.* **65**, 239.

Infrared to Centimetre Variability of the Quasar 3C 273.

Ian Robson, Simon Litchfield

Department of Physics and Astronomy
Lancashire Polytechnic
Preston PR1 2TQ, UK

Abstract: We present a series of lightcurves of the quasar 3C 273 from the infrared to centimetre wavelengths. Correlation analysis reveals that the synchrotron emission from the infrared to around 2 mm is from a single component, while the 3 mm to centimetre wavelength emission is most probably a secondary emission component from further down the jet. This is in agreement with the findings from our multifrequency blazar observations reported by Brown *et al.* (1989a).

1 Introduction

In terms of the discussion at this meeting of the variability of the central engine and the link with the relativistic jet, multifrequency observations are a key to furthering progress. However, such observations are heavily discriminated against in this modern epoch of large groundbased, common-user national facility telescopes. A few groups have managed to operate private facilities and have produced extremely valuable databases, notably in the radio, but also in the optical (see the discussion by Robson 1991).

The difficulty of continued monitoring, plus the annoying fact that the sun prevents most of the sources from being observed throughout the year for all wavelengths shortward of about 0.4 mm, means that temporal databases are extremely patchy, with the data sampling being highly irregular.

There are two approaches to attempt to determine the characteristic behaviour of the synchrotron emission due to the relativistic jet associated with the central engine: (i) take a lengthy timebase for a single source and follow the evolution of flares with time at a number of wavelengths; (ii) take a statistically large sample of objects and monitor them at a number of wavelengths. Both approaches are fraught with difficulty.

Joel Bregman and collaborators have been the master exponents of the first method, with a number of excellent papers in the literature (*e.g.* see Bregman and Hufnagel 1989 and references therein). The Preston group have tried to make headway with the second method, and results up to 1988 are presented in Brown *et al.* (1989 a,b). In this paper we describe the work from 1988 to 1991 on the best measured source in our sample, the local, luminous quasar 3C 273.

2 Observational Programme

The Preston programme of monitoring a sample of 18 blazars obtains infrared observations from the United Kingdom Infrared Telescope (UKIRT) SERVICE programme and 0.45 mm – 2.0 mm observations from a programme on the James Clerk Maxwell Telescope[7] (JCMT) which has had long term status for the past three years. We have also been assembling data from a wide variety of sources in the literature and we are also involved in programmes using the SEST millimetre telescope. The data described in this paper come, in addition to the above facilities, from infrared observations at ESO, millimetre measurements using the Berkeley-Illinois-Maryland (BIM) interferometer and IRAM and mm-cm observations from the Metsahovi monitoring programme. The data will be published in full in a paper by Robson *et al.* (1992).

In relation to the topic of this conference, one point to note from our previous work is that the single epoch infrared through centimetre lightcurve for blazar emission cannot in general be described by a single parabolic curve as proposed by Landau *et al.* (1986), rather there is good evidence that the infrared through 1 mm is a single component, and the longer millimetre through centimetre emission comes from a second component (see Brown *et al.* 1990a). This is illustrated in the paper by Joel Bregman in these conference proceedings.

The form, strength and variability of the synchrotron spectrum has clear implications for the emission processes and further campaigns were undertaken in an attempt to extend the above two-component conclusion to a wider sample of objects and to test whether there is any correlation with jet parameter or luminosity. Unfortunately, although two campaigns have been carried out, on neither were all the four telescopes operating successfully (mainly due to weather problems) and so the wavelength coverage was not as complete as planned. The complete data and analysis will be published by Gear *et al.* (1992) and a further campaign is currently planned for 1992.

3 Discussion

3.1 Overview

The goal of any multifrequency monitoring programme is to assemble a temporal database such that a number of factors can be determined: viz (i) the quiescent spectral energy distribution of the emission from the radio through to the gamma regime and hence (ii) identification of the physical emission processes producing them; (iii) the flaring behaviour at all wavelengths and hence (iv) identification of the physical processes

[7] The James Clerk Maxwell Telescope is operated on a joint basis between the United Kingdom Science and Engineering Research Council (SERC), the Netherlands Organisation for the Advancement of Pure Research (ZWO), the Canadian National Research Council (NRC) and the University of Hawaii (UH)

of the emission; (v) the relationship between the flaring behaviour and hence the causal linkage of the emission mechanisms.

We have selected 3C 273 for intensive study because it has by far the best sampled lightcurves at all wavelengths. Previous detailed studies of the behaviour of this source can be found in Courvoisier *et al.* (1987 and 1990). In this discussion we shall concentrate only on the synchrotron emission from the relativistic jet. The current best models for describing the jet behaviour are those based on shocks in relativistic jets as described by Marscher & Gear (1985), Hughes, Aller and Aller (1988) and Marscher, Gear and Travis (1991).

Synchrotron emission dominates the radio through to mid-infrared part of the energy output of 3C 273. It should be noted however that when 3C 273 is in its low, or quiescent state, the near infrared emission is not of synchrotron origin, but dominated by another component which is most probably due to dust heated by the central engine. However, as soon as the quasar flares, the synchrotron component dominates the radio through to the infrared and, for major flares, into the optical regime.

In this paper we shall discuss only one aspect, the variability of the synchrotron emission. Other aspects are discussed in Robson *et al.* (1992). The data comprise 12 relatively well sampled wavebands from 1 micron to 13.6 mm. Data at 5, 10 and 20 microns are poorly sampled and are not considered further in this paper. A representative sample of wavebands is shown in Fig. 1. Inspection and subsequent analysis (see below) reveals that all the infrared wavebands are indistinguishable, likewise the 0.8–1.3 mm are also sufficiently close that any one could be taken as a representative sample.

The overall behaviour of 3C 273 can be seen as

- fast flaring behaviour at IR wavelengths in the spring of 1988,
- near mm flaring at the same time with rapid fluctuations,
- a slower rise to flare maximum at longer mm and cm wavelengths,
- a second flare in the summer of 1988, not seen at IR wavelengths,
- a period of quiescence during 1989, the flaring of 1988 slowly decaying at the longer mm and cm wavelengths,
- a flare, well sampled at most wavelengths, during the summer of 1990,
- a further flare during the spring of 1991.

A number of features can be observed: as expected from standard synchrotron theory, events have shorter timescales at higher frequencies; clear linkage can be identified between the flaring behaviour; cooling times can be derived for the longer wavelengths.

The temporal axis of Fig. 1 is too coarse to reveal the fine details of the flaring behaviour, but some of the flares and fades occur on timescales of hours, notably in early 1988 with infrared luminosity changes of 10 million solar luminosities per second (assuming isotropic emission and a Hubble constant of 50 km/s/Mpc). Flaring of 50% in 3.5 hours was observed at 0.8 and 1.1 mm followed by a fall to the quiescent level in three days. Such fast events lead one to consider the possibility that geometrical factors may be influencing the emission, rather than an increase in energy of a relativistic population of electrons.

We will not consider the details of the emission here, but instead concentrate on the correlation of the respective wavebands.

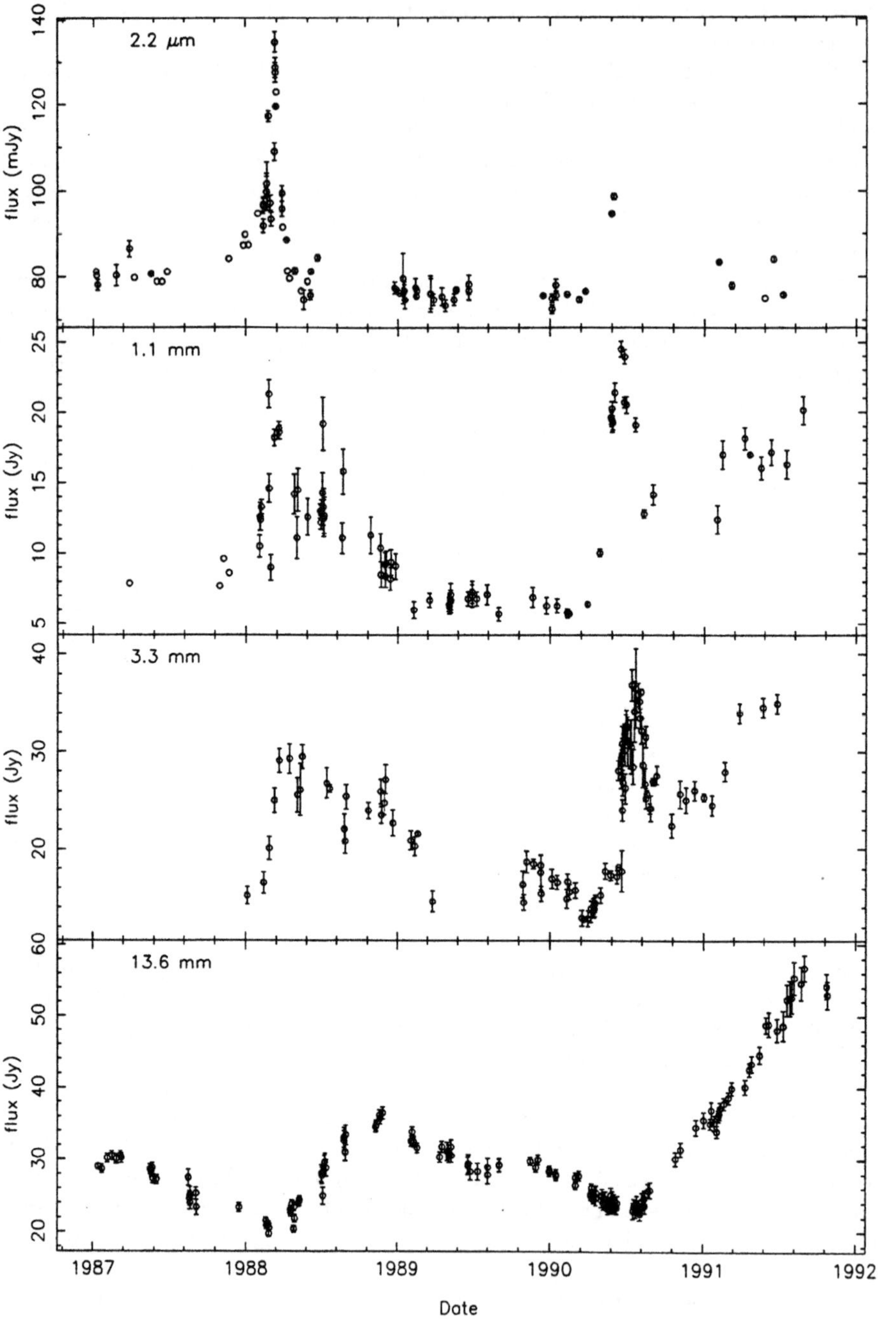

Fig. 1. A representative sample of lightcurves of 3C 273 from 1988 to 1991.

3.2 Correlation Analysis

When considering the analysis of irregularly sampled sets of data, there are a number of methods of deriving correlations. These have been well described by Joel Bregman at the Como blazar conference (Bregman & Hufnagel 1989). We have taken the discrete correlation analysis method given by Edelson and Krolick (1988). A sample of results is plotted in Fig. 2. The maximum peak in the plot indicates a correlation and a positive time lag occurs where the shorter wavelength precedes the longer wavelength (which is always the case).

It is clear that a consistent picture can be built up for the synchrotron emission from 3C 273. When flaring, the infrared through to 1.3 mm reacts either simultaneously or within about five days (the latter is consistent with zero time delay within the resolution) and then decays with a time-constant proportional to wavelength; there is a significant delay to the 3.3 mm emission from the IR and mm wavelengths; there are increased delays to longer wavelengths.

Although it is not clear that all flares should behave in the same way, there is little evidence for significant differences in the data presented. It makes insignificant difference to the correlation plots whether an individual flare or all flares are investigated, as long as there is sufficient sampling of the timespan concerned. However, this statement might be a reflection of the quality of the data sampling rather than on a unique flare behaviour.

If it is assumed that the flare is commensurate with the IR behaviour and this is taken as zero lag, then the flare propagation is given by

IR —	1.3 mm	zero – 5 day lag
—	3.3 mm	40 days
—	8.1 mm	100 days
—	13.6 mm	160 days

3.4 The IR-Near Millimetre Component

The above clearly shows the break between 1.3 mm and 3.3 mm. This is entirely consistent with the earlier observations of the snapshot mm-cm spectra of blazars as reported in Brown *et al.* (1990a). We see that the synchrotron emission from the flaring component is manifest as a single component extending from the shortest observable wavelengths (IR or optical) through to the near mm–regime. The flare emission peaks at about 1 mm and is then presumably optically thick to longer wavelengths. The flare then propagates to longer wavelengths, by a mechanism not yet understood, but presumably reflecting optical depth effects. This is under detailed investigation at the current time in terms of jet modelling.

Given the correlation link between the IR-mm flaring and that at 3 mm, it is interesting to explore the relation of the 3 mm VLBI from 3C 273 with this flaring behaviour. One would expect that major flaring would be accompanied by the appearance of a new superluminal blob in high frequency VLBI studies, the appearance lagging behind the IR/1 mm flare by the order of a month. Unfortunately, VLBI at these very high frequencies is extremely difficult and few campaigns have been attempted as yet (Bååth *et al.* 1991).

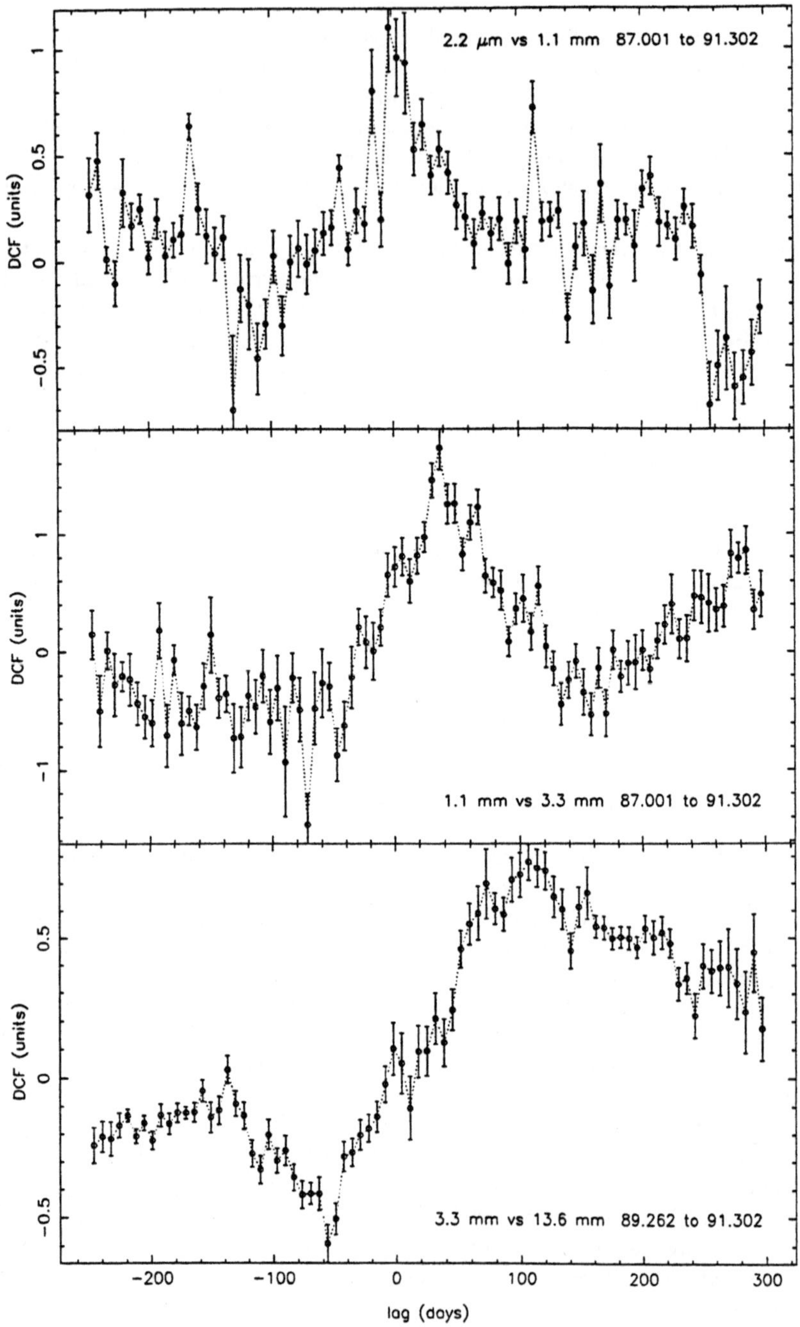

Fig. 2. Discrete correlation coefficient curves (DCF) for 3C 273 for a sample of wavelengths (a) K-band v 1.1 mm; (b) 1.1 mm v 3.3 mm; (c) 3.3 mm v 13.6 mm. The maximum value of the DCF represents the primary cross correlation time and a positive lag indicates that the shorter wavelength precedes the longer wavelength data.

3.3 The Centimetre Wavelengths

There is clearly a link between the mm and cm behaviour, but because the cm emission is effectively degraded, it is not expected to be such a good indicator of the energetics of the flaring behaviour, although one would expect the lightcurve to reveal the major flares. However, as is seen from Fig. 1, when multiple fast flaring or flickering is underway, the longer time-constants at cm wavelengths preclude observing this phenomena. Essentially the cm reveals more about the propagation of the flare down the jet and the physics therein. This is where the excellent work of Hughes, Aller and Aller (1988) becomes important and links directly with the cm VLBI spatial evolution of the blobs of emission in the jet as superluminal motion.

At the shorter wavelengths, 7 mm VLBI is becoming a wavelength at which regular campaigns are now carried out and Krichbaum *et al.* (1990) have already produced evidence of the link between an IR/mm flare and a VLBI component. Figure 3 from their paper shows the expansion diagram for the jet of 3C 273 as obtained from VLBI studies. Projection of the component C9 back to zero separation clearly fits with the spring 1988 flare (see Krichbaum *et al.* for an extensive discussion concerning this aspect).

Clearly, the further away the wavelength from the peak of flaring, the greater is the possibility that either a flare may die away before it can be observed by VLBI, or that with the increased time constant at the longer millimetre wavelengths, the flares may merge together. Hence extrapolating back from the cm regime to look for flares in the optical/IR lightcurves is fraught with difficulty due to: (a) the much faster time-constant in the short wavelength regimes, (b) the possibility that flares are missed because of relatively poor sampling compared with the cm radio monitoring campaigns, (c) the fact that 3C 273 spends about 5 months of the year in an unobservable state being too close to the Sun.

4 Conclusions

We have clearly shown that for the quasar 3C 273, the synchrotron emission associated with the relativistic jet has one component which peaks at about 1 mm, is optically thick to longer wavelengths, and is optically thin to the far and mid-infrared. This is the component which represents the most flaring aspect of the synchrotron emission and we believe directly reflects the injection of a new relativistic electron population or a re-acceleration of existing electrons by the central engine. There is a further component which peaks in the cm regime which presumably comes from regions in the jet further from the core and flare emission propagates to these wavelengths from the IR-mm regime as the optical depth changes with time.

If all flares from the IR-mm regime propagate out to 3 and 7 mm, as seems highly likely, then it is probable that all flares are associated with superluminal components. 3C 273 appears to have undergone a much more active phase over the past few years and we predict high frequency VLBI blobs to emerge associated with the summer 1990 and spring 1991 flares. We look forward to confirmation of this from our VLBI colleagues.

When VLBI moves to the highest currently achievable frequencies of the near mm regime on a more regular basis, then for the first time, the flux of the flare emission of the primary synchrotron radiating electrons of the jet can be related to the spatial size and motion of the radiating blob. An exciting prospect for the near future.

References

Bååth, L.B. *et al.*: 1991, *Astron. Astrophys. (Lett.)* **241**, L1.

Bregman, J., Hufnagel, B.R.: 1989, Lecture Notes in Physics 334, eds. L. Maraschi, T. Maccacaro, and M.-H. Ulrich, p159.

Brown, L.M.J. *et al.*: 1989, *Astrophys. J.* **340**, 129.

Brown, L.M.J., Robson, E.I., Gear, W.K., Smith, M.G.: 1989, *Astrophys. J.* **340**, 150.

Courvoisier, T.J.-L. *et al.*: 1987, *Astron. Astrophys.* **176**, 197.

Courvoisier, T.J.-L. *et al.*: 1990, *Astron. Astrophys.* **234**, 73.

Edelson, R., Krolick, J.H.: 1988, *Astrophys. J.* **333**, 646.

Gear, W.K.: 1993, in preparation.

Hughes, P.A., Aller, H.D., Aller, M.F.: 1988, *Astrophys. J.* **341**, 54.

Krichbaum, T.P *et al.*: 1990, *Astron. Astrophys.* **237**, 3.

Landau, R. *et al.*: 1986, *Astrophys. J.* **308**, 78.

Marscher, A., Gear, W.K.: 1985, *Astrophys. J.* **298**, 114.

Marscher, A., Gear, W.K., Travis, J.P.: 1992, in *Variability of Blazars*, eds. E. Valtaoja & M. Valtonen, Cambridge University Press, p.85.

Robson, E.I.: 1991, *ibidem* , p.111.

Robson, E.I. *et al.*: 1993, in preparation.

Multifrequency Observations of Intraday Variability: Implications for the Emission Processes

Stefan J. Wagner

Landessternwarte Heidelberg
Königstuhl
D-6900 Heidelberg, Germany

Abstract: Recently it has been suggested that a class of compact flat-spectrum sources varies both in the radio and optical waveband regimes in a correlated fashion on very short time-scales. In this note further properties of this rapid variability are described. The short lag observed between the various waveband regimes argues in favour of a close spatial and physical correlation of the regions dominating the emission at the corresponding frequencies. A few implications and speculations are discussed.

1 Introduction

Besides VLBI imaging, the small-scale structure of jets has often been studied from variability observations. Detailed models have been suggested on the basis of the scenario of shocks propagating within the jet plasma (e.g. Königl and Choudhuri, 1985, Marscher, 1991, Hughes *et al.* 1989). Although all of these models refer to variations on time-scales of weeks to months, more rapid variations are predicted by most models for optically thin regions if the shocks enter turbulent regions. Rapid variations at optical frequencies are easily explained within such schemes. At lower frequencies the rapid variations would be washed out completely.

Rapid variations, as described by Heeschen (1984) and others, were, therefore, ascribed to interstellar scintillation. As has been shown by Witzel *et al.* (1986), however, rapid variability also occurs at high galactic latitudes where interstellar scintillation is expected to be a rare phenomenon. The observed frequency dependence and swings in polarization angle, as observed in 0917+62 (Quirrenbach *et al.* 1989a, b), gave further strength to the assumption that intraday variability (IDV) at radio frequencies may be an *intrinsic* phenomenon. Wagner *et al.* (1990) found a statistical correlation during simultaneous observations at optical and radio frequencies where the same sources were found to be variable or quiet in both wavebands.

For an individual object, the compact radio source S5 0716+714, correlated variations were found during a four-week monitoring campaign at the VLA and the optical observatories at Calar Alto and Heidelberg-Königstuhl (Quirrenbach *et al.* 1991; Wagner, 1991).

The correlations between optical and radio variations would rule out an extrinsic, propagation-induced origin, leaving emission from compact structures in optically thin conditions as the only alternative.

2 The Multifrequency Behaviour

The most amazing behaviour is the close correspondence of optical and radio fluxes. As described by Quirrenbach, 1991, 0716+714 showed a change in modes during the four-week monitoring campaign in Feb. 1990 when flipping from intraday variability to less rapid variations on time-scales of ~ 5 days. This change occurred at radio-frequencies and in the optical regime nearly simultaneously with a time-lag of less than a day. During each of the two modes, the light-curves at 6 cm and 650 nm are very similar. Cross-correlation functions of 650 nm and 6 cm data during the two modes are shown in Figs. 1a, b. While the near-simultaneous change of modes argues for a time-lag of less than 1 day, we find a similar quasi-periodicity during the second part, when the cross-correlations are multiple-peaked with possible time-lags of about -2^d or $+4^d$.

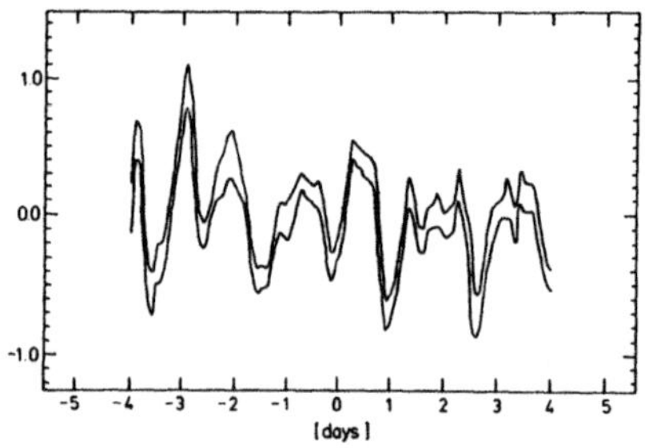

Fig. 1a. Cross-Correlations of radio and optical light-curves during the period of rapid variations of 0716+714 in February 1990.

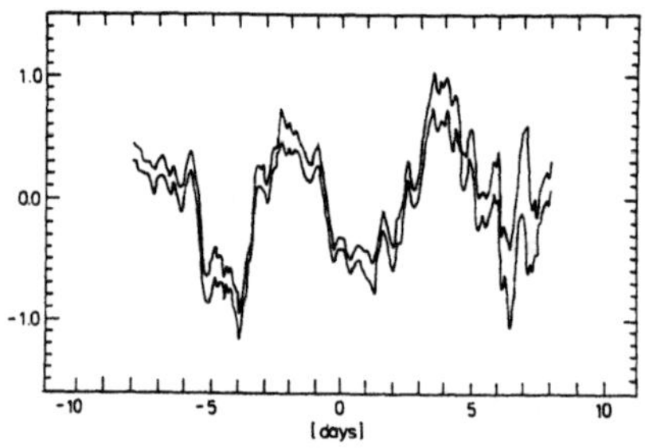

Fig. 1b. Same as 1a for the slower mode. Multiple peaks occur in the ccf. due to the quasi-periodic nature of the variation, making an exact determination of the lag difficult.

Within the optical regime, changes in spectral slope remain small on short time-scales. B, R, I colours were measured during several campaigns. No significant changes were found on intraday time-scales (Fig. 2). On longer time-scales the source slightly reddens during bright phases.

Within the radio regime significant changes of spectral indices are observed. Figure 3 shows the 3.6 cm, 6 cm and 650 nm variations of a limited part of the campaign. In spite of the small gap between 3.6 cm and 6 cm, the lag between these two frequencies is comparable to the lag displayed in Fig. 1a. In spite of the lag the amplitudes and time-scales are very similar at 3.6 cm and 6 cm. Optical depth effects do not wash out IDV.

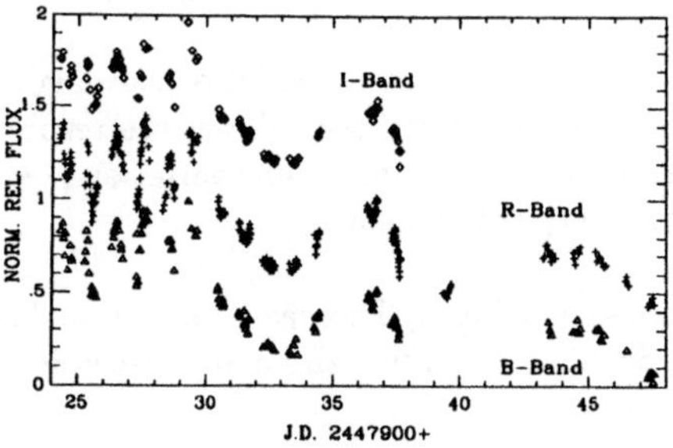

Fig. 2. Variations of 0716+714 at different optical bands during the Feb. 1990 campaign. Only small spectral changes are seen on the short time-scales.

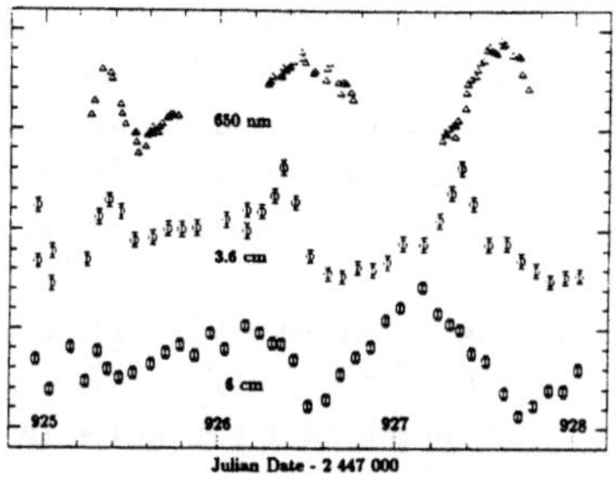

Fig. 3. A three day period of the variations of 0716+714 in Feb. 1990 at 650 nm (top), 3.6 cm (middle) and 6 cm (bottom). Clear lags can be seen in the radio regime.

In February 1990 we performed simultaneous optical-UV monitoring during 2 subsequent IUE shifts. 0716+714 varied significantly during these 32 hours. Within the measurement accuracy, the UV light-curves are identical with the R-band variations with a spectral slope R-band – UV identical to the R-B colour ($\alpha_{(650\,\mathrm{nm}-250\,\mathrm{nm})} = \alpha_{(850\,\mathrm{nm}-400\,\mathrm{nm})}$). A short simultaneous optical – X-ray campaign took place in March 1991 during which a close correspondence of optical and X-ray fluxes were found. A preliminary description is given in Wagner, 1992.

3 Emission Processes

The comparable time-scales, the simultaneous changes of modes and the statistical similarity of the light-curves of 0716+714 at 650 nm and 6 cm argue against the assumption that the variations are caused by two completely different mechanisms (such as intrinsic variations at optical frequencies and scintillation induced variations in the radio regime). Since this is a statistical argument, further simultaneous campaigns are needed to strengthen this conclusion. If the variations are intrinsic, they are not explained by the current models of shocks propagating in shocks. They do not explain the high brightness-temperatures required from the short time-scales. Even for the optical variations it is difficult to explain the rapid declines. If the variations were due to geometric effects (e.g. by emission regions moving on curved trajectories, causing the beam to sweep over the line-of-sight) shorter time-scales could be explained. Depending on the shape of the radiation characteristics at different frequencies, such a scenario may also reproduce the lags between the different frequencies in the GHz regime.

The similarity of the variations at optical, UV and X-ray frequencies indicate that the photons in these regimes are either due to the same radiation mechanism, or to *in situ* reprocessing. Classical SSC models are inconsistent with the low optical thickness in the radio regime. If the emission in the GHz regime is not due to incoherent synchrotron emission, but is emitted coherently by an anisotropic distribution of primary particles, the latter could also act as an instantaneous but anisotropic comptonizing reprocessor of the coherent synchrotron emission.

It would also be possible that the primary particles of the rapid variations are highly relativistic particles, whose decay-products and downscattered secondaries are responsible for the close correlation between X-ray and optical variations. The variations seen at radio-frequencies would then be primarily a skin-phenomenon.

Acknowledgement: It is a pleasure to thank Jochen Heidt, Thomas Krichbaum, Andreas Quirrenbach, and Arno Witzel for the enjoyable cooperation and many discussions. This work is partially supported by SFB 328.

References

Heeschen, D.S.: 1984, *Astron. J.* **89**, 1111.
Hughes, P.A., Aller, H.D., Aller, M.F.: 1989, *Astrophys. J.* **341**, 54
Königl, A., Choudhuri, A.: 1985, *Astrophys. J.* **289**, 173
Marscher, A.: 1991, in: *Physics of Active Galactic Nuclei*, ed. S.J. Wagner and W.J. Duschl, Heidelberg: Springer Verlag, 510.
Quirrenbach, A., Witzel, A., Krichbaum, T., Hummel, C., Alberdi A., Schalinski, C.: 1989a, *Nature* **337**, 442.
Quirrenbach, A., Witzel, A., Qian, S.J., Krichbaum, T., Hummel, C., Alberdi, A.: 1989b, *Astron. Astrophys. (Lett.)* **226**, L1.
Quirrenbach, A., Witzel, A., Wagner, S., Sanchez-Pons, F., Krichbaum, T., Wegner, R., Anton, K., Erkens, U., Haehnelt, M., Zensus, J., Johnston, K.: 1991, *Astrophys. J. (Lett.)* **372**, L71.
Wagner, S.J.: 1991, in *"Variability of Active Galaxies"*, Lecture Notes on Physics, Vol. 377, ed. W. Duschl, S. Wagner, M. Camenzind, Heidelberg: Springer Verlag, p.163.
Wagner, S.J.: 1992, in *"X-ray Emission from AGN and Cosmic X-ray Background"*, ed. W. Brinkmann and J. Trümper, Garching: MPE report 235, p.97.
Wagner, S.J., Sanchez-Pons, F., Quirrenbach, A., Witzel, A.: 1990, *Astron. Astrophys. (Lett.)* **235**, L1
Witzel, A., Heeschen, D.S., Schalinski, C., Krichbaum, T.: 1986, *Mitt.Astron.Ges.* **65**, 239.

Formation of Jets in Extragalactic Radio Sources

M. Camenzind

Landessternwarte Königstuhl
D-6900 Heidelberg, Germany

Abstract: There is now a fairly clear indication that not all extragalactic jets are relativistic. The fundamental parameters of the parsec-scale jets (radius, speed, density, magnetic fields, collimation) must be related in a way to the properties of the very central object and its accretion disk. Early models of jet formation were based on pure hydrodynamic acceleration in the funnels of geometrically thick tori. For sub-Eddington luminosities such flows are only accelerated to moderate jet speeds, $v_j \leq 0.3c$. The majority of jet formation models is nowadays, however, based on magnetohydrodynamic processes of geometrically thin disks. The question of an electron-positron load in the central region of the jet is of particular importance. The rapid rotation of the accretion disk around a black hole essentially forbids the use of the Newtonian approximation. It turns out that relativistic jets are naturally formed in the rapidly rotating magnetospheres of geometrically thin accretion disks around rotating black holes. Disk fields generated by dynamo effects in the boundary layer between disk and rotating hole immerse the hole in a rotating magnetosphere which is loaded by plasma from the disk. Magnetized winds ejected from the surface of the disk will be collimated by magnetic effects on scales somewhat larger than the light cylinder of the magnetosphere. Non-axisymmetric perturbations dragged along by these jets will rotate and produce quasi-periodic emission by a kind of lighthouse effect.

1 Introduction

The actual formation of jets in radio galaxies is fundamental for the understanding of the emission from these objects, as well as for the astonishing ability of these jets to stay together over a very large range of distance scales. On their way from the black hole in a galactic nucleus to the hot spot and the radio lobe, cosmic jets cover a stupendous factor of 10^8 in length scale — this is beyond any capability of modern computers.

The actual initiation of jets remains a subject of controversy despite the obvious importance of this fundamental topic. There are many reasons for this. First, detailed structural information on the subparsec-scale is still lacking. VLBI observations can now provide structural information on scales of ≤ 0.1 pc for extragalactic sources (Krichbaum, 1990; Bååth *et al.*, 1991), but the processes that govern the initiation of the jets occur on still smaller scales. The mm-VLBI is now able to resolve the parsec-scale structure and therefore to produce some constraints on possible models for the knots on this scale.

A second difficulty has to do with the complexity of the physical processes which are involved in producing jets. Even in the case of stellar jets, we have not yet reached the stage of a fully accepted model for the generation of those jets (Camenzind, 1992). We only know that somehow a central star is surrounded by an accretion disk, formed during stellar collapse, and that magnetic fields must play a major role in the initiation of the outflows. In the case of Active Galactic Nuclei, most astrophysicists assume the existence of a supermassive black hole in the center of the galaxy with a mass M_H somewhere between 10^6 and $10^{11}\,M_\odot$ also surrounded by an accretion disk which extends to a few thousand Schwarzschild radii. However, this object could also be replaced by a non-collapsed supermassive fast rotator producing a system very similar to the protostellar case. Also, the existence and influence of magnetic fields is one of the key questions for the jet formation mechanisms. This all means that general relativity, magnetohydrodynamics, plasma physics and radiation transport must be included in the various scenarios.

This latter point is the origin of a third difficulty in this field of research. It is clear that all this input physics cannot be carried out in its full extent. One has to deal with some limiting cases, *e.g.* leave aside plasma physics and general relativity in a first attempt. The rest of the physics must be carried out correctly if one wants to produce quasar jets. As an example, many people dream of relativistic jets by thinking of Newtonian MHD. As a rule, special relativity must always be included — and this already rules out many of the "models" discussed in the literature.

2 The Standard Model for AGN

Extragalactic radio sources are a subset of Active Galaxies which show, by definition, a strong nuclear activity (Seyfert galaxies, QSOs, quasars). All of these sources have most probably jets, or at least a kind of bipolar outflow in the nuclear region. But the power in the jets of Seyfert galaxies *e.g.* is many orders of magnitude lower than the overall power of the AGN, while in a typical jet source, such as 3C 273, jet power and AGN power are comparable. The difference between these two types of AGNs (historically called radio-weak and radio-loud sources) must be hidden in the physics of the nuclear region of the corresponding host galaxies. Despite many attempts to resolve this key puzzle of AGN physics, there is no convincing argument in the air.

2.1 The Various Elements of the Standard Model

The generally accepted standard model for AGN assumes the existence of a supermassive rotating black hole in the center of a galaxy surrounded by a Keplerian accretion disk with an extension of a few hundred Schwarzschild radii (≤ 0.1 lyr). This entire structure is embedded into a nuclear star cluster with mass $M_c > 10^{10}\,M_\odot$ and core radius $R_c \geq$ parsecs. This inner disk is fueled by material present in the form of molecular clouds on the scale of a few parsecs to a few hundred parsecs. In the innermost few parsecs, galactic matter is assembled in the form of a molecular torus, quite similar to structures found in our galactic centre. Direct evidence for the existence of the inner accretion disks follows from the presence of a strong optical-UVX excess in the continuum of quasars. Cold molecular gas appears as a thermal component in the far-infrared, while hot dust is the origin of the near-infrared emission in radio-weak objects (Bregman, 1990).

2.2 What are Jets made of ?

Jets contain in general four components:

(1) normal plasma consisting of ions and electrons injected from the normal disk plasma;

(2) magnetic fields, visible in the synchrotron emission and the polarization of the central core of the jet and in VLBI knots;

(3) relativistic electrons also visible in the synchrotron emission with Lorentz factors up to at least 1000;

(4) a possible electron-positron load injected from the immediate vicinity of the black hole; until now no sign of electron-positron annihilation is visible in the gamma-ray spectra of AGN.

In any jet model, normal plasma from the disk is the basic background for the jet matter. The origin of the magnetic fields visible in the synchrotron emission remains, however, mysterious in many jet models, in particular in pure hydrodynamic models. Many authors assume that these fields are completely turbulent with arbitrary turbulence scales. Jets could amplify the magnetic fields by dynamo action if the plasma streamlines are helical (Ruzmaikin *et al.*, 1988; Gvaramadze *et al.*, 1988).

2.3 Origin and Essence of Magnetic Fields in Jets

In my opinion, the most natural origin for these magnetic fields is simply the plasma itself which is injected into the jet. Magnetic fields are part of the jets, just as magnetic fields in the solar wind are also part of the solar wind. The fact that the magnetic fields of the solar wind are strongly turbulent near the Earth does not imply that they are already turbulent at the birth place of the solar wind. Similarly, the fact that magnetic fields of jets observed on the scale of a few hundred of parsecs (VLBI scale observed in the GHz regime) are turbulent does not necessarily imply a turbulent nature of these fields at the birth place of the jets, i.e. on a scale of a factor 1000 smaller. If the jet plasma has its origin in the accretion disk, it will also carry along magnetic fields. The interstellar matter which accretes onto the central objects in galaxies also brings in magnetic fields. Due to the strong differential rotation and the turbulence in the disk, these fields evolve according to the induction equation of mean field electrodynamics. Magnetic fields in the disk grow rapidly to equipartition with the turbulent pressure in the disk. Under high accretion rates, $\dot{m} = \dot{M}/\dot{M}_{\rm ED} > 0.001$, the pressure in the disk is dominated by radiation pressure, which makes the magnetic field B_D in the disk practically independent of the accretion rate $\dot{m} = \dot{M}/\dot{M}_{\rm ED}$

$$B_D \leq B_{ED}\,(M_H/R)^{1/2} \quad , \quad B_{ED} = \sqrt{L_{ED}/M_H^2 c} = 4\times 10^4\,\text{Gauss}\,M_{H,8}^{-1/2}\,. \qquad (1)$$

This implies that, in addition to the mass M_H, the angular momentum a_H of the black hole and the accretion rate $\dot{m}$, also the disk field B_D in the inner part of the accretion disk is one of the essential parameters for a classification of AGNs. In Camenzind (1990) we discuss such a classification scheme which is based on the mass M_H, the accretion rate $\dot{m}$ and the field B_D. Strong quasars such as 3C 273 and 3C 345 must have high masses and relatively high accretion rates $\dot{m} \geq 0.1$. Their magnetic fields are therefore near the upper limit given by the Eddington argument Eq. (1). Mass limits are known for M 87, $M_H \leq 10^9\, M_\odot$, implying therefore a low accretion rate, $\dot{m} \simeq 10^{-4}$. The corresponding accretion disk cannot be radiation pressure supported, so that the typical magnetic field is then about a factor 10 lower. In between these two extremes, we find the typical radio sources such as FR I and FR II type radio galaxies. In addition, the masses M_H may cover a large range, since this mass is most probably related to the bulge mass of the galaxy. One should not forget that objects like 3C 273 and 3C 345 are associated with extremely massive galaxies, their black holes could have masses exceeding $10^{10}\, M_\odot$. The mass of the black hole is, however, not the key parameter for the generation of relativistic jets, the nearest superluminal object 3C 120 is associated with a moderate galaxy with a black hole mass $M_H \simeq 10^8\, M_\odot$ (Maraschi *et al.*, 1991). Are there superluminals with even lower masses ?

The disk fields of geometrically thin disks are dominantly toroidal corresponding to a total current driven by the differential rotation of the disk

$$I_D \leq 3 \times 10^{18}\,\mathrm{Amp}\, M_{H,8}^{1/2}\, (R/M_H)^{1/4}\,. \tag{2}$$

This explains the typical magnetic fields found in the large-scale jets (3C 273: $B_j \simeq 200\,\mu$Gauss for jet radii of a few kpc, M 87 : $B_j \simeq 100\,\mu$Gauss for jet raddii of a few hundred parsecs) when part of this current is closed over the jet

$$B_j \leq 2 I_D/cR_j \simeq 70\,\mu\mathrm{Gauss}\,(I_D/10^{18}\,\mathrm{Amp})\,(\mathrm{kpc}/R_j)\,, \tag{3a}$$

or in the compact nuclear jets

$$B_j \leq 0.7\,\mathrm{Gauss}\,(I_D/10^{18}\,\mathrm{Amp})\,(0.1\mathrm{pc}/R_j)\,. \tag{3b}$$

In general, the magnetic field in the jet follows roughly from the remarkably simple scaling law $B_j \simeq B_D\, R_D/R_j$. Since the radius of the jet in M 87 is about a factor 10 smaller than the radius of the jet in 3C 273, the scale of the inner disk in M 87 must also be at least a factor 10 smaller than the corresponding radius in 3C 273. This is in agreement with the mass estimates (M 87: $M_H \leq 10^9\, M_\odot$ with $M_B = -22.2$ and 3C 273: $M_H \geq 3 \times 10^9\, M_\odot$ with $M_B \simeq -24.0$ for $H_0 = 50$ km/s/Mpc). It is not easy to find an alternative argument for the origin of the strength of the magnetic fields as observed in the jets.

The typical strengths of the magnetic fields derived from observation also show that the sound speed c_S is not the characteristic speed in the jet. Instead of this we have to use the magnetosonic speed v_{MS} defined as (Camenzind, 1990)

$$v_{MS} = \sqrt{\frac{B_j^2}{4\pi m_p n_j}} = \sqrt{\frac{B_j^2 R_j^2 c\beta_j}{4\dot{M}_j}} \simeq 0.7\, c\, \frac{I_D}{10^{18}\, A}\, \beta_j^{1/2} \sqrt{\frac{10^{-2}\, M_\odot/\mathrm{yr}}{\dot{M}_j}}\,. \tag{4}$$

This is the main argument which forces the use of MHD instead of HD for the jet manufacturing. Since v_{MS} is only determined by current and mass flux, the argument holds for the compact jets as well as for the kiloparsec-scale jets. In addition, since v_{MS} is close to the speed of light, we are also forced to use special relativity, otherwise the characteristic speeds are not correctly determined (Appl & Camenzind, 1988).

3 Gasdynamic Bipolar Outflows

Despite the fact that magnetic fields determine the characteristic speeds of jets, one large class of early models for jet formation assumes a priori that magnetic fields do not play a dominant dynamical role in the extraction of energy from the central object, nor are they important in the acceleration and collimation of beams. Although thin accretion disks obviously define a preferred direction for the ejection of matter that could then be collimated into beams, there are no well-developed models for jet production in the absence of dynamically controlling magnetic fields. Disk winds are certainly a generic phenomenon for geometrically thin disks, but the collimation mechanisms are unclear in this case.

More promising for a hydrodynamic production of jets is a geometrically thick disk with steep funnel walls. Such disks should be formed either under high accretion rates, $\dot{m} > 0.5$ (radiation pressure supported tori), or under extremely low accretion rates, $\dot{m} \leq 10^{-4}$ (ion tori). The time evolution of axisymmetric supercritical accretion disks around Newtonian quasi-black holes has been investigated by Eggum *et al.* (1988). The solutions have in general four distinct zones: (A) Centered on the disk midplane is a thick, dense region of turbulent convection. This convective core is characterized and dominated by the dynamics of convection cells. This convective core is variable. In a thick disk, both entropy and angular momentum contribute to the buoyancy of matter. (B) Above the convective core is an accretion zone in which low angular momentum matter rapidly flows onto the black hole. (C) This accretion zone is bounded by a photocone, in which matter becomes optically thin. The photocone is the source of both the mass and the radiation which drive an outflow near the rotational axis. (D) Inside the photocone, surrounding the angular momentum axis, they find a broad outflow of expelled matter. The primary source of this outflow is near $9\,GM_H/c^2$; here the radiative flux which drives the outflow becomes super-Eddington, and the mass-flow bifurcates into an inward-directed accretion flow and an outward-directed badly collimated jet flow.

Super-Eddington numerical solutions always produce radiation-pressure driven outflows. These outflows are conical in shape, with opening half-angle of $25° - 35°$. The apex of the cone, which corresponds to the point where radiation pressure balances gravity, lies at $z_0 \simeq 9\,GM_H/c^2$ along the rotational axis. The density in the outflow is low, and the optical depth to infinity is less than 1. The outflow is laterally bounded by the photosphere. The corresponding mass ejection rate $\dot{M}_j \simeq 0.01 - 0.004\,\dot{M}$. The velocity of the outflow increases from zero in the jet formation region near the central object to a maximum value of $\simeq 0.5c$ at the top of the computational grid. This speed must be corrected for relativistic effects (see below), resulting in a maximum outflow speed of $\simeq 0.3c$. *If these radiation-driven bipolar outflows are to become highly collimated, a*

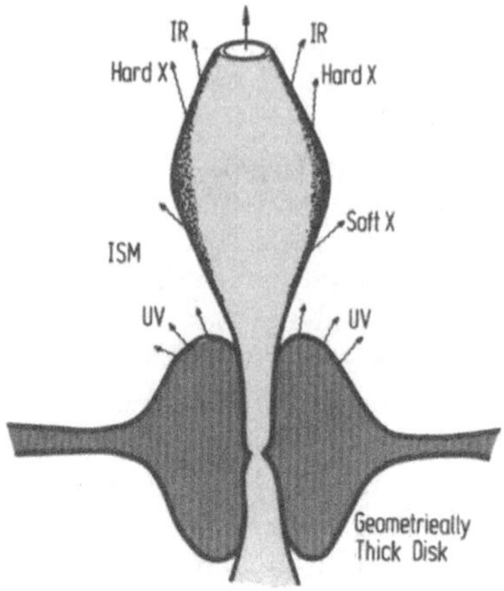

Fig. 1. The collimation of a funnel wind by the external interstellar matter ISM (Courvoisier & Camenzind, 1989). The funnel wind expands into the low pressure medium and recollimates. Shocks would appear in this region resulting in synchrotron and X-ray emission.

collimation mechanism must act externally to the radiation acceleration region — the action of magnetic stresses is the most natural mechanism (Fukue *et al.*, 1991).

These calculations also show that the total disk luminosity always remains sub-Eddington, despite the super-critical accretion. Typically, $L_D \simeq 0.8\, L_{ED}$. The accretion zone itself has no effective mechanism for converting kinetic energy to radiation, so that the gravitational potential energy of the material in the accretion zone makes little contribution to the disk luminosity. A large fraction of the gravitational potential energy of the accreted mass is swallowed by the black hole as kinetic energy. In addition, the advection of radiant energy within the accreting fluid also sweeps a large amount of the total energy released into the black hole (typically another 40%). Without these loss mechanisms, the luminosity would be super-Eddington.

The problem of funnel wind acceleration, which is crucial for the question of the jet speed, has to be formulated in terms of at least special relativistic hydrodynamics including also radiation transfer. Ferrari *et al.* (1986) developed models based upon results taken from solar wind theory (Parker winds). This model is analogous to the classical proposal of Blandford and Rees (1974), with the crucial difference that here the rapid expansion of the gas at the exit of the funnel brings the critical point inward along the funnel axis. Starting with the energy-momentum tensor for a perfect fluid, one can derive a simple equation for the evolution of the Lorentz factor of the fluid

$$\gamma^2\left(1 - \frac{\beta_S^2}{\beta^2}\right)\frac{d\beta^2}{dZ} = \left(-\frac{B}{Z^2} + \frac{2\beta_S^2}{A}\frac{dA(Z)}{dZ} + \frac{2D_1\, z_0}{\gamma^2 \rho c^2}\right), \tag{5}$$

with $Z = z/z_0$ and $B = R_S/z_0$. $A(Z)$ is the cross-sectional area of the funnel normalized to its value at base $z = z_0$, β_S is the local sound speed in units of c, $D_1(z, \beta)$ is the radiation momentum addition term. The flows generated in this way are typically transsonic with the sonic point achieved very deep in the funnel (only at a few Schwarzschild radii above the equatorial plane). Depending on the form of the funnel walls, multiple critical points can occur, which make the flows quite complicated. In general, such flows are only accelerated to $\beta \leq 0.28$ for a funnel luminosity $L_F < L_{ED}$. In the case of hot winds (temperature $\simeq 10^9$ K), $\beta > 0.9$ may be possible for highly supercritical disk luminosities, $L_F \gg L_{ED}$, which are, however, completely unphysical.

Narrow funnels containing an extremely high radiation density immediately suggested a natural way to produce beams (Lynden-Bell, 1978; Paczyński and Wiita, 1980). Early calculations were based on the particle approach which is completely irrelevant for the plasma flow in the jet. As a result of the investigations over the last ten years we come to the result that *radiation pressure supported disks always initiate mass outflows with* $\dot{M}_j \simeq 0.01\,\dot{M}$, *which can neither be accelerated to relativistic speeds, nor be collimated to small opening angles.*

4 The Hydromagnetic Origin of Jets

The other major category of jet formation models considers magnetic fields to be important. Most of these models are based on geometrically thin disks starting with the pioneering work by Blandford and Payne (1982) for extragalactic jets and Pudritz and Norman (1983) for protostellar jets.

4.1 Jets driven by Geometrically Thin Disks

Winds that originate from surfaces and coronae of accretion disks can carry off an appreciable amount of angular momentum if they are magnetized. In 1982, Blandford and Payne realized for the first time the importance of disks also for the collimation problem. They proposed that extragalactic jets could be formed and collimated if they originated from accretion disks around black holes. In the hydromagnetic limit, gas may be viewed as being centrifugally slung outward along rotating magnetic field lines, which form axisymmetric flux surfaces for the flow (Camenzind, 1989). These flux surfaces will be collimated either to asymptotically current-carrying cylindrical surfaces (perfect collimation) or to current-free paraboloidal surfaces (Heyvaerts and Norman, 1989; Chiueh *et al.*, 1991). The terminal 4-velocity on a cylindrical flux surface is then directly proportional to the current enclosed within that surface (Camenzind, 1989; Chiueh *et al.*, 1991).

4.2 Origin of the Magnetic Fields

All models of this type are based on some distribution of magnetic fields in accretion disks. Fully turbulent accretion disks are ideal set-ups for magnetic dynamos, quite similar to dynamo action in the gaseous disks of spiral galaxies. In turbulent plasma, the magnetic fields evolve according to the induction equation of *mean field electrodynamics* (Moffatt, 1978; Krause and Rädler, 1980; Pudritz, 1981)

$$\partial_t \boldsymbol{B} + (\boldsymbol{u}\cdot\nabla)\,\boldsymbol{B} = (\boldsymbol{B}\cdot\nabla)\boldsymbol{u} - (\nabla\cdot\boldsymbol{u})\,\boldsymbol{B} + \nabla\wedge\left\{\mathcal{E} - \eta_T\,\nabla\wedge\boldsymbol{B}\right\}. \tag{6}$$

$\boldsymbol{B}$ is the mean magnetic field, and $\boldsymbol{u}$ the mean velocity field in the disk, $\boldsymbol{u} = (u_R, R\Omega, u_z)$. η_T is the total magnetic diffusivity including the microscopic one and the turbulent diffusivity. $\mathcal{E} = \alpha_{\mathcal{D}}\,\mathbf{B}$ represents the electromotive force induced by correlations between turbulent motions and turbulent magnetic fields.

When the α-effect is neglected in a disk, $\alpha_D = 0$ throughout the disk, the strong differential rotation just amplifies the toroidal field. The poloidal flux Ψ is, however, only redistributed in the disk by advection and diffusion (there is no source term for Ψ). In this case, one can find stationary solutions for the distribution of the disk fields under various assumptions for the magnetic diffusivity $\eta(R,z)$ in the disk (Lovelace *et al.*, 1987; Königl, 1989; Khanna and Camenzind, 1991), provided there is a continuous magnetic flux injection at the outer edge of the disk.

When the inner disk is convective, differential rotation and convection provide a non-vanishing dynamo effect represented by $\alpha_D = L_T^2\,\Omega/H(R)$. Under this condition, a strong amplification of magnetic fields occurs near the inner edge of the disk, since shear is maximal there. The dynamo builds up axisymmetric magnetic flux tubes near the marginal stable orbit of the disk independent of the initial configuration (Camenzind, 1990). The lowest modes excited by such disk dynamos are mostly of the quadrupolar type.

4.3 Magnetic Plasma Confinement and Collimation

As in the case of the static axisymmetric plasma confinement in the lab, the magnetic surfaces of the outflowing disk wind also provide a kind of plasma confinement. If the pinching force of the toroidal field is sufficiently strong to overcome pressure gradients and centrifugal forces, the magnetic surfaces can bend towards the rotational axis resulting in a collimated outflow. Axisymmetric flux tubes are therefore stretched by the outflowing disk wind into a kind of *reverse field pinch* at large distances from the disk. The fast rotation in the disk generates electric radial fields in the disk, which drive a poloidal current system in the magnetosphere. These currents produce toroidal magnetic fields, which exert a momentum on the foot points on the disk. The magnetic structure of this hole-disk-system must then be calculated self-consistently including all the currents driven by the rotation by using Ampere's equation for the poloidal magnetic field $\boldsymbol{B}_p$

$$R\nabla\cdot\left\{\frac{1}{R^2}\,\nabla\Psi\right\} = -\frac{4\pi}{c}\,j_\phi \quad , \quad \mathbf{B}_p = \frac{1}{R}\,(\nabla\Psi\wedge\mathbf{e}_T)\,. \tag{7}$$

When plasma is moving along the magnetic flux surfaces, the general form of the current j_ϕ follows from a force balance perpendicular to the flux surfaces

$$n\,(\mathbf{u}\cdot\nabla)\,\mu\mathbf{u}_\perp + \nabla_\perp P = \rho_e\,\mathbf{E}_\perp + \frac{1}{c}\,(\mathbf{j}\wedge\mathbf{B})_\perp\,. \tag{8}$$

The projection of this equation perpendicular to the magnetic flux surfaces provides then the current j_ϕ (a derivation is given in Camenzind, 1987)

$$\frac{1}{c}B_p j_\phi(1 - M^2 - (R/R_L)^2) = -\frac{RB_\phi}{4\pi R^2}\nabla_\perp(RB_\phi) + \rho_{\rm GJ}E_\perp$$
$$- \mu n\gamma\{\nabla_\perp\gamma - \frac{1}{c^2}\Omega\nabla_\perp j\} - \nabla_\perp P + \eta B_p^2\nabla_\perp(\eta/n)\,. \qquad (9)$$

In rapidly rotating magnetospheres, the influence of the electric field $\boldsymbol{E}$ cannot be neglected, $\boldsymbol{E}_\perp = -(\Omega^F/c)\,\nabla\Psi$, $4\pi\rho_{GJ} = \nabla\cdot\boldsymbol{E}_\perp$. The first two terms on the right hand side of Eq. (9) are of electromagnetic origin and therefore important in the force-free approximation, while the last four terms are due to the inertia of the flowing plasma.

Solutions of the above cross-field force balance equation can indeed form highly collimated outflows (Camenzind, 1987; Haehnelt & Camenzind, 1991; Li *et al.*, 1992). In the force-free limit, which is a good approximation for the AGN case, the structure of a rotating magnetosphere is simply given in terms of the flux function $\Psi(R,z)$ and the integrated poloidal current $I(\Psi)$. The flow begins to collimate around the light cylinder, which is defined for each flux tube, $R_L = c/\Omega_K(R_{\rm F})$ for footpoint radii $R_{\rm F}$ near the inner last stable orbit at $r_{\rm ms}$ around a rapidly rotating Kerr black hole

$$\Omega^F \simeq \Omega_K(r_{\rm ms}) \quad , \quad P_F = 2\pi/\Omega^F \,,$$
$$R_L = c/\Omega^F \simeq 10\,GM_H/c^2 \simeq 1.5\times10^{15}\,{\rm cm}\,M_H/(10^9\,M_\odot)\,. \qquad (10)$$

Current-carrying flows bend towards asymptotically cylindrical surfaces with a typical radius of $\simeq (10-100)R_L$. This result has been proven for relativistic flows by Chiueh *et al.* (1991).

4.4 The Question of the Final Jet Speed

The plasma escaping from the disk corona is now guided by the magnetic surfaces, since ideal MHD requires the condition $\boldsymbol{u}_p \parallel \boldsymbol{B}_p$, or

$$\mathbf{u}_p = \frac{\eta}{n}\,\mathbf{B}_p \quad , \quad u^\phi = \frac{\eta}{n}\,B_\phi + \gamma\,R\,\Omega^F\,. \qquad (11)$$

n is the particle density in the flow and η is constant along the magnetic flow lines ($\mathbf{B}\cdot\nabla\eta = 0$). With these quantities we define the special relativistic Mach number M (Camenzind, 1986a)

$$M^2 = \frac{4\pi\mu\,n\,u_p^2}{B_p^2} \quad , \quad \mu = \frac{\rho + P}{n} \simeq m_p c^2\,, \qquad (12)$$

which tells us, whether the flow is strongly magnetized ($M^2 \ll 1$) or weakly magnetized ($M^2 \geq 1$). For low velocities near the disk surface, the magnetic field controls the flow, but far away from the disk the flow becomes super-Alfvénic. The exact flow properties depend on the particular form of the magnetic flux surfaces.

The wind plasma slides along a magnetic surface Ψ in such a way that the total energy E and the total angular momentum L are conserved

$$E(\Psi) = \mu\gamma - \frac{\Omega^F\,RB_\phi}{4\pi\eta} \quad , \quad L(\Psi) = \frac{\mu}{c^2}\gamma j - \frac{RB_\phi}{4\pi\eta}\,. \qquad (13)$$

These relations determine closed expressions for the Lorentz factor γ, the specific angular momentum $j = R^2\Omega$ and the toroidal magnetic field (Camenzind, 1986a,b):

$$\gamma(R,\Psi) = \frac{E}{\mu}\,\frac{1 - x_A^2 - M^2(R,\Psi)}{1 - M^2(R,\Psi) - (R/R_L)^2}\,, \tag{14}$$

$$j(R,\Psi) = \frac{R^2\Omega^F\,(1 - x_A^2) - M^2(R,\Psi)c^2 L/E}{1 - M^2(R,\Psi) - x_A^2}\,, \tag{15}$$

$$RB_\phi = -\frac{4\pi\eta(\Psi)\,E(\Psi)}{\Omega^F(\Psi)}\,\frac{x_A^2 - (R/R_L)^2}{1 - M^2(R,\Psi) - (R/R_L)^2}\,. \tag{16}$$

μ is the specific enthalpy of the plasma, $\mu = m_p c^2$ for normal plasma, $x_A^2 = \Omega^F L/E$, and $M(R,\Psi)$ increases along a given flux surface Ψ when the plasma moves outwards to larger radii R (Camenzind, 1989)

$$M^2(R,\Psi) = \frac{\mu}{m_p c^2}\,u_p \left(\frac{R}{R_L}\right)^2 \frac{\Phi(R,\Psi)}{\sigma_D(\Psi)}\,. \tag{17}$$

u_p is the poloidal component of the 4-velocity, $\Phi(R,\Psi) = B_p R^2$ the flux tube function normalized to its value at the footpoint in the disk, and $\sigma_D(\Psi)$ the magnetization parameter for the flux surface Ψ measured at the footpoint of the flux surface (Michel, 1969; Camenzind, 1991a,b)

$$\sigma_D(\Psi) \simeq \frac{(B_p R^2)_D^2}{2\dot{M}_{\rm jet} c R_L^2} \geq \pi^2\,\frac{\dot{m}}{\epsilon_H}\,(3GM_H/c^2 R_F)^{3/2}\,, \quad \dot{M}_j \leq \frac{1}{2}\left(\frac{R_F}{R_L}\right)^2 \dot{M}\,. \tag{18}$$

This magnetization is the essential parameter for the question of the speed of the plasma attained in the asymptotic collimated region. Pressure effects are of no importance for a normal plasma, they would have some influence for the outflow speeds of electron-positron plasma. The magnetization essentially determines the poloidal velocity u_p independent of the opening of the collimated jet (Camenzind, 1989). u_p is a solution of the hot wind equation along a given magnetic flux surface. Relativistic speeds can only be obtained for strongly magnetized disk plasma with $\sigma_D > 1$, $u_{p,\infty} \simeq \sigma_D(\Psi)$ (Fig. 2). The highest magnetization is then obtained for accretion rates near the Eddington accretion rate $\dot{m} = \dot{M}_{\rm acc}/\dot{M}_{\rm ED} \simeq 0.5$ with values $\sigma_D \simeq 10$. When the magnetization of the disk is low, $\sigma_D < 1$, only semi-relativistic speeds can be obtained, $v_\infty \simeq \sigma_D^{1/3}\,c$. It is therefore quite compelling that strong quasars with their high accretion rate show the highest superluminal speeds, $v_\perp \simeq 10\,c$.

5 The Knots in the Parsec-Scale Jets

One particular problem in the interpretation of observed compact jets is the identification of the jet feature which corresponds to the VLBI radio core and the moving VLBI knots. This core as the most compact feature in VLBI maps is most probably stationary and is situated at one end of the jet. In the standard model of Blandford and Königl (1979), the core is identified with the throat of the diverging conical jet, where density and magnetic fields are highest. This would occur in a HD jet if the mean energy per gas

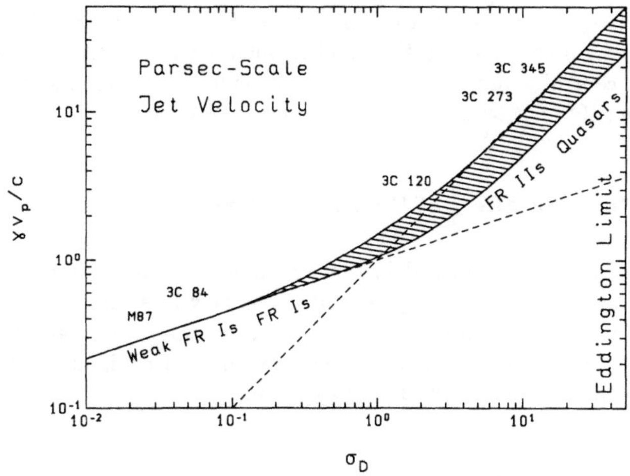

Fig. 2. The asymptotic outflow velocity $u_p = \gamma v_p/c$ as a function of the magnetization σ_D of the accretion disk corona. For high accretion rates (quasars), the corona plasma is highly magnetized resulting in a high outflow speed. For sources with extremely low accretion rates (FR I radio galaxies), the corona is weakly magnetized and the outflow is non-relativistic.

particle were to drop below the rest mass energy, such that the flow becomes ballistic within a short distance from the throat, provided the ambient pressure is lower than the jet pressure (Fig. 1; see also Daly & Marscher, 1988). Superluminal knots have been identified as clouds caught up in the jet flow (Blandford & Königl, 1979), as instabilities in the flow (Marscher, 1980; Appl & Camenzind, 1991), or as shock waves propagating down the jet (Marscher & Gear, 1985; Hughes, Aller & Aller, 1989; Marscher, 1990). In the first case, the velocity of the knot is roughly that of the local jet flow. Forward shocks will slightly decelerate with respect to the background flow.

5.1 Knots as Electron-Positron Clouds

Several models for jets also include a relativistic beam, which is formed by electrons and positrons in pair cascades on the UV photons of the accretion disk. In such a scheme of a two-fluid model (Camenzind, 1984; Sol *et al.*, 1989), the relativistic knots are considered as clouds of $e^{\pm}$ plasma pervading the ambient subrelativistic flow ejected by the accretion disk. These beams can be stable against the excitation of plasma waves for moderate bulk Lorentz factors ($\gamma < 43$) and for sufficiently strong magnetic fields $B > B_c$ ($\omega_B > \omega_p$) (Pelletier & Sol, 1991). The Alfvén turbulence in the jet can be used to heat up the pair plasma to relativistic energies by a second order Fermi process (Henri & Pelletier, 1991). The pairs are cooled by inverse Compton scattering on soft photons, which would be the origin of the X- and γ-rays in quasars. The pair production zone would be located outside the light cylinder, typically at a few hundred Schwarzschild radii, i.e. just at the position where the knots have to be injected into the jets in order to show the lighthouse effect. According to this scheme, VLBI knots would just consist of pair clouds ejected along the magnetic surfaces of the underlying jet. These $e^{\pm}$ beams just propagate until they reach a region with $B \simeq B_c$, where they loose their energy to Langmuir waves and heat up the ambient medium (typically at a distance of $\simeq 1$ kpc from the central source). $e^{\pm}$ beams with higher bulk Lorentz factors (Kundt & Gopal-Krishna, 1986), $\gamma \gg 43$, can only propagate in a vacuum.

5.2 The Lighthouse Effect of Relativistic Jets

When plasma bubbles are formed in the innermost part of relativistic jets, they will be dragged along with the underlying continuous flow in such a way that the specific angular momentum j of the bubble stays constant along the magnetic surface. The spectral flux S_ν observed from such a knot moving in a collimated jet with constant opening angle Θ_0 is strongly amplified by beaming

$$D(t) = \frac{1}{\gamma_{\rm jet}(1 - \beta_{\rm jet}\,\mu(t))}\,, \tag{19}$$

$$S_\nu(t) = \int_B d\Omega\, j_\nu(\nu)\, dl = \frac{1}{d^2}\, D^{2+\alpha}(t)\,(1+z)^{1-\alpha} \int_B dV\, j_{\nu'}(\nu)\,. \tag{20}$$

This flux is in general time-dependent, since the Doppler factor D is a function of t. The angle θ in $\mu = \cos\theta$ is the actual angle between the velocity of the bubble and the direction of the observer. Due to the rotation of the bubble, the beaming factor can strongly vary and lead to quasi-periodic flaring as the knot moves along a magnetic surface with constant opening (Camenzind & Krockenberger, 1991). Due to this rotation of the bubble around the central axis of the jet, the flux emitted by the bubble is also quasi-periodic with the fundamental period P_0 following from Eq. (15) for $R(t=0) = R_0$, $P_0 = P_F\,(R_0/R_L)^2$, $R_0 \simeq 10 R_L$. The observed period $t_{\rm obs}$ is however drastically shortened by the relativistic effect

$$t_{\rm obs} = \left(t_{\rm sys} - \int_0^{t_{\rm sys}} \beta_{\rm jet}\,\mu(t)\,dt\right)(1+z)\,. \tag{21}$$

The intrinsic period P_0 is of the order of years for quasars such as 3C 273 and 3C 345, and of the order of weeks for BL Lac objects. For 3C 273 we then find $P_{\rm obs} \simeq 15$ days (Courvoisier *et al.*, 1988) for $\gamma_{\rm jet} \simeq 15$ and an inclination of 3°, for BL Lac objects $P_{\rm obs} \simeq 1$ day (Wagner, 1992) with $\gamma_{\rm jet} \simeq 2-4$ and small inclination angles. If this interpretation of the intraday variability is correct, this would be the first evidence for rotation in the escaping jet material and for the perfect collimation of current-carrying jets. With high resolution mm-VLBI, this helical motion of young knots should be visible.

6 Conclusions

The formation of relativistic jets cannot be understood without inclusion of magnetic effects. Current-carrying jets are perfectly collimated in the sense that their magnetic surfaces bend outside the light cylinder towards asymptotically cylindrical surfaces. The typical jet radii are between 10 and 100 light cylinder radii, i.e. of the order of a few hundred Schwarzschild radii. This is in agreement with the observed jet widths on the VLBI scale. The absolute scale of the parsec-scale jets follows therefore from the central mass. According to this, powerful quasars should host masses $M_H \simeq 10^{10}\,M_\odot$, BL Lac objects and FR I radio galaxies masses $M_H \simeq 10^8\,M_\odot$.

The magnetic formation of jets predicts a correlation between the accretion rate and the outflow speed of parsec-scale jets. Quasars should have relativistic jets, FR I radio galaxies only subrelativistic jets. This is mainly a consequence of the fact that the magnetic energy of accretion disks cannot exceed the turbulent energy density. The very origin of these magnetic fields is, however, still an open problem. The angular momentum of the black hole is somehow involved in the solution of this problem.

An important consequence of this magnetized jet model is the fact that the plasma in the jet is rotating with a rotational velocity decreasing with increasing radius. But also the gas of hydrodynamic disk outflows is rotating, since disk matter always carries along angular momentum. Magnetized outflows carry along, however, a higher specific angular momentum due to the lever arm formed by Alfvén point (near the light cylinder). Non-axisymmetric structures formed in these jets can therefore explain the observed intraday variability in the optical, X-ray and radio emission of compact sources.

Acknowledgements. This work is supported by the Deutsche Forschungsgemeinschaft (SFB 328).

References

Appl, S.: 1990, Ph.D. thesis, University of Heidelberg.
Appl, S., Camenzind, M.: 1988, *Astron. Astrophys.* **206**, 258.
———: 1992, *Astron. Astrophys.* **256**, 354.
Bååth, L.B. *et al.*: 1991, *Astron. Astrophys. (Lett.)* **241**, L1.
Blandford, R.D., Rees, M.J.: 1974, *Mon. Not. R. astr. Soc.* **169**, 395.
Blandford, R.D., Königl, A.: 1979, *Astrophys. J.* **232**, 34.
Blandford, R.D., Payne, D.: 1982, *Mon. Not. R. astr. Soc.* **199**, 883.
Bregman, J.N.: 1990, *Astron. Astrophys. Rev.* **2**, 125.
Camenzind, M.: 1984, *Astron. Astrophys. (Lett.)* **130**, L9.
———: 1986a, *Astron. Astrophys.* **156**, 137.
———: 1986b, *Astron. Astrophys.* **162**, 32.
———: 1987, *Astron. Astrophys.* **184**, 341.
———: 1989, in *Accretion Disks and Magnetic Fields in Astrophysics*, ed. G. Belvedere, Dordrecht: Kluwer, p. 129.
———: 1990, in *Reviews of Modern Astronomy* **3**, 234.
———: 1991a, in *Variability of Active Galactic Nuclei*, ed. W.J. Duschl, S. Wagner, M. Camenzind; Lecture Notes in Physics **377**, Heidelberg: Springer, p. 201.

——: 1991b, in Proc. of *TEXAS/ESO-CERN Symposium on Relativistic Cosmology, and Fundamental Physics, Ann.N.Y.Acad.Sci.* **647**, 610.
——: 1993, in *Stellar Jets and Bipolar Outflows*, 6th Int. Workshop of the OAC, ed. A. Vittone, Dordrecht: Kluwer, in press.
Camenzind, M., Krockenberger, M.: 1992, *Astron.Astrophys.* **255**, 59.
Chiueh, T., Li, Z.Y., Begelman, M.C.: 1991, *Astrophys.J.* **377**, 462.
Courvoisier, T.J.-L., Camenzind, M.: 1989, *Astron.Astrophys.* **224**, 10.
Courvoisier, T.J.-L., Robson, E.I., Blecha, A., Bouchet, P., Hughes, D.H., Krisciunas, K., Schwarz, H.E.: 1988, *Nature* **335**, 330.
Daly, R.A., Marscher, A.P.: 1988, *Astrophys.J.* **334**, 539.
Eggum, G.E., Coroniti, F.V., Katz, J.I.: 1987, *Astrophys.J.* **323**, 634.
——: 1988, *Astrophys.J.* **330**, 142.
Ferrari, A., Trussoni, E., Rosner, R., Tsinganos, K.: 1986, *Astrophys.J.* **294**, 397.
——: 1986, *Astrophys.J.* **300**, 577.
Fukue, J., Shibata, K., Okada, R.: 1991, *Publ.Astron.Soc.Japan* **43**, 131.
Gvaramadze, V.V., Lominadze, J.G., Ruzmaikin, A., Sokoloff, D., Shukurov, A.M.: 1988, *Astrophys.Space Sci.* **140**, 165.
Haehnelt, M., Camenzind, M.: 1993, *Astron.Astrophys.*, in press.
Henri, G., Pelletier, G.: 1991, ApJL 383 7.
Heyvaerts, J., Norman, C.A.: 1989, ApJ 347 1055.
Hughes, P.A., Aller, H.D., Aller, M.F.: 1989, *Astrophys.J.* **341**, 68.
Khanna, R., Camenzind, M.: 1992, *Astron.Astrophys.* **263**, 401.
Königl, A.: 1989, *Astrophys.J.* **342**, 208.
Krause, F., Rädler, K.-H.: 1980, *Mean-Field Magnetohydrodynamics and Dynamo Theory*, Pergamon, Oxford
Krichbaum, T.: 1990a, Ph. D. thesis, University of Bonn.
——: 1990b, in *Parsec-scale radio jets*, ed. J.A. Zensus and T.J. Pearson, Cambridge: Cambridge University Press, p. 83.
Kundt, W., Gopal-Krishna: 1986, *J.Astrophys.Astr.* **7**, 225.
Li, Z.Y., Chiueh, T., Begelman, M.C.: 1992, *Astrophys.J.*, in press.
Lovelace, R.V.E., Wang, J.C.L., Sulkanen, M.E.: 1987, *Astrophys.J.* **315**, 504.
Lynden-Bell, D.: 1978, *Phys.Scripta* **17**, 185.
Maraschi, L., Chiappetti, L., Famolo, R., Garilli, B., Malkan, M., Tagliaferri, G., Treves, A.: 1991, *Astrophys.J.* **368**, 138.
Marscher, A.P.: 1980, *Astrophys.J.* **235**, 386.
——: 1990, in *Parsec-scale radio jets*, ed. J.A. Zensus and T.J. Pearson, Cambridge: Cambridge University Press, p. 236.
Marscher, A.P., Gear, W.: 1985, *Astrophys.J.* **298**, 114.
Michel, F.C.: 1969, *Astrophys.J.* **158**, 727.
Moffatt, H.K.: 1978, *Magnetic Field Generation in Electrically Conducting Fluids*, Cambridge: Cambridge University Press.
Paczyński, B., Wiita, P.J.: 1980, *Astron.Astrophys.* **88**, 23.
Pelletier, G., Sol, H.: 1991, *Mon.Not.R.astr.Soc.*, in press.
Pudritz, R.: 1981, *Mon.Not.R.astr.Soc.* **195**, 881.
Pudritz, R., Norman, C.A.: 1983, *Astrophys.J.* **274**, 677.
Ruzmaikin, A., Sokoloff, D., Shukurov, A.: 1988, *J.FluidMech.* **197**, 39.
Sol, H., Pelletier, G., Asséo, E.: 1989, *Mon.Not.R.astr.Soc.* **227**, 411.
Wagner, S.: 1992, these Proceedings.

MHD-Stability of Jets

S. Appl, M. Camenzind

Landessternwarte Königstuhl
D-6900 Heidelberg, Germany

1. Introduction

Models that collimate the jets in rotating magnetospheres automatically produce a toroidal magnetic field, which by virtue of Ampère's law is equivalent to a poloidal electric current. Or, conversely, the electric current, being the source term in the Grad-Shafranov equation for the flux function, determines the collimation and the magnetic profile of the jet (Camenzind, 1992). Based on the experience with fusion plasmas there is a major concern that jets generated this way are liable to destructive nonaxisymmetric MHD instabilities. We are addressing the stability of such configurations.

Observationally, there exists a morphological dichotomy in the large scale jets. Fanaroff-Riley type I (FR I) objects have turbulent looking jets, which appear to continually dissipate energy and entrain ambient material. They finally fade out into a plume. The FR II sources on the other hand, have well collimated and faint jets that propagate out to large distances until they terminate in a strong shock. Apparently they are much more stable than the jets of weak radio galaxies. The two classes hence exhibit widely different stability properties. From the absence of a shock in FR Is it may be inferred that their jets propagate at sub- or transmagnetosonic velocities, while those of FR IIs are clearly supermagnetosonic. The latter also possess magnetic fields that are about an order of magnitude stronger than in the FR Is. Mach number, magnetic field strength and configuration are crucial parameters and may well determine the distinct stability behaviour of the two classes of jets.

Various studies that address this point have been performed, and we shall start by briefly summarizing the essential findings of these investigations, including our own work. Only the idealized situation of an infinitely long cylindrical jet confined either by thermal pressure or magnetic tension is considered. Velocity and density are usually constant across the jet, additional effects caused by shear are therefore disregarded. This contribution differs from earlier proceedings (Appl and Camenzind, 1992b,c) in that it tries to summarize the essential points, interrelates the stability properties of current-carrying jets with other equilibria that were studied, and that it includes current-carrying configurations with a Mach number of the order of unity.

Hydrodynamic Jets. The relevant instability is the Kelvin-Helmholtz instability driven by the relative kinetic energy of the jet and the ambient medium. The jet is unstable at all wavelengths. For all Mach numbers the fundamental mode is excited. This is a growing perturbation which travels backward w.r.t. the frame of the jet with approximately the speed of sound. For supersonic jets there are additionally reflected modes, which are growing by resonant reflection at the boundary of the jet. They only exist for wavelengths sufficiently small to fit into the jet diameter. These modes also travel backwards, however somewhat slower due to their non straight propagation. In a supersonic jet this results in a net forward motion w.r.t. the observer, the instabilities are hence convected along the flow in a highly supersonic jet. This makes the growth length a more appropriate quantity than the growth rate when comparing stability results. A high Mach number naturally tends to stabilize by the mere fact that a perturbation has been convected further away until it has grown by a factor of e in amplitude, compared to a low Mach number jet. In the nonlinear regime the reflection modes will form weak oblique shocks, while the fundamental modes may destroy the beam (Norman and Hardee, 1988).

Jets with homogeneous longitudinal field. An analogous dispersion relation as in the hydrodynamic case can be obtained (Ferrari et al., 1981; Ray, 1981; Bodo et al. 1989). Hence, the mode structure is the same. However, the *sonic* Mach number loses its meaning and the *fast magnetosonic* Mach number has to be taken instead. This is important when comparing magnetized and hydrodynamic equilibria or equilibria at different plasma-β, $\beta = 8\pi p/B^2$. A magnetically dominated jet with a moderate sonic Mach number of, say $M = 3$, is in fact a subcritical flow, in which neither reflection modes nor shocks can develop. In the limit of zero and dominant magnetic field the phase velocity reduces to the sound and Alfvén velocity, respectively. Generally, it was found that the axial field has a stabilizing effect. In particular, subalfvénic flows were shown to be entirely stabilized when the magnetic field becomes dominant. The magnetic field affects the stability properties at long wavelengths only very little, short wavelength instabilities, on the other hand, are suppressed very effectively at low β (Appl and Camenzind, 1992a,c).

Jets with currents. Relatively little has been known for this case since an analytic dispersion relation is usually no longer available. Generally, both Kelvin-Helmholtz and current-driven instabilities will have to be considered. Cohn (1983) studied the axisymmetric instabilities of a hydrodynamic jet that is confined by an azimuthal field. The stability properties were reported to be essentially unaffected by the confinement mechanism. Pietrini and Torricelli-Ciamponi (1989) and Corbelli and Torricelli-Ciamponi (1990) considered a jet that carries an electrical current and return current. They investigated the $\beta \gtrsim 1$ case and studied the influence of a varying pitch. They also found a stabilization for sufficiently longitudinal field at subalfvénic velocities. Finally, our work addresses the stability of low-β super- and transmagnetosonic jets that carry a poloidal electric current, and compares the results to jets with longitudinal field and hydrodynamic jets. It is shown that supermagnetosonic current-carrying jets are more stable than hydrodynamic jets and even jets with a homogeneous axial field (Appl and Camenzind, 1992a). At transmagnetosonic speeds, on the other hand, the azimuthal field acts destabilizing w.r.t. the longitudinal field, but the configuration remains more stable than the hydrodynamic counterpart (Appl, 1992). As in the current-free MHD-jets, a low plasma-β strongly suppresses short wavelength instabilities while the large scale modes are much less affected. Pinch modes of low-β jets, with and without current, are stabilized.

2. Equilibria

The following types of equilibria were investigated:

Reversed field pinch (RFP). The jet leaves the collimation zone with an almost force-free configuration, its current distribution being determined by the conditions given at its foot point. Most likely, the jet will develop into a state minimal energy. In a low-β configuration this happens through magnetic relaxation (Taylor, 1986). It results in a linear force-free field $\nabla \wedge \mathbf{B} = \alpha \, \mathbf{B}$, with $\alpha = const.$ The magnetic profile is then given by the Bessel functions $B_\phi = B_0 J_1(\alpha \, r)$ and $B_z = B_0 J_0(\alpha \, r)$. Requiring that the current is closed within the jet fixes the constant to the value of $\alpha = 3.83/R_{\rm jet}$. Being a state of minimum energy the configuration is widely independent of the detailed conditions at the footpoint of the jet.

The static RFP (see Bodin and Newton, 1980) was studied for fusion purposes as an alternative to the Tokamak. It possesses high magnetic shear which is by virtue of Suydam's criterion favourable w.r.t. current driven instabilities. Both components of the magnetic field are of the same order of magnitude, unlike in the Tokamak, and the axial field reverses its sign in the outer part. The fact that with minor modifications a considerable plasma-β can be supported without changing the stability properties makes it even more attractive as a model for extragalactic radio jets.

Cocoon Jet. Motivated by axisymmetric MHD simulations with a helical field at the orifice (Kössl et al., 1990), we choose as equilibrium for the kpc-scale a beam with a homogeneous longitudinal field, surrounded by a non-moving cocoon with a toroidal field, extending out to three jet radii. Current and return current hence flow on the interfaces between beam and cocoon, and cocoon and IGM, respectively.

Jet with homogeneous longitudinal field. This equilibrium differs from the cocoon jet only in that it has no electric current and azimuthal field associated with it. It can be considered as the asymptotic limit of a (non-moving) Tokamak configuration. For the stability of the Tokamak it is essential that the current driven through the plasma column does not exceed the value allowed by Kruskal's criterion. Stabilization is therefore obtained by a longitudinal field which is much stronger than the azimuthal field, in contrast to the RFP.

Hydrodynamic Jet. This is a cylindrical jet supported by thermal pressure with an adiabatic index of $\gamma = \frac{4}{3}$ against the ambient medium.

3. Stability

The linearized MHD equations for a general cylindrical equilibrium can be reduced to a nonlinear eigenvalue problem for two first order differential equations using a normal mode expansion $\propto \exp i(m\phi + kz - \omega t)$ for the perturbed quantites. In the case of the two discussed current-free equilibria it reduces to Bessel's equation and a transcendent dispersion relation can be obtained. In the static limit, which was extensively studied in fusion research, the eigenvalue problem is self-adjoint, and many more mathematical tools are available to probe for and classify instabilities. A current profile introduces a radial inhomogeneity that is responsible for the appearance of a stable continuous spectrum of highly localized modes. The equilibria for which results will be given and compared consist of two current-carrying (RFP and cocoon jet) and two current-free (longitudinal field and hydrodynamic jet) configurations. The three magnetic equilibria are discussed in the low-β limit.

Static Jets. Both current-free equilibria are absolutely stable in the static limit, while the current-carrying configurations are liable to current driven instabilities (Voslamber and Callebaut, 1962; for the RFP). The pinch ($m = 0$) is entirely stabilized in low-β configurations, leaving the nonaxisymmetric instablilities as the most dangerous ones.

Supermagnetosonic Jets. By definition of the Mach number the kinetic energy of the flow exceeds the sum of thermal and magnetic energy. The relevant instability is therefore the Kelvin-Helmholtz instability modified by the magnetic and current profile. We start from the known mode structure of hydrodynamic jets, namely the fundamental and first reflection mode, and study how the modes change in the presence of the current. In the axisymmetric case only the reflection modes remain unstable.

Transmagnetosonic Jets. These are jets with a fast magnetosonic Mach number not much different from $M = 1$. Here both Kelvin-Helmholtz and current driven modes determine the stability. The method applied to supermagnetosonic jets is unable to find current driven modes that might become more dangerous. The method of eigenvalleys (Scarton, 1973) is appropriate to find all eigenvalues in the interesting domain of the complex plane. It turned out that the modified fundamental mode remains indeed the most dangerous one, even at Mach numbers below unity. Reflection modes are stabilized.

4. Results

Complex eigenvalues $\omega(k, m)$ are calculated for the most dangerous modes. From this the inverse growth length $k_g = \Im\omega / v_{\mathbf{group}}$ is derived to account for the convective nature of the instabilities. Both axisymmetric ($m = 0$) and nonaxisymmetric ($m = \pm 1$) modes were considered, the focus, however, is on the nonaxisymmetric instabilities being most dangerous. The instabilities are discussed in three groups that lead to different manifestations. As an example for the super- and transmagnetosonic jets the results for the nonaxisymmetric fundamental mode of a jet with Mach number $M = 3$ and $M = 1.5$ are displayed (Fig. 1).

Instabilities with $k \gtrsim 1$. Instabilities with wavelength larger than or of the order of the jet radius are most dangerous to the integrity of the jet. In particular the $|m| = 1$ mode will produce a helical sideways displacement of the beam. This can finally disrupt it when grown to nonlinear amplitude. At moderate and large wavelengths there is not much difference between the four equilibria, nevertheless, both supermagnetosonic current-carrying jets are more stable than the hydrodynamic and even than the jet with $B_{\|}$, i.e. the current also stabilizes w.r.t. the axial field. Only at very large wavelengths is there a tendency for the cocoon jet to become less stable. With this only exception, at all k and for all modes, the current-carrying jets are more stable than their current-free counterparts. At low Mach number the jet with axial field stabilizes (Fig. 1b) and

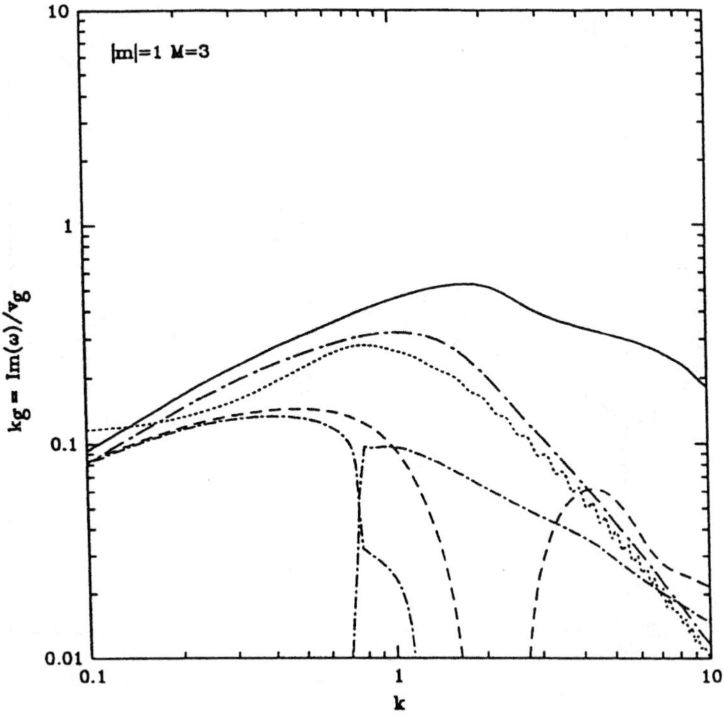

Fig. 1a. Inverse growth length of the fastest growing nonaxisymmetric instabilities for a jet with Mach number $M = 3$ for various equilibria: hydrodynamic (solid), longitudinal magnetic field (long dash-dotted), RFP ($m = 1$ dashed; $m = -1$ short dash-dotted), cocoon jet (short dashed)

hence the current has a destabilizing effect w.r.t. the latter. Now the toroidal field shows its destabilizing nature that it generally exhibits in the static case. Nevertheless, the equilibria carrying a current remain more stable than the hydrodynamic jet, again with the exception of the cocoon jet at large wavelength.

Instabilities with $k < 1$. Instabilities with wavelengths much smaller than the jet radius are usually thought to cascade down to even smaller scales and are a source for MHD turbulence. The turbulence in turn can lead to dissipation of kinetic energy, facilitate entrainment of ambient material, and accelerate relativistic particles. In all three magnetic dominated jets short wavelength instabilities are very effectively suppressed compared to the hydrodynamic case, in particular at higher Mach numbers. Generally, the wavelengths with the shortest growth length are typically a factor of two larger, if the jet carries a current. In the case of the reflection modes (see Appl and Camenzind, 1992a,b,c) this is even more striking. They also set in at longer wavelengths. An increasing Mach number has a similar effect.

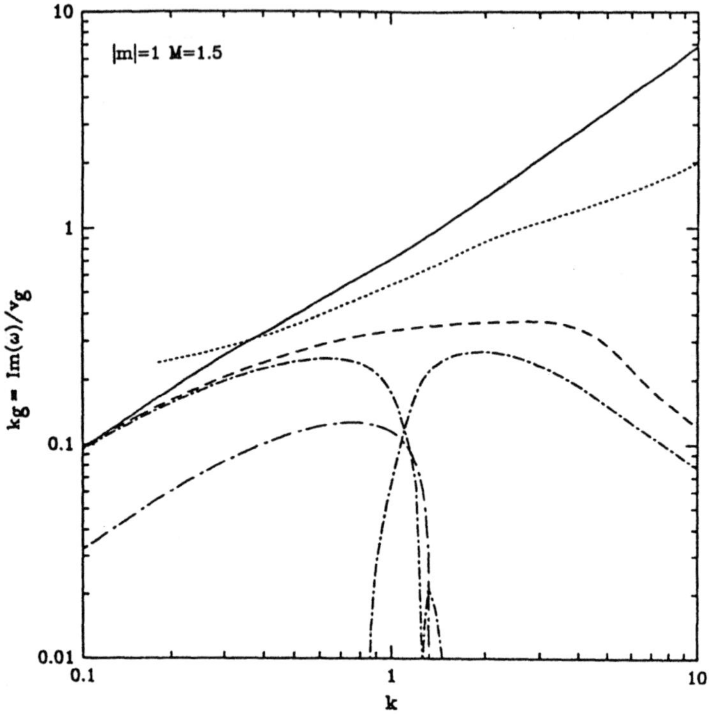

Fig. 1b. as Fig. 1a, but for $M = 1.5$

Localized instabilities. Both the $m = 1$ and $m = -1$ dispersion branch for the RFP separate into two curves that vanish at some value of k. This is due to the existence of the continuous part of the spectrum. Apart from the stable singular modes there are also unstable localized modes (Appl and Camenzind, 1992a,c). Growing in amplitude, such modes will produce very high gradients, and *e.g.* the Alfvén mode will finally convert into a kinetic Alfvén wave. The latter can dissipate and lead to a sheath ($m = 0$) or a helical strand ($m = 1$) with enhanced brightness within the beam.

5. Conclusion

We showed that for supermagnetosonic velocities the current-carrying equilibria become more stable than their current-free counterparts. Extrapolating stability properties from static equilibria to jets, where the situation is opposite, is therefore very dangerous. At transmagnetosonic flow velocities the current destabilizes w.r.t. the axial field. Together with the estimated values for the magnetic fields, the presence or absence of hot spots suggests that FR Is have transmagnetosonic jets with moderate currents, and the jets of FR IIs are supermagetosonic low-β configurations with a high current. This also nicely accounts for the fact that FR IIs possess a low surface brightness and propagate very effectively while FR Is suffer from larger losses and entrainment due to turbulence produced by moderate plasma-β. The inhomogeneous magnetic profile causing localized instabilities naturally accounts for the helical filaments that are observed in the jet of M87. Having demonstrated the enhanced stability of current-carrying jets strongly supports models that collimate them through the rotation of a magnetic field anchored in the central engine (Camenzind, 1990). It is also in accordance with the hypothesis that FR II jets arise from AGNs with a rapidly rotating black hole, and FR Is from a nucleus with a slow rotator. In the former case, the accretion disc can approach closer to the hole, can carry more magnetic flux and rotates faster, resulting in higher electric currents.

Acknowledgement: This work was supported by the Deutsche Forschungsgemeinschaft (SFB 328).

Appl, S., Camenzind, M.: 1992a, *Astron. Astrophys.* **256**, 354.
——: 1992b, in *Physics of AGN*, eds. W. Duschl, S. Wagner, Heidelberg: Springer Verlag.
——: 1992c, in *Extragalactic Radio Sources — from Beams to Jets*, eds. G. Pelletier, J. Roland, H. Sol, Cambridge: Cambridge University Press.
Appl, S.: 1992, in preparation.
Bodin, H.A.B., Newton, A.A.: 1980, *Nucl.Fusion* **20**, 1255.
Bodo, G., Rosner, R., Ferrari, A., Knobloch, E.: 1989, *Astrophys. J.* **341**, 631.
Camenzind, M.: 1990, in *Rev. Mod. Astron.* **3**, 234, ed. G. Klare, Heidelberg: Springer Verlag.
Cohn, H.: 1983, *Astrophys. J.* **269**, 500.
Corbelli, E., Torricelli-Ciamponi, G.: 1990, *Phys.Fluids B* **2**, 828.
Ferrari, A., Trussoni, E., Zaninetti, L.: 1981, *Mon. Not. R. astr. Soc.* **196**, 1051.
Kössl, D., Müller, E., Hillebrandt, W.: 1990, *Astron. Astrophys.* **229**, 378.
Pietrini, P., Torricelli-Ciamponi, G.: 1989, *Phys.Fluids B* **1**, 923.
Ray, T.P.: 1981, *Mon. Not. R. astr. Soc.* **196**, 195.
Scarton, H.A.: 1973, *J.Comp.Phys.* **11**, 1.
Taylor, J.B.: 1986, *Rev. Mod. Phys.* **58**, 741.
Voslamber, D., Callebaut, D.K.: 1962, *Phys.Rev.* **128**, 489.

Dynamo-Supported Magnetic Fields in Radio Jets

Jean A. Eilek

New Mexico Institute of Mining & Technology
Socorro NM 87801, USA

Abstract: Turbulence in radio jets will amplify any initial magnetic field to a level comparable to the energy density in the turbulence. If the turbulence is helical, as could be the case in a jet arising from a rotating accretion flow, magnetic flux as well as energy will be generated, and magnetic fields will be maintained which are ordered on scales comparable to the jet radius. This paper presents large-scale dynamo solutions for magnetic fields in turbulent, expanding jets.

1. Turbulent Dynamos in Radio Jets

The problem of magnetic flux generation in radio sources is well known. Flux freezing during the expansion of the plasma from the galactic nucleus to the extended lobes will reduce an initial magnetic field to extremely low values. In addition, it is likely that the jet plasma is turbulent. Transonic, turbulent flows have been suggested as models for type I sources (*e.g.*, Bicknell, these proceedings). Turbulence is not as fundamental a part of the supersonic-flow models of type II sources, but one would expect small-scale turbulence to be generated from the internal shocks and/or instabilities in these flows.

Turbulence can destroy magnetic flux through driven, small-scale reconnection. However, if the turbulence is helical, it will also act as a *source* of magnetic flux. Helicity is likely to be injected in turbulent radio jets from the base, where the flow arises from a rotating plasma. Helical turbulence will generate magnetic flux on a wide range of scales; in particular, it will amplify the field on scales large compared to the turbulence scale. In jets, this suggests that a field can be maintained with power on scales comparable to the jet radius – such as is suggested by the helical structures seen in M 87 (Owen *et al.*, 1989) or in Cygnus A (Perley *et al.*, 1984), for example.

In this paper I present analytic solutions for magnetic fields which can be supported by turbulent, helical dynamos in a steady flow. My starting point is the induction equation with a macroscopic dynamo term included:

$$\frac{\partial \mathbf{B}}{\partial t} = \nabla \times (\mathbf{V} \times \mathbf{B} + \alpha \mathbf{B} - \eta \nabla \times \mathbf{B}) \tag{1}$$

Here, $\mathbf{V}$ is the bulk flow field (non-turbulent) in the jet; the dynamo coefficient, α, can be related to the turbulent velocity as $\alpha \sim v_t$; and the turbulent diffusivity can be scaled as $\eta \sim \lambda_t v_t$, in terms of the turbulent velocity and length scale. This equation can be used to predict the geometry of the magnetic field, for instance in the steady-state limit, $\partial\mathbf{B}/\partial t = 0$. It cannot be used to predict the strength of the dynamo-supported field. For this, one must turn to numerical simulations, which find that the magnetic energy density will be maintained at a level comparable to the turbulent energy density, $E_B \lesssim \rho v_t^2/2$ (*e.g.* De Young 1980).

2. Solutions for Non-Expanding Jets

Two of the terms in equation (1) act as loss terms in a jet. The $\mathbf{V}\times\mathbf{B}$ term describes flux freezing, which is an effective loss term in an expanding flow. In addition, the $\eta\nabla\times\mathbf{B}$ term describes turbulent destruction of magnetic fields, which is always a loss term. In jets where the expansion is very slow, the last term will dominate the losses. Then, the steady-state dynamo-supported field will be described by $\alpha\mathbf{B} \simeq \eta\nabla\times\mathbf{B}$. This is just the equation describing a force-free magnetic field; it has been applied to radio jets (with a related but somewhat different justification) by Königl and Choudhuri (1985). Axisymmetric solutions of this equation in cylindrical geometry are well known: the magnetic field lines are helices, with pitch angle increasing away from the center of the jet. In the limit of no expansion, the magnetic field will be independent of distance along the jet; the solutions describe the behavior of the field as a function of jet radius. The scaling arguments for the α and η parameters predict the physical scale of the solutions: field reversals will occur on a scale $\sim \eta/\alpha \sim \lambda_t$.

3. Solutions for Expanding Jets

In jets which expand rapidly, flux-freezing losses may dominate turbulent, resistive losses. The steady-state equation in this limit becomes

$$\nabla\times(\mathbf{V}\times\mathbf{B}+\alpha\mathbf{B})=0. \tag{2}$$

I have found axisymmetric solutions for the magnetic field in an expanding, decelerating jet. These solutions will be described in more detail in an upcoming paper. In this paper I highlight the nature of the solutions and their relation to observable features of the jets.

I picked a jet velocity field which allows deceleration along the jet axis, as $V_r \propto r^k$ (if $\hat{\mathbf{r}}$ is the direction along the flow), and which allows the flow to have a small azimuthal velocity, $V_\phi \ll V_r$. In addition, I assumed the ratio α/V_r is constant; this is essentially assuming the turbulent velocity, $v_t \propto V_r$. These assumptions allowed the solution of (2) to be expressed as a sum of modal solutions, $\mathbf{B} = \sum_n \mathbf{B}_n$, where the nth mode depends on the distance along the jet, r, as $B_n \propto r^m$ and depends on the distance across the jet through Legendre polynomials and their derivatives. The geometry of the field is again helical, with a pitch angle increasing outwards across the jet; as in the non-expanding case, the pitch angle goes to zero on the jet axis. For higher mode numbers, n, the pitch angle increases faster away from the jet axis. The exponent m, describing the evolution of the field strength going out along the jet, depends on the mode number, and on the deceleration index k. All modes are found to decay more slowly than predicted by flux freezing. Higher-n modes decay more slowly than lower-n modes; if a mixture of modes is excited, the higher-n ones will come to dominate as one goes out along the jet.

A modal solution to a differential equation is usually completed by reference to the boundary conditions, which specify which modes can exist in the solution. In the case of jets, however, the boundary is unlikely to be rigid, and is likely to carry a thin current sheet. Thus, arbitrary jumps between the magnetic fields in the jet and in the surrounding medium are possible, and textbook boundary conditions may not be operative. Thus, I suspect the solution can contain a mixture of modes, and this mixture may well be specified by conditions at the origin of the jet. The "real" mixture would require knowing the initial conditions for the jet; in this work I will stop with a description of individual modes.

4. Consequences and Conclusions

This work shows that turbulent dynamos will support helical magnetic fields in radio jets, even in the presence of dissipative or expansion losses. I have considered the field structure that will be maintained in two limits: well-collimated jets with little lateral expansion, and jets where expansion losses dominate turbulent dissipation of the field. It is tempting to identify these two types of solutions with the two types of jets, corresponding to type II jets (only slowly expanding) and type I jets (more rapidly expanding; *e.g.*, Leahy, these proceedings).

For the type II jet solutions, the dynamo equation predicts force-free solutions in the steady-state, axisymmetric case. These magnetic fields are maintained at a steady level, going out along the jet, even in the presence of internal turbulence and consequent dissipation. The helical field predicted by these solutions is consistent with the observed helical patterns in resolved type II jets (for instance, M 87 or Cygnus A). The details of the emission patterns in these jets, however, require more physics than the present model contains. These real jets are limb brightened. However, simulations of the total intensity profile produced by the magnetic fields in this model do not find noticable limb brightening unless the electron density is also assumed to be enhanced towards the edge of the jet. In addition, the helical emission patterns apparent in both M 87 and Cynus A cannot be reproduced by solutions which are independent of distance along the jet. These solutions may describe an equilibrium state which is perturbed by some instability, such as a tearing mode (*e.g.*, Owen *et al.*, 1989), which might lead to field-aligned flux ropes, and local particle acceleration, both of which would enhance the local synchrotron emissivity.

The Type I jet solutions also predict helical magnetic fields, which are maintained against expansion losses by the dynamo. These solutions can be applied to expanding, decelerating flows, such as in type I jets, or possibly in the tails of some type I sources (*e.g.*, O'Donoghue *et al.*, 1990). These solutions make several interesting observational predictions. For poorly resolved jets, the main predictions are as follows. Jets with a larger opening angle are more likely to be dominated by a magnetic field with a large

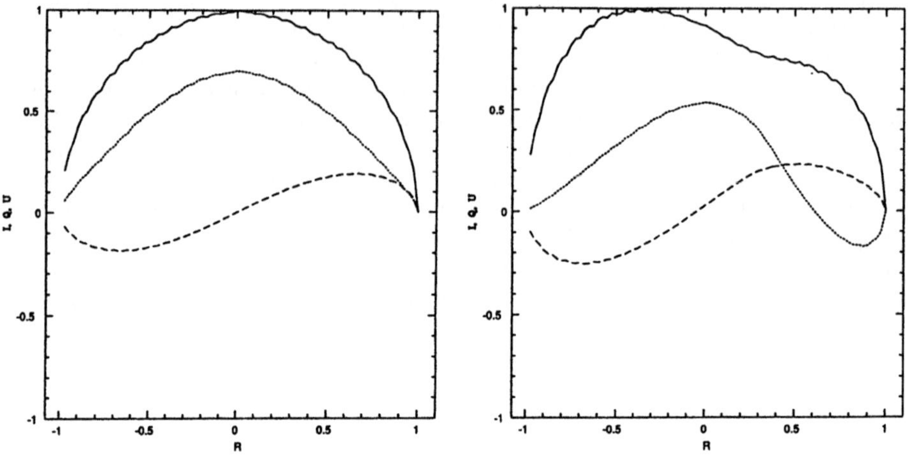

Fig. 1. Total and polarized intensity from a uniformly filled type I jet with mode $n = 3$ solutions. The jet opening angle is 10°. The solid line is the Stokes I parameter, the dotted line is Q and the dashed line is U; the sign of Q is chosen so that $Q > 0$ corresponds to $B_{\|}$, and $Q < 0$ to $B_{\perp}$. The left-hand figure shows a jet lying in the plane of the sky, and the right-hand figure shows a jet at 60° to the plane of the sky.

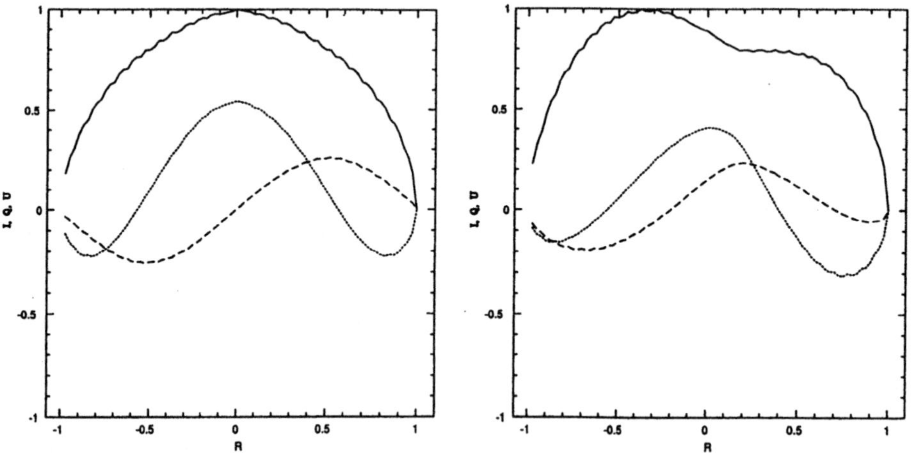

Fig. 2. The same as Fig. 1, for mode $n = 10$.

pitch angle; if these jets lie close to the plane of the sky, their projected field is likely to be perpendicular to the jet axis ($B_\perp$). Going out along the jet, lower mode numbers decay faster, so that higher-n modes will come to dominate; since higher mode numbers have larger pitch angles, this can lead to a transition from $B_\parallel$, close to the origin of the jet, to $B_\perp$, further out along the jet. (It is worth noting that standard flux-frozen models predict the same qualitative behavior, but for quite different reasons). The field strength is likely to decay, going out along the jet, but more slowly than in the simple flux-frozen case. Thus, the synchrotron emissivity will not decay as rapidly as simply models predict, and less deceleration of the flow is needed to account for the observed luminosity (*e.g.*, O'Donoghue *et al.* 1991).

For well-resolved jets, the present model makes detailed predictions for the total and polarized synchrotron intensity. Figures 1 and 2 show representative cases, from a code which evaluates the I, Q and U Stokes parameters integrated along a line of sight through the jet.

In these calculations, the electron density is uniform within the jet. We find little limb brightening in total intensity; however, normal polarization transfer effects can lead to asymmetric and limb brightened Q and U Stokes parameters. The total intensity distribution is sensitive to the electron distribution throughout the jet. Simulations in which the electron density rises linearly, or faster, with jet radius show appreciable limb brightening in total intensity. The polarized intensity is sensitive to the electron distribution, and also to the particular mode chosen. Low-n modes are dominantly $B_\parallel$, while high-n modes are dominantly $B_\perp$. The angle at which the jet is viewed is also important: jets close to the plane of the sky are approximately symmetric about their central line, while jets lying at a large angle to the plane of the sky can have strongly asymmetric total and polarized intensity distributions, while being physically axisymmteric. (Laing 1982 has pointed out such effects for other models of the magnetic field).

Unlike the type II jets, the models for the type I jets are not force-free. The Lorentz force, $\mathbf{j} \times \mathbf{B}$, has components both across the jet and around the jet. The former will contribute to jet confinement, while the latter will exert a torque on the flow (which must arise from the tying of the field lines to something at the base of the jet, say an accretion flow). The magnitude of these forces cannot be predicted from this model, which is independent of the strength of the field. We may note, however, that turbulent dynamos can reach approximate equipartion with the turbulent energy density in the plasma; this suggests these forces might be significant in the overall jet structure.

References

De Young, D.S.: 1980, *Astrophys. J.* **241**, 81.
Königl, A., Choudhuri, A.R.: 1985, *Astrophys. J.* **289**, 173.
Laing, R.A.: 1982, *Astrophys. J.* **248**, 87.
O'Donoghue, A.A., Owen, F.N., Eilek, J.A.: 1990, *Astrophys. J. Suppl.* **72**, 75.
O'Donoghue, A.A., Eilek, J.A., Owen, F.N.: 1991, *Astrophys. J.*, in press.
Owen, F.N., Hardee, P.E., Cornwell, T.C.: 1989, *Astrophys. J.* **340**, 698.
Perley, R.A., Dreher, J.W., Cowan, J.J.: 1984, *Astrophys. J. (Lett.)* **285**, L53.

Theory and Simulation of Jet Stability

Philip E. Hardee [1], Michael L. Norman [2]

[1]Department of Physics & Astronomy
University of Alabama
Tuscaloosa, AL 35487, USA
[2]Beckman Institute & Department of Astronomy
University of Illinois
Urbana, IL 61820, USA

1 Introduction

Observations show many extragalactic jets with large scale oscillation in the plane of the sky, *e.g.*, 3C 449 (Perley, Willis, & Scott 1979), NGC 6251 (Perley, Bridle, & Willis 1984), Cygnus A (Dreher, Carilli, & Perley 1987), and other structures consistent with the helical structures predicted from analysis of the Kelvin-Helmholtz instability of extragalactic jets (see Birkinshaw 1991 for a review and references). Over the past few years we have tested the predictions of stability theory in two dimensions through numerical simulations of a two dimensional slab jet. A slab jet is spatially resolved along two Cartesian axes and is infinite in extent in the third dimension. Stability analysis of cylindrical jets and slab jets has shown that these jets share both qualitative and quantitative similarities (Hardee & Norman 1988). The wave dispersion relation describing wave propagation and growth in slab or cylindrical geometries allows a symmetric pinching and an asymmetric sinusoidal mode or helical modes, respectively. Each mode contains a single fundamental surface wave solution and multiple reflection body wave solutions.

2 Stability of the Fluid Slab Jet

Stability analysis and numerical simulations of equilibrium jets in constant atmospheres and in atmospheres containing density and temperature gradients were performed by Norman & Hardee (1988), and Hardee *et al.* (1991). Solution of the dispersion relation for the sinusoidal mode on a jet with internal Mach number $M_j = 3$ and density ratio $\eta = \rho_j/\rho_{ext} = 0.1$ and a density image of a corresponding simulation are shown in Fig. 1.

Solution of the dispersion relation illustrates the basic behavior of growing sinusoidal waves on a supersonic jet. There is a frequency and real part of the wavenumber (dotted

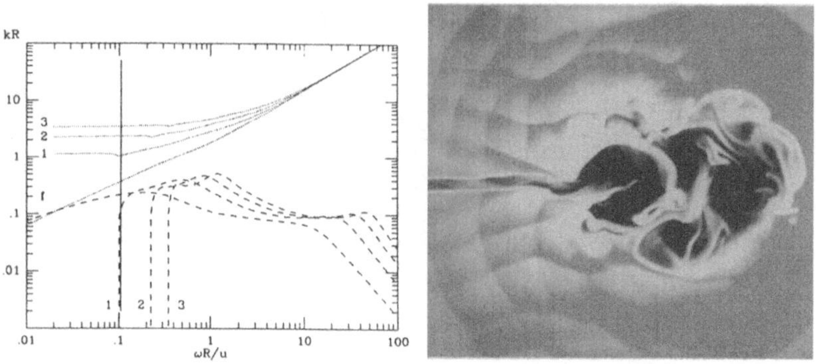

Fig. 1. Complex wavenumber normalized by the jet's radius as a function of the frequency normalized by the jet's radius and velocity, and a grey scale density image from a corresponding numerical simulation of a supersonic jet initialized in equilibrium with a constant atmosphere ($M_j = 3, \eta = 0.1$).

lines) at which the imaginary part of the wavenumber or growth rate (dashed lines) is a maximum for fundamental and for reflection solutions. The growth length and associated wavelength at maximum growth vary approximately proportional to the jet's Mach number and radius with a lesser dependence on conditions in the external medium. In the simulation, a jet established initially in equilibrium with the external medium was perturbed at the origin to break the symmetry at a frequency indicated by the vertical solid line in Fig. 1. The jet was disrupted initially by sinusoidal oscillation near the resonant wavelength, i.e., fastest spatially growing wavelength. Ultimately, by the time shown in the image, the jet flaps within a lobe at about the wavelength corresponding to the perturbing frequency and at a distance on the order of 10 M_jR. In addition to the fundamental sinusoidal oscillation clearly evident in the image, weak oblique shocks inside the jet can be identified with reflection wave solutions.

Subsequently, simulations were performed with jets of constant opening angle established in equilibrium with appropriate pressure gradient atmospheres containing various density and temperature gradients. The effect of jet expansion is predicted to partially stabilize a jet because the jet's Mach number increases as an adiabatic jet expands. On the other hand, a decrease in the external density relative to the jet's density, and increase in the external temperature can destabilize a jet. Fig. 2 contains density images from simulations of a jet with constant opening angle of 0.04 *radian*, and with conditions at the inlet equal to those used in the simulation in Fig. 1, in two different atmospheres.

In the isothermal atmosphere the pressure gradient consistent with constant jet expansion is the result of a density gradient only. In the rising temperature atmosphere, temperature rises by about a factor of ten across the grid while the density falls more rapidly to maintain the pressure gradient consistent with constant jet expansion. The jet in the isothermal atmosphere illustrates the stabilizing effect of jet expansion predicted

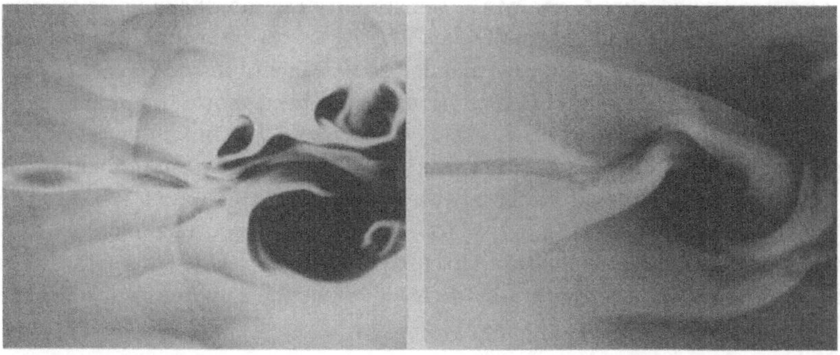

Fig. 2. Grey scale density images of a supersonic jet initialized with constant opening angle in equilibrium with (left) an isothermal atmosphere and (right) a rising temperature atmosphere.

by stability theory (compare this image with that shown in Fig. 1). The jet in the rising temperature atmosphere illustrates a destabilization which is largely the result of the rapid decrease in the external density required to maintain the appropriate pressure gradient. This rapid decrease in the external density results in reduced resistance of the external medium to sideways jet motion.

One of the most important results of the simulations has been to show that the wavelength and amplitude growth along a jet are properly computed from the dispersion relation by conserving the wave frequency along the jet (Hardee *et al.* 1991). Additionally, results obtained in the equilibrium jet simulations were confirmed in propagating jet simulations (Hardee & Norman 1990; Hardee *et al.* 1992).

3 Stability of the Magnetized Slab Jet

Fluid models may be able to account for the dynamics and morphology of most extended jets and radio sources but several sources have emission properties that imply pressures considerably in excess of an external thermal pressure, *e.g.*, M 87 (Owen, Hardee & Cornwell 1989) and 4C 32.69 (Potash & Wardle 1980). These observations and theoretical models for jet production, *e.g.*, Lovelace *et al.* (1987) and Königl (1989), imply the existence of dynamically significant magnetic fields.

The effect of magnetic fields on Kelvin-Helmholtz instability has been examined in cylindrical geometry for axial magnetic fields (Ray 1981; Ferrari, Trussoni, & Zaninetti 1981; Fiedler & Jones 1984; Bodo *et al.* 1989) which were found to stabilize sub-Alfvenic flows to fundamental surface modes, and were found to admit a new region of unstable slow reflection body modes (Bodo *et al.* 1989) which corresponded to slow magnetosonic waves in the jet. Since magnetic pressure and tension effects are similar for axial magnetic fields in slab and cylindrical geometries, we have recently extended stability theory in slab geometry to incorporate dynamically significant magnetic fields.

For magnetized jet and external medium where the magnetic fields are axial, values of Alfvenic, and magnetosonic Mach numbers defined by $M_{A(in,ex)} = u/V_{A(in,ex)}$, and $M_{ms(in,ex)} = u/(a^2_{in,ex} + V^2_{A(in,ex)})^{1/2}$ where V_A is the Alfven speed, a is the sound speed, and u is the jet speed, determine the stability regime. If the jet is highly super-magnetosonic internally and externally, i.e., $M_{ms(in,ex)} \gg 1$, the stability properties are similar to a fluid jet with scaling by magnetosonic Mach number replacing scaling by sonic Mach number. A similar scaling relationship has been shown to apply to cylindrical jets by Appl and Camenzind (1992 & these proceedings). The fundamental surface wave can be identified with Alfven waves propagating in the jet or external medium modified by the flow boundary. The reflection body waves can be identified with fast magnetosonic waves inside the jet. When $M_{A(in,ex)} = u/V_{A(in,ex)} < 1$ the surface mode is stabilized. Reflection waves still exist on the sub-Alfvenic jet when the jet speed is slightly less than the slow magnetosonic speed, and this corresponds to the slow magnetosonic mode on the cylindrical jet found by Bodo *et al.* (1989).

If the internal or external magnetic field is perpendicular to the flow velocity, then the magnetic field has no effect on surface waves or reflection waves which propagate parallel to the flow. For arbitrary angles, θ, of wave propagation and arbitrary angles of magnetic field orientation, ϕ, the slab jet is unstable to surface sinusoidal waves when $[V^2_{A(in,ex)}/u^2][\cos^2(\theta - \phi_{in,ex})/\cos^2\theta] \leq 1$. Thus, on a super-Alfvenic jet when $\phi_{in,ex} > 0$ there will be some range $\pi/2 \geq \theta \geq \theta_{min}$ for which surface waves are stabilized but the instability remains for propagation angles $\theta < \theta_{min}$. In general, when there are magnetic fields inside and outside a jet, a surface wave will be stabilized by the magnetic field if the jet's velocity projected on the wavevector, u $\cos\theta$, is less than the largest Alfven velocity projected on the wavevector, $V_{A(in,ex)}\cos(\theta - \phi_{in,ex})$. For the reflection modes, the condition for instability is that the jet's velocity projected on the wavevector must be greater than the fast magnetosonic speed or slightly less than the slow magnetosonic speed.

On a cylindrical jet fundamental surface waves associated with the helical mode propagate around the jet with angle $\theta = tan^{-1}(1/k_z R)$ where k_z is the axial wavenumber and R is the jet's radius. Provided the internal fields are axial, equivalent to $\phi_{in} = 0$ on the slab jet, then the slab jet results imply that the helical surface mode is stabilized only when the jet is sub-Alfvenic, and this is the result found in cylindrical geometry for axial fields. On the other hand, our present results suggest that small wavenumber, long wavelength helical twisting of a cylindrical jet can be partially stabilized by helically wrapped magnetic fields, equivalent to $\phi_{in} > 0$ on the slab jet, even when the jet is super-Alfvenic. This prediction agrees with work performed by Appl and Camenzind (1992 & these proceedings) who find that helical fields on cylindrical jets partially stabilize the long wavelengths.

For magnetosonic Mach numbers greater than three, solutions to the dispersion relation are similar to those shown in Fig. 1 with some suppression of the growth rate at high frequencies. Simulations confirm this result and reveal only small changes in stability for dynamically significant magnetic fields. This changes when the magnetosonic Mach number approaches one. Solution of the dispersion relation for sinusoidal modes on a jet with $M_{ms} \sim M_A \sim 1.5$, and density one-tenth that of an unmagnetized external medium, and a density image of a corresponding simulation are shown in Fig. 3.

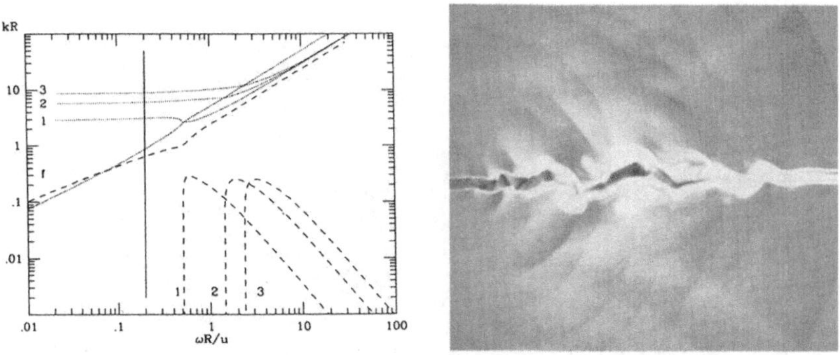

Fig. 3. Complex wavenumber normalized by the jet's radius as a function of the frequency normalized by the jet's radius and velocity, and a grey scale density image from a corresponding numerical simulation of a trans-magnetosonic jet initialized in equilibrium with a constant atmosphere.

Solution of the dispersion relation shows that the reflection modes have been suppressed relative to the surface wave whose resonance has disappeared, and whose growth rate has increased at all frequencies (compare to Fig. 1). The simulation shows the predicted decrease in stability of this trans-magnetosonic jet. Continuous flow is disrupted at about 10 M_{ms}R and this is about half of the distance reached by the Mach three jet shown in Fig. 1. The strong magnetic field has prevented the formation of a lobe. This provides confirmation of the prediction of a scaling of stability properties with the magnetosonic Mach number for super-Alfvenic magnetized jets. An interesting implication is that a sub-Alfvenic jet which becomes unstable if the jet goes through the Alfven point to become super-Alfenic will be doomed to disruption by the low magnetosonic Mach number.

4 Stability of the Cylindrical Jet

A difference between slab and cylindrical geometries lies in the different propagation angles of waves excited by a perturbation relative to the axial or flow direction. For example, on a cylindrical jet, surface waves associated with the helical mode are constrained to propagate around the jet with angle given by $\theta = tan^{-1}(1/k_z R)$. On the other hand, sinusoidal surface waves on a slab jet can propagate at any angle. However, waves on slab jets which propagate at an angle appropriate to their wavelength in cylindrical geometry behave very nearly identically to similar growing waves in cylindrical geometry. Thus, this difference can be easily accounted for, and slab jet results can be applied to cylindrical jets.

A difference between slab and cylindrical geometries that cannot be determined from slab jet results is the effect of helical fluting modes in cylindrical geometry that

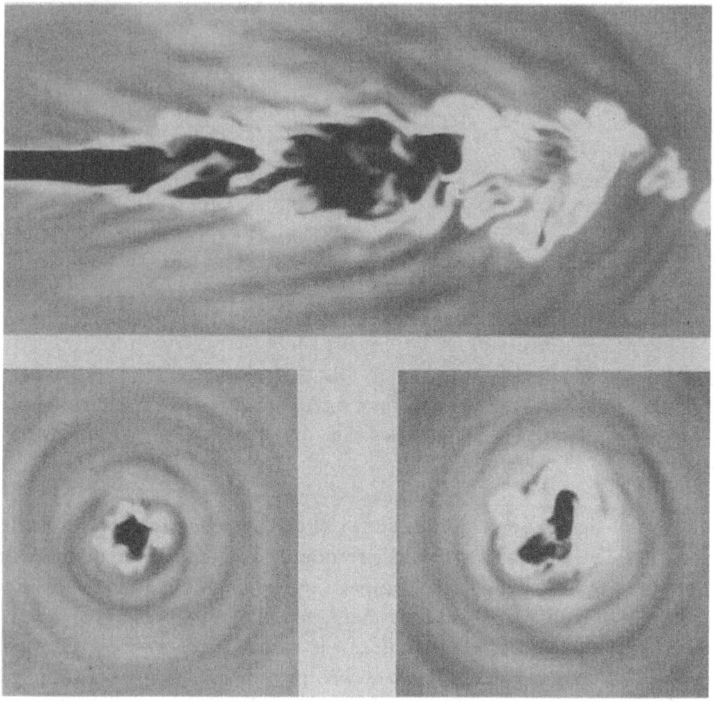

Fig. 4. Grey scale density images of one 2D slice along the 3D jet's axis (top) and two 2D slices perpendicular to the 3D jet's axis (bottom) at ten and twenty jet radii from the origin, respectively.

do not have a counterpart in slab geometry. These fluting modes have large growth rates and lead to helically twisted filamentation of a cylindrical jet. On a supersonic jet the fluting modes appear at sucessively higher resonant frequency, with sucessively shorter resonant wavelength, and with sucessively higher resonant growth rate as the mode number (equal to the number of flutes around a jet circumference) increases. At any fixed frequency the highest growth rate is associated with one of the surface fluting modes. Clearly it is of utmost importance to ascertain the importance of the fluting modes to jet dynamics.

Recently, in collaboration with David Clarke, a fully three dimensional simulation of the equilibrium jet in a constant atmosphere has been performed using ZEUS-3D (see Clarke these proceedings). Results of a simulation of a Mach three cylindrical jet with density one-tenth that of the external medium are shown in Fig. 4. The parameters chosen were identical to those used in the slab jet simulation in constant atmosphere shown in Fig. 1.

The top panel is a two dimensional density slice along the axial direction. The jet is disrupted at fifteen to twenty jet radii from the origin. This is much closer to the origin than the slab jet shown in Fig. 1. The cylindrical jet has become unstable to disruption by helical and filamentary modes with a disruption length of about $5M_jR$ where the slab jet had a disruption length of about $10M_jR$. The lower panel contains two slices perpendicular to the jet's axis at a distance of ten and twenty jet radii from the origin. The images reveal the importance of filamentation. At ten jet radii higher order fluting modes are apparent, and at twenty jet radii the jet has been broken into at least two filaments. From this result we conclude that surface fluting modes are important and have an important effect on jet dynamics.

Acknowledgements

P. Hardee acknowledges support from the National Science Foundation through grant AST-8919180 and EPSCoR grant RII-8996152 to the University of Alabama. Supercomputer time was provided by the National Center for Supercomputing Applications.

References

Appl, S., Camenzind, M.: 1992, *Astron. Astrophys.* **256**, 354.

Birkenshaw, M.: 1991, in *Beams and Jets in Astrophysics*, ed. P.A. Hughes, Cambridge: Cambridge University Press, p.278.

Bodo, G., Rosner, R., Ferrari, A., Knobloch, E.: 1989, *Astrophys. J.* **341**, 631.

Dreher, J.W., Carilli, C.L., Perley, R.A.: 1987, *Astrophys. J.* **316**, 611.

Fiedler, R., Jones, T.W.: 1984, *Astrophys. J.* **283**, 532.

Ferrari, A., Trussoni, E., Zaninetti, L.: 1981, *Mon. Not. R. astr. Soc.* **196**, 1051.

Hardee, P.E., Norman, M.L.: 1988, *Astrophys. J.* **334**, 70.

———: 1990, *Astrophys. J.* **365**, 134.

Hardee, P.E., Norman, M.L., Koupelis, T., Clarke, D.A.: 1991, *Astrophys. J.* **373**, 8.

Hardee, P.E., White, R.E. III, Norman, M.L., Cooper, M.A., Clarke, D.A.: 1992, *Astrophys. J.* **387**, 460.

Königl, A.: 1989, *Astrophys. J.* **342**, 208.

Lovelace, R.V.E., Wang, J.C.L., Sulkanen, M.E.: 1987, *Astrophys. J.* **315**, 504.

Norman, M.L., Hardee, P.E.: 1988, *Astrophys. J.* **334**, 80.

Owen, F.N., Hardee, P.E., Cornwell, T.J.: 1989, *Astrophys. J.* **340**, 698.

Perley, R.A., Bridle, A.H., Willis, A.G.: 1984, *Astrophys. J. Suppl.* **54**, 291.

Perley, R.A., Willis, A.G., Scott, J.S.: 1979, *Nature* **281**, 437.

Potash, R.I., Wardle, J.F.C.: 1980, *Astrophys. J.* **239**, 42.

Ray, T.P.: 1981, *Mon. Not. R. astr. Soc.* **196**, 195.

Velocity and Confinement of Large-Scale Jets

Mitchell C. Begelman

Joint Institute for Laboratory Astrophysics,
University of Colorado and National Institute of Standards and Technology,
Boulder, CO 80309-0440
USA

Abstract: I consider several issues connected with the velocities and confinement properties of kpc-scale jets in FR II sources. There is considerable evidence that large-scale jets are at least mildly relativistic. I discuss whether the Doppler-beaming interpretation of one-sidedness can be reconciled with Barthel's unified scheme, and consider modifications to the Doppler beaming formula due to the directionality of the magnetic field. The synchrotron-emitting regions of many FR II jets appear to be overpressured with respect to their surroundings, and yet seem to be effectively confined. I consider several possible confinement mechanisms, and discuss the implications of a small volume filling factor in the synchrotron emissivity. Finally, I argue that most lobe asymmetry could be due to environmental effects, and show that even a small asymmetry in the ambient intergalactic medium can lead to a large asymmetry in lobe/core separation.

1 Introduction

The arguments used to estimate speeds and internal pressures in large-scale extragalactic jets are indirect, and the conclusions drawn from them are often inconclusive. Taken together, however, these arguments suggest that 1) many jets have relativistic speeds, even on kiloparsec scales and larger; and 2) the synchrotron-emitting regions in some large-scale jets have surprisingly high energy densities. Since the inferred velocities and minimum pressures are most extreme in the powerful jets associated with FR II radio galaxies and quasars, I will focus on these objects. Indeed, the inferred conditions in some of these sources are so extreme that they stress the assumptions behind our diagnostic techniques.

2 How Relativistic are Large-Scale Jets?

The case for bulk relativistic flow in extragalactic jets depends on establishing evidence for one or more of the following phenomena — Doppler boosting, apparent superluminal motion, and the relativistic aberration of light. The second of these has been used to argue forcefully that many VLBI jets have highly relativistic speeds, although one is never quite sure that the superluminal changes on the sky reflect motion of matter rather than pattern. Since there is no direct evidence (other than superluminal motion) that one-sided VLBI jets are on the near side of the nucleus, as required by the Doppler boosting interpretation, one-sidedness has played a corroborating role in establishing relativistic flow on VLBI scales. Also corroborating the relativistic interpretation are the sharp bends often observed in VLBI jets, which would result from foreshortening of a weakly curved jet, viewed end-on.

Continuity of large-scale structure with VLBI structure, exemplified by the spectacular montages of NGC 6251 (Bridle & Perley 1984) and 3C 120 (Walker *et al.* 1987), provides strong evidence that whatever makes VLBI jets one-sided is also responsible for the one-sidedness of large scale jets. Other studies have yielded independent evidence for relativistic bulk motion on large scales, and it now appears that all three of the relativistic phenomena listed above can be associated with large-scale jets. As is the case with VLBI jets, one-sidedness or *brightness asymmetry* between jet and (often putative) counterjet is readily explained by "Doppler favoritism," but in the large-scale case it is possible to argue independently that the visible jet is indeed the one that is pointing towards the observer, by measuring the *depolarization asymmetry* between the jetted and unjetted radio lobes (Laing 1988; Garrington *et al.* 1988). That there is a similar, "Doppler enfeebled" (in the words of Peter Scheuer) jet pointing away from the observer is suggested by the observation that the two diffuse radio lobes are generally quite similar in appearance. Attempts to explain the simultaneous jet asymmetry and lobe symmetry without resorting to Doppler boosting require either subtle differences in the dissipative properties of the two jets (which must be correlated with the depolarization of the lobe), or a "flip-flopping" jet (*e.g.* Rudnick & Edgar 1984).

Evidence for superluminal motion and relativistic aberration in large-scale jets has been developed more recently. Although superluminal motion in 3C 120 on 2 kpc scales (Walker *et al.* 1988) has not been confirmed by Muxlow & Wilkinson (1991), the latter do detect motion at an apparent speed of $3.7h^{-1}c$ about 100 pc from the nucleus. Knots in the M 87 jet also appear to show at least mildly relativistic proper motion on moderately large scales (Biretta, these proceedings). The interpretation of the polarization vectors of hot spots in terms of relativistic aberration (Kollgaard *et al.* 1990) is more model-dependent, but intriguing nonetheless. Note that compact hot spots, which are seen only on the jetted side (Laing 1989), may exhibit Doppler boosting as well, which suggests that they could represent relativistic flow downstream of an oblique shock (Begelman and Kirk 1990).

Despite the evidence favoring relativistic bulk motion for jets in FR II sources, some nagging doubts remain. For example, there are persistent claims that the diffuse lobe emission, which is supposed to be isotropic, is systematically different on the jetted side than on the unjetted side (Leahy *et al.* 1986). Liu & Pooley (1991) have found a significant correlation between lobe spectral index and depolarization in a sample consisting of 9 FR II radio galaxies (RGs) and 4 quasars (QSRs). Garrington *et al.* (1991) suggest that a similar correlation in their sample could be explained by beaming of hot spot radiation, however, Pooley (these proceedings) doubts that this explanation can account for his result. It should be noted that the correlation of jettedness with depolarization asymmetry is established only for a sample rich in QSRs which, if Barthel's (1989) unified model is correct, corresponds to small mean angles between the jet axis and the line of sight (see § 3.1). It is in just such a sample that one would expect relativistic effects to stand out most clearly. At the same time, *some* intrinsic jet and lobe asymmetries are bound to occur, caused by asymmetries in the environment if not by the jet production mechanism (see *e.g.* § 5). It could be that intrinsic jet asymmetries are correlated with lobe properties, and that the bias shows up mainly in samples which include many sources oriented close to the plane of the sky, *i.e.*, a sample rich in RGs. Alternatively, we could be naive in our assumption that lobes emit isotropically in our frame of reference (*e.g.* Browne & Perley 1986).

Given these caveats, it seems likely that *most* brightness asymmetries and one-sidedness in kpc-scale jets of FR II radio galaxies and quasars are due to Doppler favoritism. The question is then, how relativistic is relativistic enough? For VLBI jets, superluminal motion and the statistics of "unified schemes" (*e.g.* Orr & Browne 1982) suggest that Lorentz factors $\gtrsim 5$ are typical, but such high velocities are certainly not required to explain most existing observations of large-scale jets. Although counterjets are often not seen, the dynamic range constraints in most cases are not particularly strong ($\mathcal{R} \equiv S_{jet}/S_{counterjet} \gtrsim 4 - 10$), leading to minimum velocities $\beta \gtrsim 0.5 - 0.8$ ($\gamma \gtrsim 1.2 - 1.7$). Nevertheless, a few exceptions, such as 3C 120 ($v_{app} \sim 3.7h^{-1}c$; Muxlow & Wilkinson 1991) and M 87 ($\mathcal{R} > 340$; Biretta, these proceedings, confirmed in the optical by Stiavelli *et al.* 1991), give cause for worry. If most jets had extremely relativistic speeds, we would expect Doppler enfeeblement to render *both* jets invisible in the majority of sources. In those jets which were visible, small fluctuations in the velocity vector along the jet, perhaps indicated by wiggles, would lead to huge changes in surface brightness, contrary to what appears to be the norm (van Groningen *et al.* 1980).

3 Relativistic Beaming and Unified Models

3.1 Beaming Ratios and Unification à la Barthel

The association of FR II jets with well-defined lobes, which are presumably isotropic emitters to a reasonably good approximation, affords us the possibility of testing the effects of orientation on jet visibility. Since we are concerned with large-scale jets, I will cast the discussion in terms of Barthel's (1989) model, which seeks to unify FR II radio galaxies with steep spectrum radio quasars in terms of a single parameter, the orientation of the jet axis with respect to the line of sight. By choosing an angle $\theta_{crit} \approx 44.4°$ as the demarcation between RGs ($\theta > \theta_{crit}$) and QSRs ($\theta < \theta_{crit}$), Barthel is able to: 1) explain the relative numbers of the two classes; 2) unify their projected linear size distributions; and 3) explain why the tendency of QSRs to exhibit one-sidedness and superluminal motion on VLBI scales is insensitive to their apparent linear size. Padovani & Urry (1992) have shown that flat spectrum QSRs can also be incorporated within this scheme.

What does Barthel's scheme imply for the Doppler boosting interpretation of one-sidedness? For an intrinsically symmetric source, the jet–to–counterjet surface brightness ratio is expected to be

$$\mathcal{R} \equiv \frac{S_j}{S_{cj}} = \left(\frac{1+\beta\cos\theta}{1-\beta\cos\theta}\right)^{\delta}, \tag{1}$$

where $\beta \equiv v_j/c$ is the jet speed and $\delta = 2 - \alpha$ for a jet spectral index α (typically in the range -0.5 to -0.8). The expression for δ is valid if the jet emissivity is isotropic in the fluid rest frame, which is generally thought to be the case. However, I leave δ as a variable for reasons which will become clear shortly.

In principle, it is possible to test Barthel's conjecture that FR II radio galaxies and quasars are similar objects with disjoint orientation distributions by using measured jet–counterjet ratios. Specifically, if we form the statistic

$$\frac{\left\langle(\mathcal{R}^{1/\delta}-1)/(\mathcal{R}^{1/\delta}+1)\right\rangle_{RG}}{\left\langle(\mathcal{R}^{1/\delta}-1)/(\mathcal{R}^{1/\delta}+1)\right\rangle_{QSR}} = \frac{\langle\beta\cos\theta\rangle_{RG}}{\langle\beta\cos\theta\rangle_{QSR}} = \frac{\cos\theta_{crit}(1-\cos\theta_{crit})}{\sin^2\theta_{crit}}, \tag{2}$$

where $\langle\ldots\rangle_{RG}$ and $\langle\ldots\rangle_{QSR}$ represent means over radio galaxies and radio quasars, respectively, then the final equality holds only if Barthel's model is correct. If the distributions of jet speed and jet orientation are statistically independent of one another *and* the orientation distributions are disjoint, then the statistic approaches a maximum value of 0.5 when $\theta_{crit} \to 0$, vanishes when $\theta_{crit} = 90°$, and has intermediate values of 0.42, 0.33, and 0.15 for $\theta_{crit} = 44.4°$ (Barthel's value), 60°, and 80°, respectively. Any value of this statistic exceeding 0.5 would imply that the orientation distributions of FR II RGs and QSRs overlap, or that the value of θ_{crit} depends on β, or that the jet velocity distributions of QSRs and RGs are different.

It is probably not feasible to apply this test with currently available data, since most counterjets are not detected and quoted values of $\mathcal{R}$ are often lower limits. However, it may be possible to rule out simple versions of Barthel's model on the basis of non-detections alone. Table 1 shows the mean jet–counterjet ratios predicted by Barthel's model for $\theta_{crit} = 44.4°$ and various fixed values of the jet speed. For the expected beaming ratio exponent $\delta = 2.5 - 2.75$, we see that the *mean* jet–counterjet ratio for RGs should not exceed 22, even if all jets are ultrarelativistic. Note that in this limit, the mean beaming ratios for QSR jets may be arbitrarily large. Under the more plausible assumption that mean jet speeds fall in the range 0.5c–0.8c, Barthel's model predicts that the mean RG jet–counterjet ratio should not exceed 4–9. These mean beaming ratios come tantalizingly close to existing observational lower limits.

Should Barthel's model fail this test in its current form, there are a number of ways in which it could be modified without losing its unifying features, *i.e.*, without assuming a population of RGs which is intrinsically different from the QSR population. For example, θ_{crit} could be a decreasing function of β. Another possible modification which I have never heard discussed seriously is that the standard formula used to calculate beaming ratios, eq. (1) with $\delta = 2 - \alpha$, might not be correct. From Tab. 1 it is clear that a larger value of δ would ease the constraints on the asymmetry of RG jets tremendously. But is there any physical basis for questioning such a time-worn formula? Below I discuss

Table 1. Barthel model predictions for mean beaming ratios, assuming $\theta_{crit} = 44.4°$ and β fixed.

β \ δ	2.5	2.75	3.5	4.0	4.5
1	∞ / 15.29	∞ / 21.49	∞ / 62.46	∞ / 131.17	∞ / 280.71
0.8	86.45 / 6.87	137.99 / 8.67	573.06 / 17.96	1504.36 / 29.87	3994.27 / 50.42
0.6	18.38 / 3.73	24.74 / 4.35	60.75 / 7.01	111.16 / 9.79	204.21 / 13.82
0.4	6.09 / 2.25	7.31 / 2.47	12.66 / 3.27	18.29 / 3.97	26.47 / 4.85

Key (upper-right / lower-left of each cell): $\langle R\rangle_{QSR}$ / $\langle R\rangle_{RG}$

one modification of the formula which is clearly applicable. However, I show that it is probably of little quantitative importance for models of real jets.

3.2 Effects of Magnetic Anisotropy on the Beaming Formula

As noted earlier, the validity of the standard beaming formula depends on the isotropy of the intrinsic synchrotron emissivity in the jet. Are there any reasons to suspect that the emissivity might be significantly anisotropic? Strong anisotropy of the electron pitch angle distribution would do the trick, but leads to too much circular polarization. However, an effect which will certainly lead to *some* anisotropy in the observed radiation is the directionality of the magnetic field. We know that the field has a preferred orientation parallel to FR II jets because of the high observed polarization (typically, $\gtrsim 15\%$) oriented perpendicular to the jet axis on the sky. In a uniform field, the synchrotron emissivity of an isotropic, power-law electron distribution has the angular dependence $j'_{\nu'} \propto |\sin\theta'|^{1-\alpha}$, where θ is the angle with respect to the magnetic field (which equals the jet orientation angle in this case) and primes denote quantities measured in the rest frame of the jet fluid. From the formula for the aberration of light, we have $\sin\theta' = \sin\theta/[\gamma(1-\beta\cos\theta)]$, therefore we arrive at a new beaming formula for a *perfectly unidirectional field aligned with the jet:*

$$\mathcal{R} = \left(\frac{\sin\theta'_j}{\sin\theta'_{cj}}\right)^{1-\alpha}\left(\frac{1+\beta\cos\theta}{1-\beta\cos\theta}\right)^{2-\alpha} = \left(\frac{1+\beta\cos\theta}{1-\beta\cos\theta}\right)^{3-2\alpha}. \tag{3}$$

In this case, the effect of the anisotropy would be to increase the exponent in eq. (1) to $\delta = 3 - 2\alpha \approx 4 - 4.5$. From Tab. 1, it is clear that such a change would have a dramatic effect on the beaming ratios predicted by the Barthel model. But it is also clear that the anisotropy caused by the magnetic field cannot possibly be this large in real jets, because the field is not unidirectional. If it were, then the polarization would have the maximum possible value $p_{max} = (3-3\alpha)/(5-3\alpha) \approx 0.7$ instead of the observed value, which is typically only a quarter to a third of p_{max}. To estimate the true degree of anisotropy, and hence to determine how the beaming formula should be modified, it is necessary to average over a distribution of field orientations, chosen to give the observed degree of polarization. Laing (1980, 1981) describes how to do this calculation using Stokes parameters. Assume that the field orientation is distributed axisymmetrically about the jet axis according to the probability function $P(\theta)$, with the normalization $\int_{-1}^{1} P(\theta) d\cos\theta = 1$. As Laing points out, the results are not very sensitive to the value of α, and the integrals are particularly simple for the case $\alpha = -1$. For this case, both the anisotropy and the polarization properties depend on the intrinsic field structure through a single moment of P:

$$A \equiv \int_{-1}^{1} P(\theta)\cos^2\theta \; d\cos\theta. \tag{4}$$

To within a multiplicative constant, the angular dependence of the intensity is given by

$$I = 2(1-A) + (3A-1)\sin^2\theta' \tag{5}$$

while the degree of polarization is given by

$$\tilde{p} \equiv \frac{p}{p_{max}} = \frac{(3A-1)\sin^2\theta'}{2(1-A)+(3A-1)\sin^2\theta'}. \tag{6}$$

By applying the aberration formula, one can derive an approximate expression for the beaming ratio of a polarized jet,

$$\mathcal{R} = \frac{\left(\frac{1+\beta\cos\theta}{1-\beta\cos\theta}\right)^{2-\alpha}}{1-\tilde{p}+\tilde{p}\left(\frac{1+\beta\cos\theta}{1-\beta\cos\theta}\right)^{-(1-\alpha)}}, \tag{7}$$

which is exact for $\alpha = -1$. Since $\tilde{p} \sim 0.25 - 0.35$ in typical jets, the anisotropy will increase the beaming ratio by less than 30–50 percent, in general. However, the effects can be much more important (a factor 2–3 increase) in those regions of jets where very high polarizations ($\sim 40 - 50\%$) are observed.

4 How Much Confinement is Necessary?

Most quoted values for the "minimum pressure" in luminous extragalactic jets, as high as they are, may well be underestimates of the true minimum pressure in the synchrotron emitting regions. The estimates are usually obtained by assuming a "filled" source, *i.e.*, one with a uniform emissivity within a volume which is deduced from a jet's cross section on the sky. However, high contrast maps of M 87 (Owen *et al.* 1989) and some other sources show that much of the synchrotron emissivity may be filamentary, *i.e.*, concentrated in a small fraction of the total volume. Even the pressures obtained assuming a filled source are often quite large by interstellar and intergalactic standards, ranging upwards of 10^{-9} dyne cm^{-2} in some cases (and higher in hot spots). In a few cases (*e.g.* 4C 32.69 [Potash & Wardle 1980], M 87, and Cygnus A [Perley *et al.* 1984]), the minimum jet pressures are demonstrably higher than upper limits on the pressure of the surrounding intergalactic or intracluster medium, derived from X-ray observations (Arnaud *et al.* 1987). In other cases, thermal confinement is marginally consistent at best (*e.g.* Burns *et al.* 1984). Therefore, it seems unlikely that luminous FR II jets are routinely confined by the thermal pressure of the quiescent ambient medium. The confinement problem remains even if one corrects the minimum pressure estimate downward to account for Doppler boosting (Bridle 1985).

Nevertheless, a number of arguments suggest that jets are confined. Some of the arguments are morphological, and therefore require faith that the contours of synchrotron surface brightness actually reflect the flow pattern. First, the most powerful jets often have very narrow opening angles, $\theta_j \lesssim$ a few degrees, suggesting that they are highly collimated to begin with. Second, they often appear to become increasingly collimated with distance, and the surface brightness contours exhibit "wiggles" which would be hard to understand if the jets were freely expanding. Third, if the jets were freely expanding, then the internal Mach numbers would have to be enormous, $\mathcal{M} \sim \theta_j^{-1} \sim 30 - 60$ (Bridle 1984). Given the minimum pressure, a large Mach number implies an enormous momentum flux (or "thrust"), which is difficult to reconcile with the estimated minimum pressures in hot spots and the speed of advance of the head of the jet (Potash & Wardle 1980).

In view of the apparent need for confinement at a pressure exceeding that of the ambient medium, a variety of alternative confinement mechanisms have been proposed. These include:

- *Magnetic tension.* There is little observational evidence for or against this mechanism. Magnetic confinement requires a toroidal field component, while the polarization data suggest that the dominant field direction is poloidal (parallel to the jet). However, as Laing (1981) showed, a helical field can lead to either perpendicular or parallel polarization, depending on the pitch of the helix and the viewing direction. Moreover, there is no reason why the synchrotron emitting region need coincide with the region responsible for confinement. Magnetic confinement involves an increase in the magnetic field strength towards the jet axis, and one might reasonably expect the synchrotron emissivity to peak near where the field strength peaks. This could well be a region dominated by poloidal field. In the limit where most of the toroidal field is contained outside the visible jet, one might hope to detect it by observing a gradient in the Faraday rotation across the jet (Bridle *et al.* 1986). However, no convincing example of this effect has yet been discovered (see, *e.g.* Clarke *et al.* 1992).

- *Overpressured cocoon.* A light, supersonic jet boring into the intergalactic medium (IGM) is enveloped by a "cocoon" consisting of shocked jet material (Scheuer 1974). Begelman & Cioffi (1989) showed that it could take a long time ($\gtrsim 10^8$ yr) for the cocoon to come into pressure balance with the IGM. Thus, a jet can be in pressure balance with its cocoon but remain overpressured with respect to the local IGM. Recent numerical simulations (Cioffi & Blondin 1992; Loken *et al.* 1992) have borne out the basic elements of this mechanism. Since cocoons are observed to have much lower surface brightness than the enclosed jet, a corollary of this picture is that the relativistic electron pressure inside the cocoon must either be far out of equipartition with the magnetic field, or the cocoon must be strongly dominated by thermal pressure.
- *Supersonic turbulent backflow in cocoon.* This mechanism was first suggested by 2–D numerical simulations of jet propagation (Norman *et al.* 1982; Norman & Winkler 1985; Kössl & Müller 1988). Supersonic vortices pummel the jet for a considerable distance upstream of the hot spot. These create local regions of overpressure of an order of magnitude or more. The drawback of this mechanism is that the overpressured regions tend to be localized and rather widely spaced along the jet. Therefore, this process might be more suitable for producing synchrotron knots in the jet rather than relatively smooth emission.

To this list, let me add a fourth possibility, which is distantly related to the third.

- *Locally overpressured regions in shear layer.* Another source of free energy which might be used for creating regions of overpressure is the shear layer between the jet and the cocoon. The energy could be tapped if the magnetic field in the jet and the cocoon became linked through reconnection. Suppose the shear layer, of width Δx and velocity difference Δv, were initially threaded by a field of B_0, as shown in Fig. 1. The Alfvén speed is $v_{A0} = B_0/(4\pi\rho)^{1/2}$. After a time t, both the field and the Alfvén speed are amplified by a factor $(\Delta v/\Delta x)t$. If unchecked by reconnection, this process would saturate when the Alfvén speed reached a value $v_{A1} \sim (\Delta v\ v_{A0})^{1/2}$, corresponding to an enhancement in the magnetic energy density by a factor $\Delta v/v_{A0}$. An overpressured synchrotron-emitting sheet could result if a fraction of the magnetic energy were dissipated via acceleration of relativistic electrons. Localized dissipation in a (highly overpressured) surface layer is indicated by recent observations of the jet in M 87 (Owen *et al.* 1989; Stiavelli *et al.* 1991). This mechanism has the added advantage that it naturally produces field aligned with the jet in the synchrotron-emitting regions, as required by polarization measurements.

The latter two mechanisms are capable of producing *localized*, *transient* regions of overpressure, and would be consistent with a small filling factor $f \ll 1$ of synchrotron-emitting material. The minimum pressures required in the overpressured regions would therefore have to be considerably higher than the minimum pressures estimated assuming a "filled source". However, it is interesting to note that the *mean* energy density required in a low filling-factor synchrotron source is actually lower than in a filled source. To see this, let j_{loc} denote the local emissivity of the synchrotron emitting regions. The

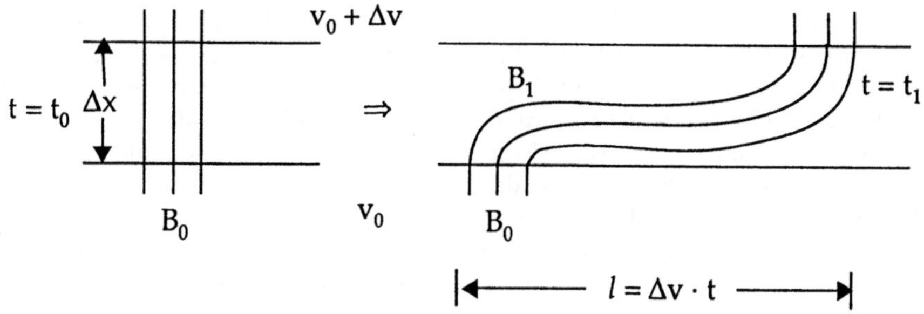

Fig. 1. Schematic illustration showing how local, transient regions of overpressure can be created by shearing the magnetic field at the boundary of a jet. See text for details.

mean synchrotron emissivity is then $\bar{j} = f j_{loc}$, leading to a minimum pressure estimate $(p_{min})_{filled} \propto \bar{j}^{4/7}$ if one assumes a filled source. The *local* minimum pressure in the emitting regions is $(p_{min})_{loc} \propto j_{loc}^{4/7} = f^{-4/7}\bar{j}^{4/7} \gg (p_{min})_{filled}$. However, the *mean* minimum pressure is

$$\bar{p}_{min} = f(p_{min})_{loc} \propto f^{3/7}\bar{j}^{4/7} \ll (p_{min})_{filled}. \tag{8}$$

The effect is analogous to decreasing the total energy requirements of a bremsstrahlung source by clumping the emitting gas. The difference is that the maximum synchrotron emissivity depends on a high power of the pressure, rather than the density.

5 Lobe Asymmetries: Illusion or Environment?

Given the evidence for relativistic velocities in large-scale jets, it is relevant to ask whether radio lobes also exhibit any manifestations of relativistic motion. In a perfectly symmetric source expanding into a symmetric medium, the lobe on the near side would appear to be both brighter and farther from the core than the lobe on the far side. The statistics of lobe/core separation asymmetries would then provide information about lobe velocities. Early studies (*e.g.* Longair & Riley 1979; Banhatti 1980; Katgert-Miekelijn *et al.* 1980) used this technique to place upper limits of $\lesssim 0.3c$ on the speed of advance of the hot spots in FR II sources. If this explanation for asymmetry is correct, then one would expect the asymmetry to become more pronounced with decreasing projected linear size, since the axes of smaller sources would tend to lie closer to the line of sight. This prediction has not been borne out observationally (Fokker 1986), although Teerikorpi (1986) finds a trend in this direction for a *subset* of sources in which the brighter lobe is the further one from the core.

Teerikorpi's finding probably reflects the crucial importance of environmental effects in determining the asymmetries of lobes in most sources. In Teerkorpi's sample, the sources which violate the expected trend between size and asymmetry are those which satisfy "the Mackay (1971) rule", which states that the brighter lobe tends to lie closer to the central object. This is the opposite of what one would expect from relativistic effects, but fits in with the idea that the IGM is denser on the side with the brighter lobe. Observations of distant radio galaxies appear to corroborate this interpretation. For example, Pedelty *et al.* (1989) find that the closer lobe tends to be more depolarized, while McCarthy *et al.* (1991) find a strong correlation between the closer lobe and the side with the brightest extended optical line emission. In addition, the typical degree of asymmetry seems to increase with redshift (Zięba & Chyży 1991), as would be expected if the more distant sources were expanding into a "messier" IGM.

It does not require a very high degree of asymmetry in the expansion of a radio source to yield a large asymmetry in the apparent arm-length ratio. To see this, consider the situation shown in Fig. 2. We suppose that arms 1 and 2 lengthen at the rates $\beta_1(x)$ and $\beta_2(x)$, respectively, where x is the distance from the core. The arm length ratio $Q \equiv \ell_1/\ell_2$ is obtained by solving the equation

$$\int_0^{\ell_1} \frac{dx}{\beta_1(x)} - \ell_1 \cos\theta = \int_0^{\ell_2} \frac{dx}{\beta_2(x)} + \ell_2 \cos\theta. \tag{9}$$

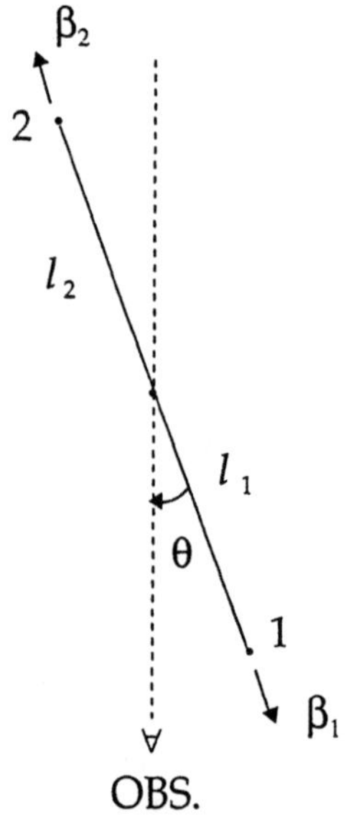

Fig. 2. Geometry used to illustrate effects of intrinsic asymmetries on observed arm-length ratios.

The first term on each side of eq. (9) represents the intrinsic time required for the growth of the arm, while the second term corrects for light-travel-time effects. If

$$\beta_2(x) = \eta\beta_1(x) = \beta\left(\frac{x}{\ell_2}\right)^{-\alpha}, \tag{10}$$

where $\eta = \text{const.}$ and $\alpha > -1$, then Q can be obtained by solving the implicit equation

$$Q = \frac{1 + (\alpha + 1)\beta \cos\theta}{\eta Q^{\alpha} - (\alpha + 1)\beta\cos\theta}. \tag{11}$$

For $\beta\cos\theta \ll 1$ the intrinsic asymmetry dominates over light-travel-time effects when $|\eta - 1| > 2(\alpha + 1)\beta\cos\theta$. The effect of even a small asymmetry ($\eta \approx 1$) becomes large when α approaches -1, corresponding to $\beta(x) \propto x$. In this limit, we have $Q \sim \eta^{-\frac{1}{\alpha+1}}$. This behavior is not too far-fetched: it corresponds to a jet with constant ram pressure propagating into a medium with $\rho(x) \propto x^{-2}$.

It would be very interesting to know whether jet asymmetries are correlated with lobe asymmetries for any subset of FR II sources. For example, it is conceivable that *intrinsic* brightness asymmetries of jets are sometimes correlated with environmentally induced lobe asymmetries. If so, we would expect the environmental effects to show most clearly in sources close to the plane of the sky (*e.g.* radio galaxies in Barthel's scheme), while relativistic effects dominate in the sources pointing at us (*i.e.*, radio quasars).

My work on extragalactic jets is partially supported by NSF grant AST88-16140 and NASA Astrophysical Theory Center grant NAGW-766.

References

Arnaud, K.A., Johnstone, R.M., Fabian, A.C., Crawford, C.S., Nulsen, P.E., Shafer, R.A., Mushotzky, R.F.: 1987, *Mon. Not. R. astr. Soc.* **227**, 241.
Banhatti, D.G.: 1980, *Astron. Astrophys.* **84**, 112.
Barthel, P.D.: 1989, *Astrophys. J.* **336**, 606.
Begelman, M.C., Cioffi, D.F.: 1989, *Astrophys. J. (Lett.)* **345**, L21.
Begelman, M.C., Kirk, J.G.: 1990, *Astrophys. J.* **353**, 66.
Bridle, A.H.: 1984, *Astron. J.* **89**, 979.
——: 1985, in *Physics of Energy Transport in Extragalactic Radio Sources*, ed. A.H. Bridle, J.A. Eilek (Green Bank: NRAO), p. 1.
Bridle, A.H., Perley, R.A.: 1984, *Ann. Rev. Astron. Astrophys.* **22**, 319.
Bridle, A.H., Perley, R.A., Henriksen, R.N.: 1986, *Astron. J.* **92**, 534.
Browne, I.W.A., Perley, R.A.: 1986, *Mon. Not. R. astr. Soc.* **222**, 149.
Burns, J.O., Basart, J.P., De Young, D.S., Ghiglia, D.C.: 1984, *Astrophys. J.* **283**, 515.
Cioffi, D.F., Blondin, J.M.: 1992, *Astrophys. J.* **392**, 458.
Clarke, D.A., Bridle, A.H., Burns, J.O., Perley, R.A., Norman, M.L.: 1992, *Astrophys. J.* **385**, 173.
Fokker, A.D.: 1986, *Astron. Astrophys.* **156**, 315.
Garrington, S.T., Conway, R.G., Leahy, J.P.: 1991, *Mon. Not. R. astr. Soc.* **250**, 171.
Garrington, S.T., Leahy, J.P., Conway, R.G., Laing, R.A.: 1988, *Nature* **331**, 147.
Katgert-Meikelijn, J., Levi, C., Padrielli, L.: 1980, *Astron. Astrophys. Suppl. Ser.* **40**, 91.
Kollgaard, R.I., Wardle, J.F.C., Roberts, D.H.: 1990, *Astron. J.* **100**, 1057.
Kössl, D., Müller, E.: 1988, *Astron. Astrophys.* **206**, 204.
Laing, R.A.: 1980, *Mon. Not. R. astr. Soc.* **193**, 439.
——: 1981, *Astrophys. J.* **248**, 87.
——: 1988, *Nature* **331**, 149.

——: 1989, in *Hot Spots in Extragalactic Radio Sources*, eds. K. Meisenheimer, H.-J. Röser (Heidelberg: Springer), p. 27.
Leahy, J.P., Stephens, P.W., Muxlow, T.W.B. 1986, in *Quasars*, IAU Symp. 119, ed. G. Swarup, V.K. Kapahi (Dordrecht: Reidel), p. 189.
Liu, R., Pooley, G.: 1991, *Mon. Not. R. astr. Soc.* **249**, 343.
Loken, C., Burns, J.O., Clarke, D.A., Norman, M.L.: 1991, *Astrophys. J.* **392**, 54.
Longair, M.S., Riley, J.M.: 1979, *Mon. Not. R. astr. Soc.* **183**, 625.
Mackay, C.D.: 1971, *Mon. Not. R. astr. Soc.* **154**, 209.
McCarthy, P.J., van Breugel, W., Kapahi, V.K.: 1991, *Astrophys. J.* **371**, 478.
Muxlow, T.W.B., Wilkinson, P.N.: 1991, *Mon. Not. R. astr. Soc.* **251**, 54.
Norman, M.L., Smarr, L.L., Winkler, K.-H.A., Smith, M.D.: 1982, *Astron. Astrophys.* **113**, 285.
Norman, M.L., Winkler, K.-H.A. 1985, *Los Alamos Science* **12**, 38.
Orr, M.J.L., Browne, I.W.A.: 1982, *Mon. Not. R. astr. Soc.* **200**, 1067.
Owen, F.N., Hardee, P.E., Cornwell, T.J.: 1989, *Astrophys. J.* **340**, 698.
Padovani, P., Urry, C.M.: 1992, *Astrophys. J.* **387**, 449.
Pedelty, J.A., Rudnick, L., McCarthy, P.J., Spinrad,: H. 1989, *Astron. J.* **97**, 647.
Perley, R.A., Dreher, J.W., Cowan, J.J: 1984, *Astrophys. J. (Lett.)* **285**, L35.
Potash, R.I., Wardle, J.F.C.: 1980, *Astrophys. J.* **239**, 42.
Rudnick, L., Edgar, B.K.: 1984, *Astrophys. J.* **279**, 74.
Scheuer, P.A.G.: 1974, *Mon. Not. R. astr. Soc.* **166**, 513.
Stiavelli, M., Møller, P., Zeilinger, W.W.: 1991, *Nature* **354**, 132.
Teerikorpi, P.: 1986, *Astron. Astrophys. (Lett.)* **164**, L11.
van Groningen, E., Miley, G., Norman, C.A.: 1980, *Astron. Astrophys. (Lett.)* **90**, L7.
Walker, R.C., Benson, J.M., Unwin, S.C.: 1987, *Astrophys. J.* **316**, 546.
Walker, R.C., Walker, M.A., Benson, J.M.: 1988, *Astrophys. J.* **335**, 668.
Zięba, S., Chyży, K.T.: 1991, *Astron. Astrophys.* **241**, 22.

The Velocity of Large Scale Jets in Quasars

John Wardle

Brandeis University — Physics Department
South Street
Waltham, MA 02254, USA

Abstract: We present an analysis of the jet/counter-jet ratios measured from the very deep VLA images of quasars made by Bridle and his collaborators. A technique is developed for including 'largest linear size' information as a *statistical* indicator of source orientation. Nine of these sources form a complete and well defined sample. If the difference in jet and counter jet brightness are attributed to Doppler beaming, then jet velocities less than $0.8c$ are rejected at the 99% confidence level, and the data are most consistent with values of v/c very close to 1.

Additional new evidence that the velocities of large scale jets are relativistic also comes from total intensity and polarized images of the extended structure surrounding core-dominated quasars.

Reference

Wardle *et al.*, in preparation.

The Jet of M 87

John A. Biretta [1] and Klaus Meisenheimer [2]

[1]National Radio Astronomy Observatory
Socorro, New Mexico, 87801, USA

[2]Max-Planck-Institut für Astronomie
Königstuhl 17
D–6900 Heidelberg 1
Germany

Abstract: Due to its outstanding surface brightness, the jet of M 87 is the earliest known and best investigated extragalactic radio jet. Thus, many contributions to our current picture of extragalactic radio sources originated from observations of M 87 (= Virgo A = 3C 274). We highlight some of these earlier results. Then we present a concise summary of the observed properties in all currently accessible wavelength regions between 1 GHz (radio) and 3×10^{17} Hz (X-rays). They show that the jet of M 87 is well collimated on all (projected) scales between 0.1 and 2000 pc. The radio images show filamentary structure on scales < 25 pc. The superposition of these filaments together with local enhancements of their surface brightness leads to the knotty appearence of the jet observed at medium resolution ($1''$, i.e. 80 pc). The magnetic field lines closely follow the filaments. Both from very high signal-to-noise observations with groundbased optical telescopes ($1''$ resolution) and recent HST results ($0\overset{''}{.}1$ resolution) one concludes that the morphology and polarization is virtually identical over the entire frequency range between 10^9 and 10^{14} Hz – at least on scales $\gtrsim 10$ pc.
Proper motions with a speed of 0.1 to 0.7 c are seen both in the parsec and the kiloparsec jet, although several small condensations may move with even larger velocities. We point out the most important implications for the physics of the jet in M 87 following from these observations.

1. Introduction

One morning session of the Ringberg workshop was devoted to the prototypes of extragalactic radio jets – M 87, 3C 273 and Cygnus A. The speakers only had a limited amount of time in which they had to concentrate on the most recent results and their interpretation, and they consequently assumed that most participants were familiar with earlier results. Since this book is aimed at a much wider audience the editors asked the authors to broaden the scope of their contributions and to write self-contained review articles which include a more complete account of earlier work.

1.1 The Prototype of Extragalactic Jets

As early as in 1918 Curtis described a curious "ray" which he had detected on a photographic plate of M 87 (= NGC 4486) – one of the dominant elliptical galaxies in the center of the Virgo cluster. However, it was the exploration of the radio sky in the late forties which attracted to M 87 and to its strong radio source Virgo A (= 3C 274; see Bolton *et al.* 1949). This identification was confirmed by Baade & Minkowski (1954) who correctly proposed that the "jet" which seems to be "ejected by the nucleus" (as they interpreted the "ray") could cause the radio source. But in their words: "No possibility exists at this time of forming any hypothesis on the formation of the jet, the physical state of its material and the mechanism which connects the existence of the jet with the observed radio emission."

One year earlier Shklovsky (1953) had suggested that the synchrotron emission process could provide a common origin for both radio and optical continuum radiation from cosmic sources. The observational confirmation that the Crab nebula does indeed emit – highly polarized – synchrotron light (Vashakidze 1954, Oort & Walraven 1956) was discussed at a symposium in Manchester (1955) and it was suggested that the blue continuum of the jet in M 87 may also be caused by synchrotron radiation. Walter Baade (1956) verified this impressively by taking photographs through a polaroid orientated at four different angles. The first quantitative polarization measurement of the brightest knots was obtained by Hiltner (1959).

In the sixties it became more and more obvious that the jet itself contributes a significant fraction of the total flux from Virgo A – at least at $\nu > 10^9$ Hz. So some idea about its spectrum could be developed and the standard formulae for synchrotron sources were applied (Burbidge 1956 and many others). Felten (1968) pointed out that the lifetime of the electrons responsible for the optical light is very short (20 years) if the optical synchrotron radiation does in fact originate from regions in which the magnetic field has roughly its equiparition value, while the time required for light to travel along the jet is 4000 years. He also showed that easy ways around this problem (propagation along the field lines of weaker fields) would probably not work since plasma instabilities would destroy the jet. He concluded that either *bulk relativistic motions* or an *in situ* acceleration process have to be considered. However, known acceleration mechanisms (second order Fermi acceleration) seemed inappropriate to balance the synchrotron losses. In fact, as one will see below (section 6), this problem has remained unsolved in the 25 years since then.

The improvement of radio interferometers in the early seventies made radio maps with resolutions comparable to optical images possible for the first time. On the maps taken with the Cambridge 5 km telescope, it is obvious that the brightest optical knots (A,B,C) have radio counterparts of similar flux ratios (Turland 1975). From these maps, together with optical (Kaneko *et al.* 1973) and near-infrared photometry (Wisniewski & Kleinmann 1968), Turland was able to derive an overall spectrum of the brightest knots which is characterized by a powerlaw $S_\nu \sim \nu^{-0.55}$ between radio frequencies and 10^{13} Hz steepening to $\sim \nu^{-1.7}$ towards optical frequencies – a finding which is rather close to the modern result (Fig. 17).

At about the same time, the very high surface brightness found in the hot spots of Cygnus A with the 5 km telescope (Hargrave & Ryle 1974) led Scheuer (1974) and Blandford & Rees (1974) to the suggestion that "beams" or "jets" have to provide a continous energy supply in all bright radio sources. This turned the curiosity of M 87 into the prototype of extragalactic radio sources in general.

The first maps of M 87 made with the Very Large Array (VLA, $\lambda = 6$ and 2 cm, resolution = 1″and 0.″6, Owen *et al.* 1980) showed that the morphology is indeed rather similar at radio and optical frequencies. Furthermore, the existence of low surface brightness inter-knot regions became obvious (see also Laing 1980). In the last 10 years, X-rays have been detected (Schreier *et al.* 1982), and the resolution of the best optical images was improved to $\simeq 0.7''$ (Nieto & Lelièvre 1982).

A matter of debate is still the shape of the spectrum in the near-infrared to optical regime. While early infrared photometry (Wisniewski & Kleinmann 1968, Stocke *et al.* 1981) combined with optical photometry indicates that there is a spectral "break" or "cutoff" around $\lambda = 1\ \mu$m, this could not be confirmed by subsequent infrared observations (Smith *et al.* 1983, Killeen *et al.* 1984). The latter groups argue for a straight powerlaw between 10^{14} and 10^{15} Hz.

Many numerical simulations were carried out to model extragalactic jets by supersonic hydrodynamical flows (Norman *et al.* 1982, Falle & Wilson 1985). Most of these early numerical experiments were tuned in such a way that the most prominent features of the jet in M 87 could be explained.

In this contribution we review the present status (end 1991) of observations in all wavebands and their interpretation. We assume a distance of 16 Mpc to the Virgo cluster (Mould *et al.* 1980), so 1″corresponds to a (projected) linear scale of 78 pc (redshift $z = 0.0043$). For greater distances, most of the arguments presented in chapters 6 and 7 become even stronger.

1.2 The Entire Radio Source Virgo A

Although the radio jet is the most conspicious part of the radio source around M 87, it emits only a small fraction of the total radio power released by the entire radio source Virgo A. The jet is embedded in a fairly symmetric double radio structure with bright filamentary lobes extending about 4 kpc towards the East and West of the core (see Fig. 2 in Owen *et al.* 1989a and our Fig. 9). In spite of the rather sharp edges which seem to confine this inner high surface brightness source, the radio emission from Virgo A can be traced tenfold further out. In Fig. 1 we display a low resolution map of Virgo A. The jet and the double lobe structure are completely contained in the unresolved central region of Virgo A which at this resolution appears as a typical core-halo source with rather elliptical outer contours of low surface brightness. However, even in this halo there still remains some signature of the preferential east–west orientation of the inner double source: a prominent "plume" or outflow stretches out to the eastern boundary of the halo (about 20 kpc from the core), and a narrower plume starts towards the West before bending sharply southwards where it can be traced to more than 30 kpc from the core. The connection between this radio halo of Virgo A and the inner double-lobe structure is not clear: the halo may either represent a relic of an earlier activity period of M 87, or it could be interpreted as the outermost region of the current outflow. Thus, it is uncertain whether the apparent North-South elongation of the outermost parts of the halo is due to a change in the radio source orientation, or to interactions between the radio plasma and the huge ($> 1°$) halo of hot X–ray emitting plasma surrounding M 87 – the central galaxy of the Virgo cluster (White & Sarazin 1988).

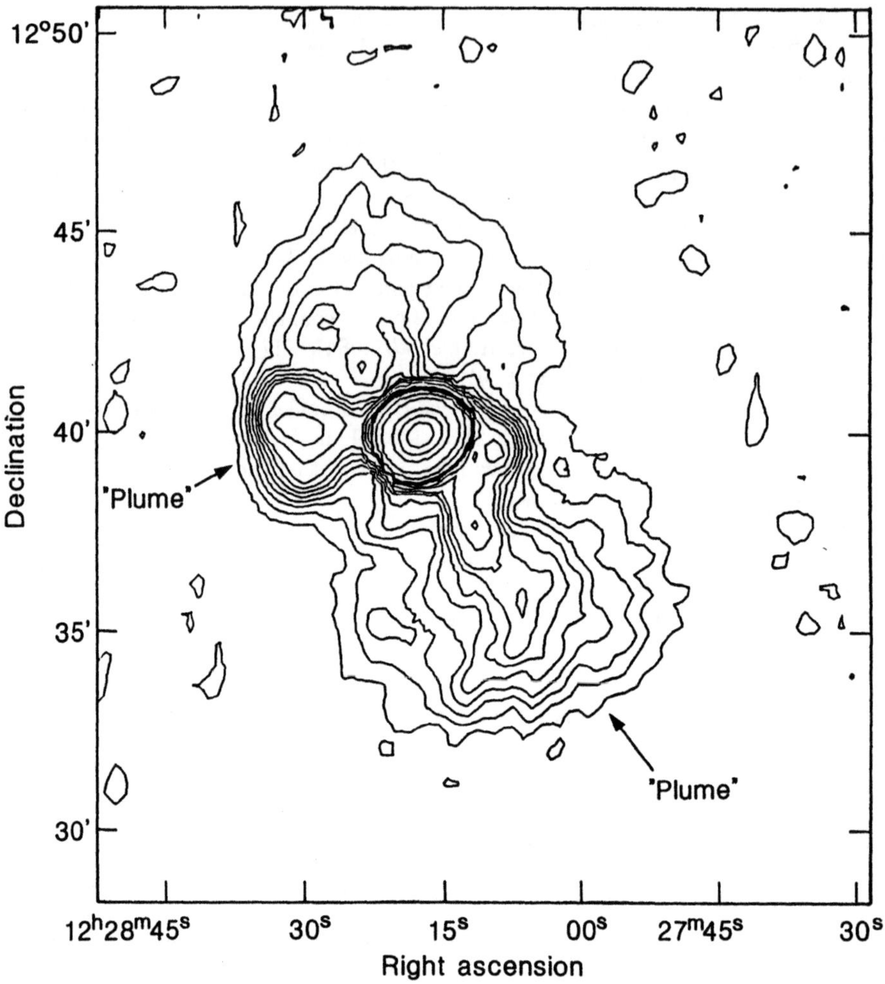

Fig. 1. Image of M 87 made with at 1.5 GHz with the VLA in the D configuration. The resolution is 100″ and the central region has a brightness of 90 Jy beam^{-1} (from Owen 1992).

2. Radio Properties

(John A. Biretta)

2.1 Morphology

Evidence of collimated outflow has been seen in M87 on scales from 0.01 pc to 35,000 pc. We will begin with the smallest structures at the nucleus and work outward.

The highest resolution observations showing evidence of a collimated jet are the 22 GHz VLBI observations of Spencer and Junor (1986, Fig. 2). This image clearly shows that the nucleus is extended in approximately the same direction as the jet seen at larger scales, and thus gives strong evidence for a collimated jet on scales $\simeq$ 0.01 pc. Future observations at 22 GHz and above with better dynamic range may reveal structural details on this scale. We note that Baath *et al.* have produced higher resolution images at 100 GHz, but the relationship of the "features" they see to structures seen in other maps is unclear.

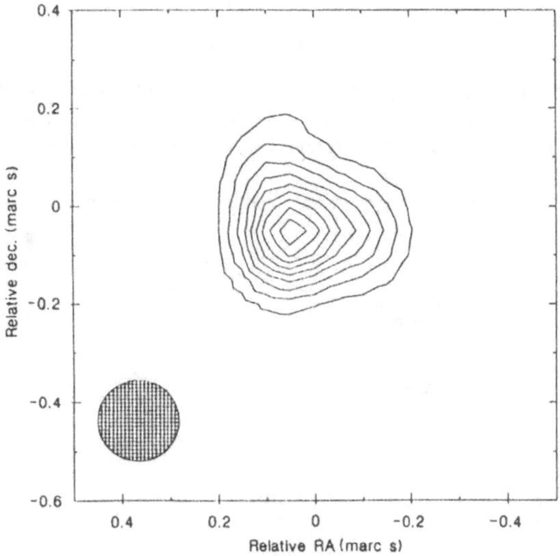

Fig. 2. VLBI image of the nucleus of M 87 at 22 GHz for epoch 1985.1. The resolution is 0.15 mas; the peak brightness is 0.34 Jy beam^{-1} and the lowest contour is at 0.03 Jy beam^{-1}. From Spencer and Junor (1986).

Moving to slightly larger scales, 18 cm World Array VLBI observations have revealed a well collimated jet 200 mas (16 pc) long (Fig. 3; Reid *et al.* 1989). This jet is approximately aligned with the northern limb of the 20 arcsecond jet, and has an opening angle (full-width at quarter-maximum) of 9.8°.

The jet appears limb-brightened over much of its length. At several places both the northern and southern limbs appear equally bright (Fig. 4). At other places one limb

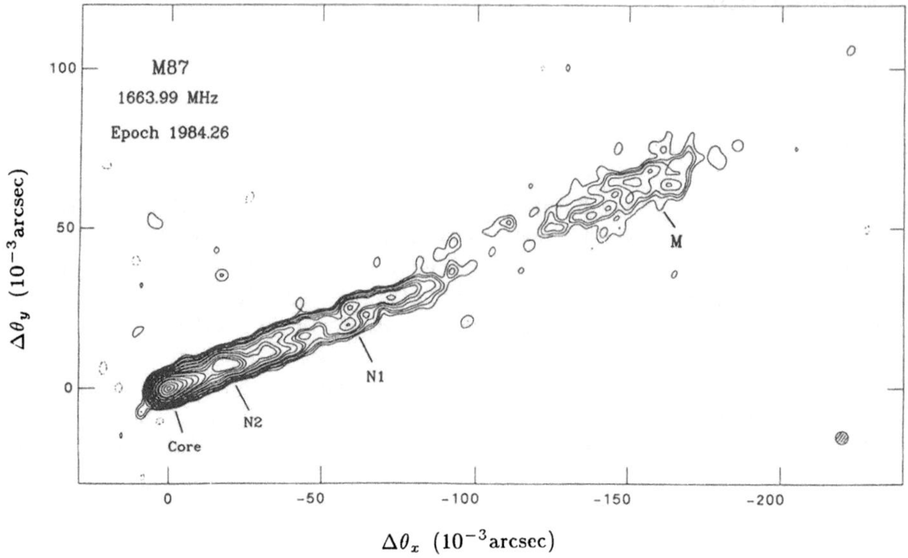

Fig. 3. Image of M87 nucleus made at 1.66 GHz with World Array VLBI at epoch 1984.26. The resolution is 4 mas (shaded circle at the lower right); the peak at the nucleus reaches 0.97 Jy beam^{-1} while the lowest contour is at 0.002 Jy beam^{-1} (from Reid *et al.* 1989).

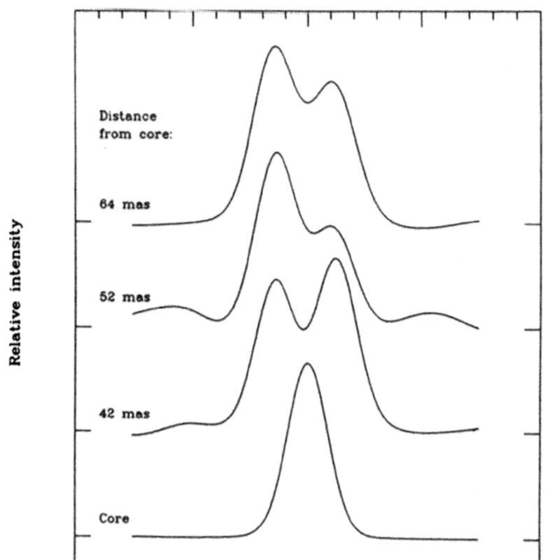

Fig. 4. Brightness profiles across the jet at core distances of 42, 52, and 64 mas showing limb-brightening (from Reid *et al.* 1989.)

dominates the brightness; at core distances between $\simeq$45 and 55 mas the northern limb is brightest, between $\simeq$55 and 60 mas the southern limb is brightest, and at distances $>$ 65 mas the nothern limb is brightest again. This oscillation of the brightness ridgeline from one side of the jet to the other might be interpreted as continuous filamentary structure which may represent a 3-dimensional helical pattern seen in projection.

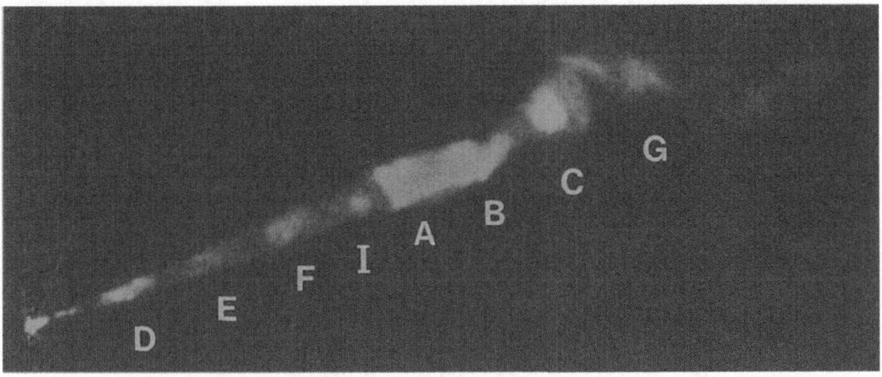

Fig. 5. Image of the M 87 jet made with the VLA at 15 GHz. The prominent knots are labeled with their usual designations. The resolution is 0.″10 and white corresponds roughly to an intensity of 3 mJy beam^{-1}. The brightest region on the map is at the nucleus, and has a brightness of about 2.5 Jy beam^{-1} (from Owen, Hardee, and Cornwell (1989)).

VLA observations at 15 GHz show that the jet continues to be well collimated out to a distance of about 20″ (1600 pc) from the nucleus (Fig. 5; Owen, Hardee, and Cornwell 1989, hereinafter OHC89). The entire jet seen in the 18 cm VLBI image (see Fig. 3) is now merely seen as a narrow 0.″2 long extension from the nucleus. The "inner" part of the kpc-scale jet, from the nucleus to knot A, appears quite straight and has an FWQM opening angle of 6.5°, which is slightly narrower than seen at the parsec-scale. As with the parsec-scale jet, there is evidence for limb-brightening along much of the inner jet. Various regions of knot D appear brightest at either the northern or southern limb, and knots E and F show a number of regions where both limbs are equally bright, the center of the jet being much dimmer (Fig. 6). However, there is no unambigous evidence that the entire inner jet is limb-brightened. One finds several regions where the jet appears center-brightened, and indeed the *mean* profile across the entire inner jet (knots D, E, F, and I) is brightest at the center of the jet.

Again, some of the features can be interpreted as filamentary structures corresponding to a 3-dimensional helix seen in projection; this is especially true of knots D and F. We also want to point out that the nucleus and the knots of the inner jet (D, E, F, and I) have a roughly uniform spacing.

At knot A we see a very bright feature which runs across the entire width of the visible jet ($\simeq$1.″5), but which is quite narrow in the direction along the jet axis ($\simeq$ 0.″2;

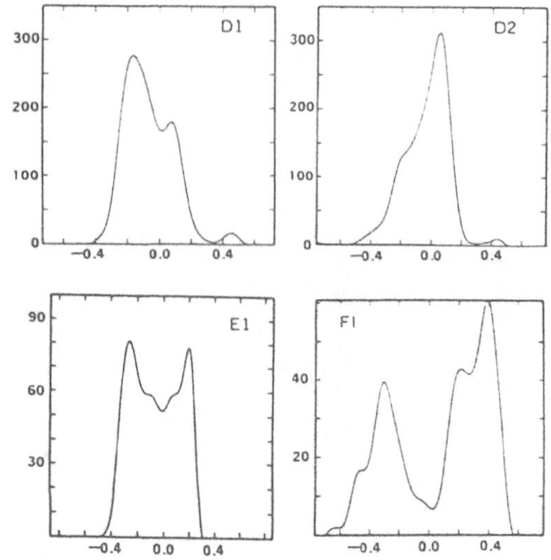

Fig. 6. A Brightness profiles taken across the jet within the indicated knots showing evidence for limb-brightening (from Owen, Hardee, and Cornwell (1989)).

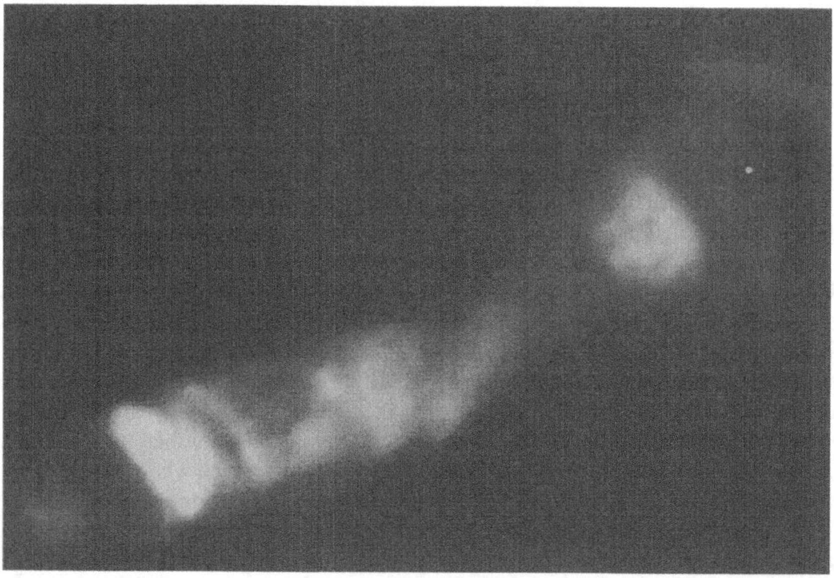

Fig. 7. Detail of knots A, B, C of the M 87 jet at 0.″10 resolution. The displayed intensities have been histogram equalized, and white corresponds roughly to 5 mJy beam^{-1} (from Owen, Hardee, and Cornwell, 1989).

Fig. 7). Knot C contains a similar feature. The jet appears to become less straight after each of these "bright transverse features." While the inner jet is quite straight, it starts oscillating from side-to-side immediately after knot A. Here the jet bends slightly to the south, and then to the north in knot B. After the second bright transverse feature in knot C, the jet appears to bend sharply to the south, and the side-to-side oscillations become stronger. An interesting feature in this region of the jet is the apparent narrow "dark lane" running down the center of the jet between knots A and B, into knot B, and perhaps as far as knot C.

Unlike the inner jet, which appears as a well defined cone, after knot C the jet has ill-defined boundaries and gradually merges into the western radio lobe (see Fig. 9). The eastern lobe shows a similar morphology, with a broad arc of emission running roughly north-south. This arc of emission is centered directly opposite the jet, and is at the same distance from the nucleus as the outermost parts of the jet on the western side ($\simeq$1850 pc from the core). We note that if one ignores the high-brightness emission from the jet, the eastern and western sides of the source appear to be quite similar (see also millimetre wave observations, section 3.3 below). Both the eastern and western lobes show filamentary and edge-brightened structures, similar to those seen at various places in the jet.

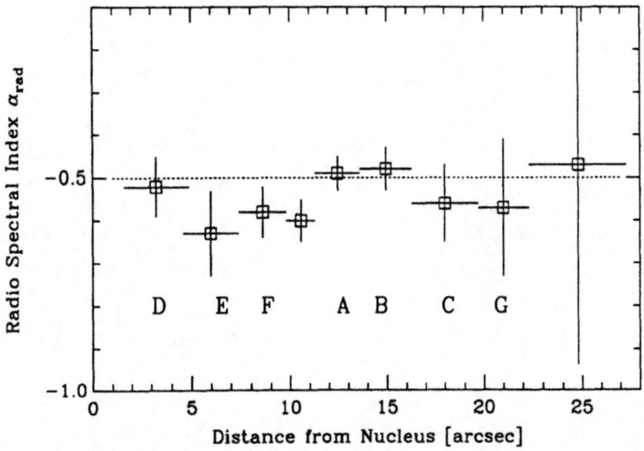

Fig. 8. Plot of radio spectral index α_{rad} *versus* distance along the jet for the nucleus and prominent knots (from Biretta, Stern & Harris 1991). Horizontal bars indicate the regions in which fluxes have been averaged.

2.2 Spectral Index

VLBI observations of the nucleus find the usual situation with the core having a flat radio spectrum ($\alpha \simeq 0$ defined in the sense $S_\nu \propto \nu^\alpha$) and the jet having a steeper spectrum ($\alpha \simeq -0.5$; Biretta and Reid 1992). Prominent knots on the kpc-scale jet all have spectral indicies roughly consistent with $\alpha = -0.5$, which is similar to the pc-scale jet (Fig. 8). There is some evidence for the weaker knots of the inner jet (E, F, and I) to have slightly steeper spectra with $\alpha \simeq 0.6$, but this is only at the $\simeq 1\sigma$ confidence level

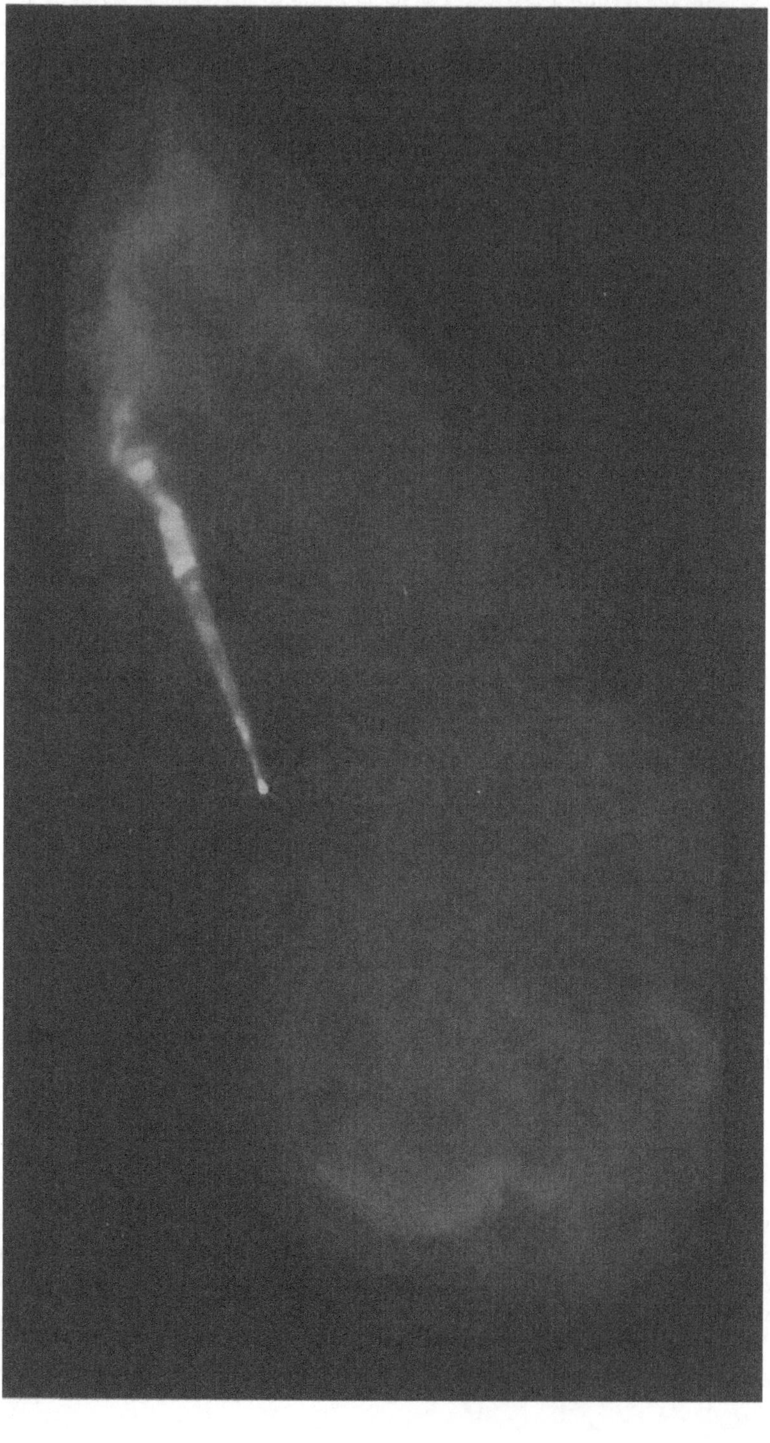

Fig. 9. High dynamic range image of the M 87 jet and lobes made with the VLA at 15 GHz. The total extent of the visible source is 75″ or about 6 kpc. The resolution is 0.″15, and intensity scale is logrithmic with white corresponding to 20 mJy beam^{-1}. Approximately 40, 6, and 2 hours of observation were used in the A, B, and C arrays, respectively.

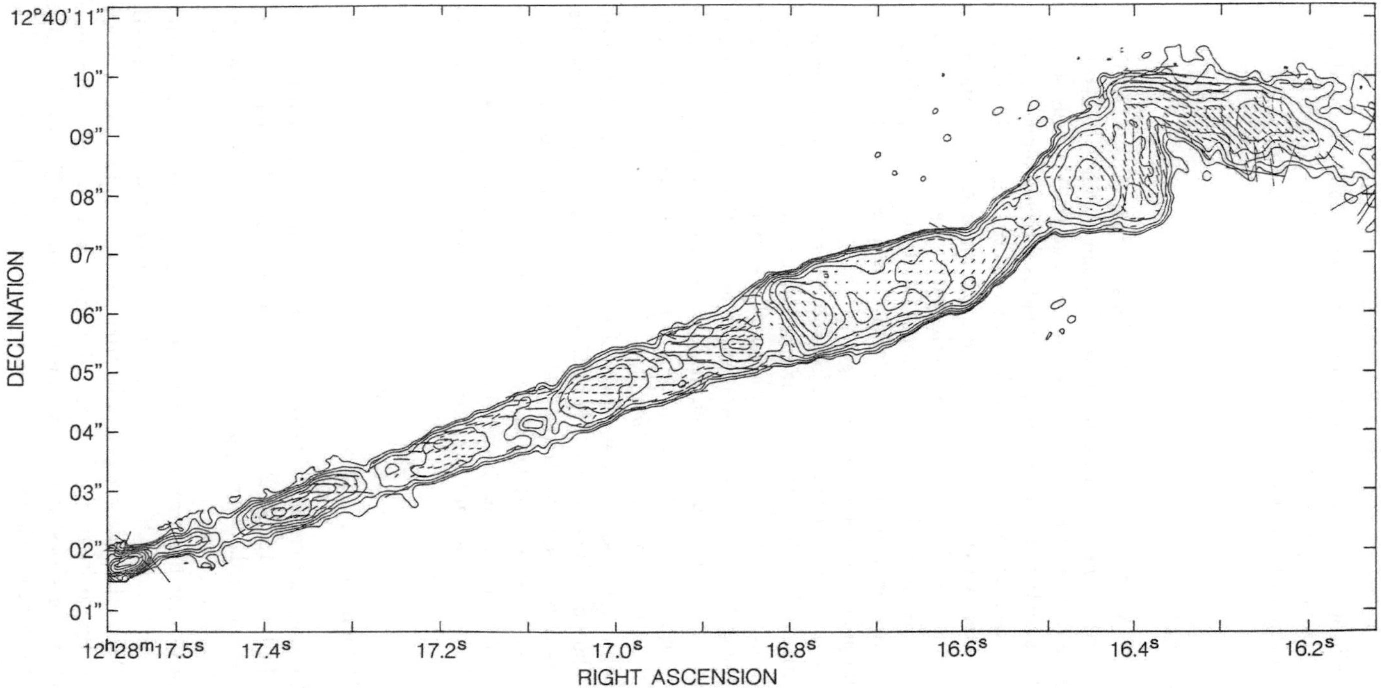

Fig. 10. VLA 2cm map of the jet of M 87 with fractional polarization superimposed (**B**-vectors, corrected for Faraday rotation). A vector length of 1″ corresponds to 216% polarization. Contour levels are at (1,2,4,...,512)×1.85 mJy/beam.

(BSH91). There is also evidence for the emission from the inter-knot regions having slightly steeper radio spectra ($\Delta\alpha \simeq -0.1$; Biretta and Owen 1993). Beyond the jet, the spectra appear to steepen appreciably with values $\alpha = -0.65$ reported in the western lobe, and -0.85 in the un-jetted eastern lobe (Forster 1980, but compare also Salter *et al.* 1989).

2.3 Polarization Structure

There are as yet no VLBI polarization observations of the nucleus, so the detailed polarization structure on these scales is unknown. Evidence from $0''.15$ resolution VLA observations (OHC89), however, suggests that the average polarization of the nuclear jet (scale size $0''.2$) is high and that the magnetic field is directed along the jet.

From the same observations (Fig. 10), it is clear that the magnetic field is roughly parallel to the jet axis everywhere in the $20''$–jet, except in knots A and C. In both these knots the field is roughly perpendicular to the jet axis, and runs along the "bright transverse features" discussed above. The fractional polarization is typically 20 to 30% in the knots. There is a tendency for higher polarizations to be seen at the edges of the jet, with fractional polarizations of 50% or larger.

In many regions of the jet elongated or "filamentary" structures are seen, and in all those places the magnetic field vectors are oriented along the morphological features, suggesting some connection between elongation and magnetic field. The bright transverse features in knots A and C are examples of this, as are other features seen in knots in D, F, and I, as well as an elongated feature between knots B and C. Filamentary structures are also seen in the eastern radio lobe (Fig. 9), and again the magnetic field is running along these filaments (Owen 1992).

The Faraday rotation across the jet and lobes of M87 has been mapped at 5 GHz with 0.4 arcsecond resolution by Owen, Eilek, and Keel (1990). Rotations in the direction of the jet are typically between -300 and +300 radians m^2. Rotations in the extended lobes are much larger (+1000 to +2000 radians m^{-2}), which suggests that the jet is in front of the extended lobes. While these values are typical for the lobes, an elongated region ($\simeq 1'' \times 10''$) in the eastern lobe shows rotations around +6000, with values up to +8500 radians m^{-2}.

2.4 Apparent Velocities

Multi-epoch VLBI observations at 1.6 GHz have detected motion of feature N2 in the parsec-scale jet (Reid *et al.* 1989). This feature is located about 20 mas (1.6 pc) from the core, and moves outwards at an apparent speed of $(0.28 \pm 0.08)\ c$, where c is the velocity of light (Fig. 11).

Apparent velocities for features in the kiloparsec-scale jet have been measured by comparing high dynamic range 15 GHz VLA images observed at several epochs (Biretta and Owen 1990, 1992). The data are primarily from two epochs in 1985, with another two in 1989. A *fifth* epoch from 1982 has poor dynamic range, but provides some useful information on the brightest knot. After an initial "cross calibration" against the 1985 Feb. 26th data, each epoch was reduced independently and in the same manner. Final images were made using data from the A, B, and C configurations of the VLA

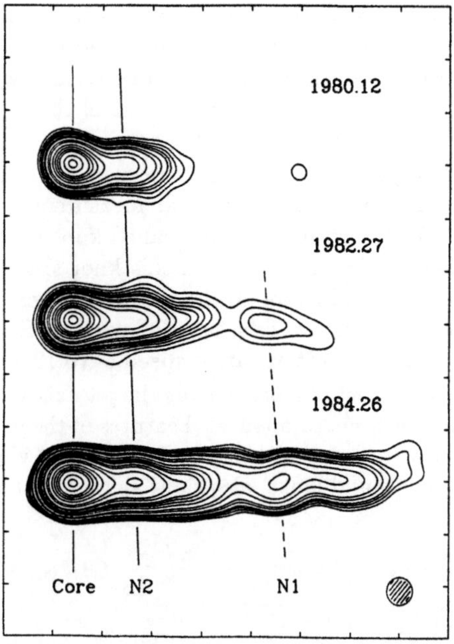

Fig. 11. Images of M 87 nucleus as observed in 1980, 1982, and 1984 with 1.6 GHz VLBI. The feature labeled N2 appears to move away from the core at (1.1 ± 0.3) mas yr^{-1} $(0.28c \pm 0.08c)$. The resolution (shown by a shaded circle) is 8 mas (from Reid *et al.* 1989).

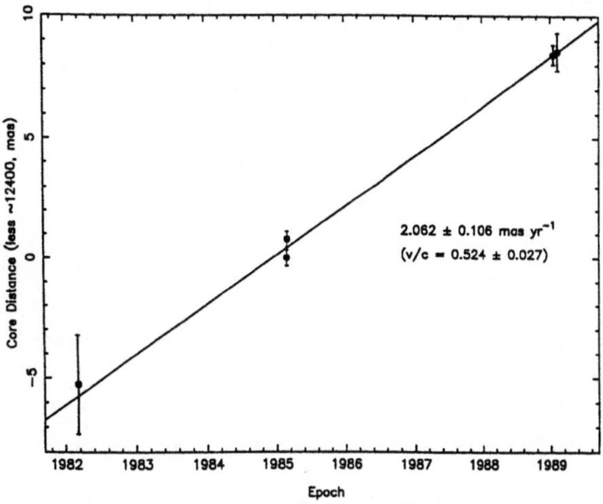

Fig. 12. Positions of knot A relative to the nucleus measured at several epochs. The 1985 Feb. 26th position is used as a reference, and is defined to be zero. Knot A is approximately 12400 mas from the nucleus. The best fit line indicates a motion away from the nucleus at (2.06 ± 0.10) mas yr^{-1} $(0.52c \pm 0.03c$; from Biretta and Owen 1993).

employing a Maximum Entropy Method. Small changes in the positions of features were then measured using a cross-correlation technique (Biretta, Owen, and Cornwell 1989). Because of the high sensitivity of the cross-correlation technique, and the high dynamic ranges of the images, it is possible to measure positional changes to an accuracy better than 1 percent of the $0\rlap{.}''15$ resolution of the images.

The results for the brightest knot (A) are shown in Fig. 12. This feature appears to move away from the nucleus at $0.52c$. Similar measurements have also been made for knots D, B, and C. For the entire structure called knot D, we find a speed of $(0.66 \pm 0.15)c$. However, small condensations within this knot show speeds of up to $(3.0 \pm 0.4)c$. A similar situation is seen in knot B; the speed of the entire knot is $(0.6 \pm 0.2)c$, while speeds of up to $(1.3 \pm 0.2)c$ are seen in small regions. Knot C, which is near the end of the 20″ jet, shows the slowest apparent speed of $(0.12 \pm 0.07)c$.

In addition to this component of motion *along* the jet axis away from the nucleus, we have also measured the *transverse* component. Features of the inner jet (knots D and A) show motion directly away from the nucleus. Knots B and C, which are located near the end of the jet where it bends and kinks, also show significant transverse components.

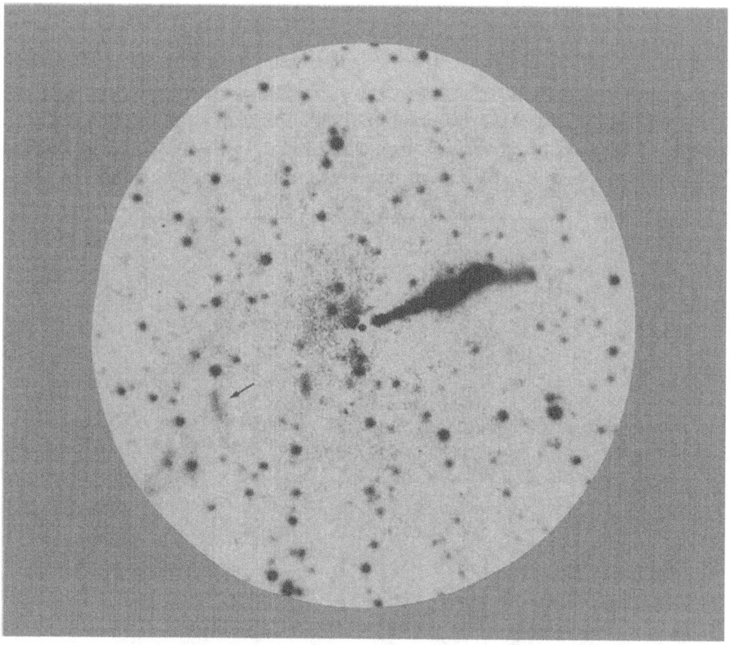

Fig. 13. Red image of the central arcminute of M 87. The stellar light from the galaxy has been removed by a model fit. Besides the jet, many globular clusters and some patches of low surface brightness are seen. Most of these are emission line filaments seen in H_α + [NII]. The continuum emitting patch which coincides with the eastern radio peak is marked by an arrow.

3. Optical Properties.

(Klaus Meisenheimer)

3.1 Morphology

On optical images the jet is embedded in the light of the giant elliptical galaxy M 87. The steep brightness gradient of the stellar light distribution requires careful subtraction of this stellar background in order to obtain undisturbed jet images between 0.3 and 2.2 μm. The stellar flux peaks around 0.7 μm. Thus, the contrast of the jet is at its minimum in the R and I bands (650, 850 nm), while in the UV ($\lambda < 400\,nm$) and near-IR ($\lambda > 2\,\mu m$) the stellar background is less severe. Observations with modern CCD detectors (Keel 1988, Perez-Fournon *et al.* 1988, Biretta *et al.* 1991, Meisenheimer *et al.* 1993) allow a rather straightforward subtraction of the stellar background which is well described by elliptical isophots. However, in the wavelength region of minimum contrast (R, I, J), details of the background following procedure employed may affect the resulting jet morphology and – more severely – the flux values of faint features.[8] After the removal of the stellar background, the jet image (Fig. 13) looks almost indistinguishably similar to a radio image convolved to the same resolution (cf. section 3.3). The most obvious differences are the globular clusters surrounding M 87 and some low brightness patches or filaments seen mainly to the East of the core. Most of these low brightness parches are due to emission lines from the extended emission line region surrounding M 87 (see section 3.4), but it has recently been pointed out (Stiavelli *et al.* 1992, Sparks *et al.* 1992) that one of these patches (20.″9 east, 10.″6 south from the core) coincides with the brightest part of the eastern radio lobe and may thus be of non-thermal origin.

The optical jet can be followed out to 25″ from the core although the brightness peak at its very edge is presumably a globular cluster. It is obvious from Fig. 13 that the optical jet is not just a chain of knots, but that it also emits light from the inter-knot regions. The best impression of how closely the optical morphology of the M 87 jet matches the radio structure is given by the deconvolved HST images (Boksenberg *et al.* 1992).

3.2 Spectral Index

As far as quantitative results are concerned, the most accurate and reliable data still come from groundbased telescopes. The best resolved spectral index map with high signal-to-noise ratio was obtained at the MPIA/ESO 2.2m telescope on La Silla (Meisenheimer *et al.* 1993) parallel to similar observations of the jet of 3C 273 (see Röser & Meisenheimer 1991). After aligning and convolving the 15 CCD images taken at B,R,I (450, 648, 863 nm, resp.), we reached an *effective* resolution of 1.″3 (FWHM). Due to the very high S/N ratio of these data, both the alignment procedure and the measurement of the seeing PSF on individual frames has to be accurate to $\lesssim$ 30 mas in order

[8] Older, photographic observations of the M 87 jet (*e.g.* Warren-Smith *et al.* 1984) have to be treated with great caution since no reliable background subtraction is possible on uncalibrated photographic plates.

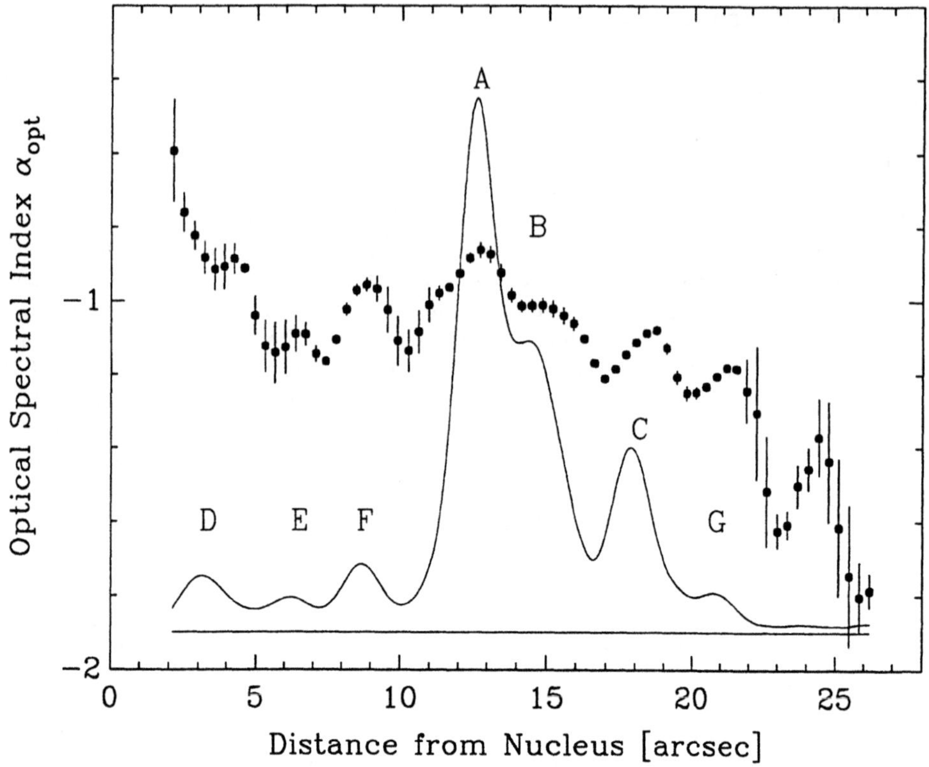

Fig. 14. Optical spectral index α_{opt} along the jet of M 87. The effective resolution is 1.″3 (FWHM). For positional reference a thin line gives the brightness profile of the jet. Prominent knots are marked.

to avoid spurious oscillations of the spectral index near steep brightness gradients. This has been achieved by a cross-correlation between the images of the brightest globular clusters on various frames.[9] Since no significant variations are found *across* the jet we display only the run of the optical spectral index *along* the jet in Fig. 14. Our most important findings are:

- The optical spectral index exhibits very smooth variations without dramatic differences between the knots and the inter-knot regions.
- In general, the spectral index steepens gradually from $\alpha_{opt} = -0.6$ in the inner jet (knots D,E) to −1.8 in the outermost parts (beyond knot G).
- There is a remarkable correspondence between the location of brightness maxima (knots) and the maxima of α_{opt}.

[9] An alignment to better than 30 mas requires that the differential refraction Δz_r between the globular cluster system (mean color assumed: G3) and the blue synchrotron jet has to be taken into account (we find $\Delta z_r = 70$ mas at airmass 1.4 !).

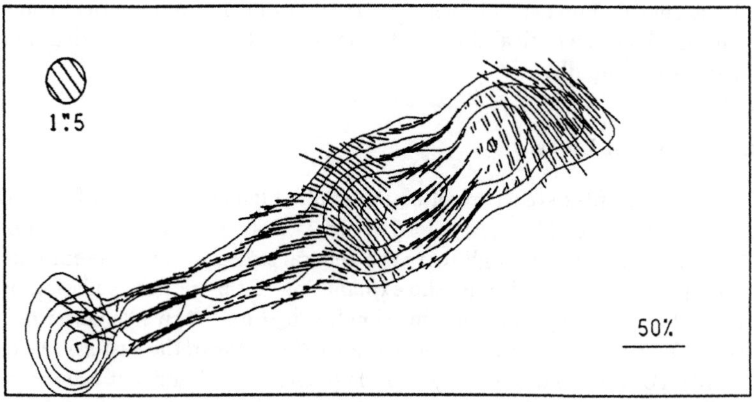

Fig. 15. Comparison of radio and optical polarization (**B** vectors) in the jet of M 87 at a common resolution of 1."5 (FWHM). The optical vectors have been shifted by 0."3 towards NW to make both polarizations visible.

3.3 Optical Polarization and Comparison with 2 cm

The best resolved optical polarization map (Schlötelburg *et al.* 1988) reaches an *effective* resolution of 1."5.[10] Although this is significantly inferior to the resolution of the 2 cm polarization map (Fig. 10), one is able to perform an accurate radio-optical comparison by convolving the I,Q,U radio maps to the optical resolution. From this comparison (Fig. 15), it is obvious that the agreement in amount and orientation of the polarization is almost perfect between 2cm and 0.65 μm. By misaligning the measuring grid on one polarization map with respect to the other, one finds that around knots A and C the observed 1:1 correspondence between radio and optical polarization is only possible if the polarization structure is, in fact, identical on scales of 0."1 or less. This result has been confirmed qualitativly by the optical images taken with the HST. The magnetic field orientation is predominantly parallel to the jet axis in the inner jet (out to knot F, 10″ from the core), the "flip" to **B** ⊥ jet occurs gradually between knot I and A, without any radio-optical offset. Small radio-optical discrepancies in between knots A and B point to a rather abrupt change back to **B** || jet. The polarization flip at the downstream edge of C also seems rather sharp. The largest discrepencies between radio and optical polarization are present around knot G. This may indicate some spatial separation between the optical emission region (i.e. the place of effective particle acceleration, see

[10] Fraix-Burnett & Nieto (1988) claim that their polarization measurements have a resolution of 1″(FWHM). This value refers to the seeing on individual exposures rather than to the effective resolution of their polarization map. Apparently, their resolution is quite similar to that of Schlötelburg *et al.*, while the S/N ratio of their map is much inferior.

section 6.2, below) and the maximum of the radio emission (i.e. region of high magnetic field strength). Further out, the constraints from the radio-optical comparison are less stringent since the optical flux, and thus the S/N ratio of the optical polarization map, is decreasing rapidly.

3.4 Emission Lines

There is a complex structure of emission line filaments within a few tens of arcseconds around the core of M 87 (Arp 1967, Ford & Butcher 1979, Meisenheimer & Hippelein 1989, Jarvis 1990). Although there is a general tendency for these emission line filaments to run parallel to the jet axis, the explanation of this radio – emission line alignment is still unclear.[11] One prominent emission line feature – Arp's "counterjet" – is located at $\Delta\alpha = 29''$ $\Delta\delta = -22''$, that is more than $10''$ outside of the eastern radio maximum (see Fig. 13). For the studies of the jet, perhaps the most significant property of the emission line filaments is that they seem to avoid those places (projected on the sky) from which we recieve the strongest synchrotron radiation. This lack of direct interaction between the emission line gas and the synchrotron emitting plasma weakens any hope of inferring the kinematics of the jet expansion from studies of the emission lines.

4. Observations at other Wavelengths

(Klaus Meisenheimer)

At present, the telescopes operating at wavelengths between the radio and optical window or in the X–ray regime do not provide a spatial resolution of $\lesssim 1''$. Thus, only rather global statements about the flux can be made. However, in this wavelength regime, dramatic improvements are soon to be expected (section 4.5).

4.1 X–ray Emission

X–ray emission from the brightest parts of the jet (knots A,B and C) was detected with the EINSTEIN HRI (Schreier *et al.* 1982). A recent re-analysis of a subset of these data[12] allowed Biretta *et al.* (1991) to produce an X–ray map with $5\overset{''}{.}5$ resolution (FWHM). Figure 16a shows that the X–ray image is dominated by two point sources which are identified with the core of M 87 and knot A. After subtraction of two point source models (see Biretta *et al.* 1991 for details), the presence of two weaker X–ray sources become evident (Fig. 16b). They can be identified with knots B and D. There might even be a marginal detection of knot C. The optical-to-X–ray spectral indices calculated from these data and our R-band photometry (Meisenheimer *et al.* 1993) are $\alpha_{OX} = -1.05$, 1.27 and -1.44 for knots D, A and B, respectively. These values are

[11] Compare the work by Baum *et al.* (1990, 1992) on extended emission line regions in nearby ($z < 0.2$) radio galaxies.

[12] Only those 14% of the data with startrackers locked on guidestars and the jet avoiding the $1\overset{'}{.}67$ gaps on the HRI were used.

significantly steeper than the radio-to-optical spectral index. So no straight power-law spectrum connects the flux values at 10^{10}, $10^{14.5}$ and $10^{17.5}$ Hz. In the EINSTEIN data, there is no indication that the X–ray emission is resolved perpendicular to the jet axis.

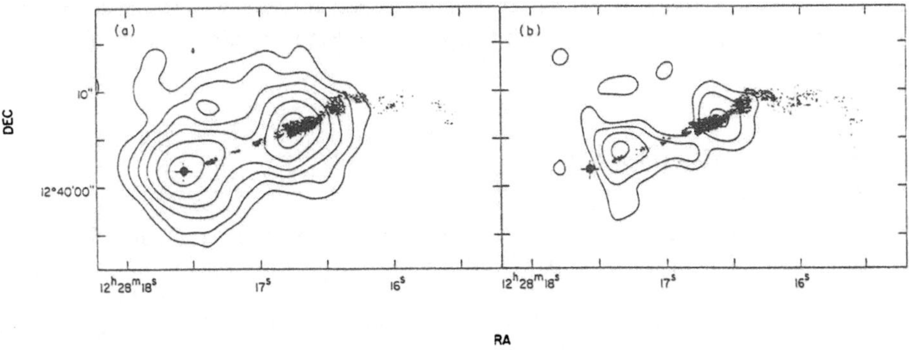

Fig. 16. X–ray contour map of M 87 overlayed on a radio image (greyscale). **(a)** Best image. **(b)** Two point sources removed at the positions of the core and knot A (from Biretta *et al.* 1991).

4.2 Ultraviolet Emission

Observations in the UV have been carried out by Perola & Tarenghi (1980) with the IUE satellite. Its entrance aperture ($10'' \times 20''$) covered knots A,B,C and the inner knot F. No background subtraction was possible. To derive the overall spectrum of the brightest part of the jet (section 5.1), we averaged their spectrophotometric flux values in two wavelength bins: $130\,\text{nm} < \lambda < 180\,\text{nm}$ and $220\,\text{nm} < \lambda < 300\,\text{nm}$.

4.3 Near-Infrared Emission

The first attempt to measure the near-IR emission of the M 87 jet dates back to the early days of extragalactic infrared astronomy (Wisniewski & Kleinmann 1968). Ten years later, Stocke *et al.* (1981) still used a single channel photometer and centered a $5''.9$ aperture on knot A, thus including also Knot B and the inner knot I (compare Fig. 5). Three filters, H, K and L were used. The background due to the starlight of M 87 was determined by measuring the position opposite to knot A with respect to the core. Due to the steep gradient of the brightness profile, this background subtraction may be rather uncertain – especially for the shortest wavelength (H), where the background contribution is $> 40\%$ of the flux in the $5''.9$ aperture centered on A. Better resolved JHK photometry was carried out by an Australian group (Smith *et al.* 1983) using a $3''.6$ aperture centered on the brightest knots A, B and C, respectively. The background subtraction was carried out in a similar manner to Stocke *et al.*. Since their results differed by more than 30% from that of Stocke *et al.*, the Australians repeated the observation and employed a raster scan across the jet to determine the background contribution more accurately (Killeen *et al.* 1984). They essentially reproduced their earlier results – the disrepancy with the Stocke *et al.* measurements remain. In spite of the inferior method of background determination used by Stocke *et al.* (1981), a smooth interpolation between the radio and optical data indicates that their flux values are more reliable than those of Killeen *et al.* (see section 5.1, below). Astonishingly, the arrival of large infrared arrays has not triggered a widespread interest in resolving this discrepancy.

4.4 Millimetre Wave Emission

A map of M 87 at $\lambda = 1.3\,\mathrm{mm}$ has been presented by Salter *et al.* (1990). Since it is based on a raster scan with the IRAM 30 m telescope on Pico Veleta, its resolution is much inferior to that available at other frequencies. To determine the jet flux, the 2 cm data of OHC89 have been convolved with a Gaussian beam of 11″ FWHM and rebinned to the grid of the IRAM scan. This "jet model" was scaled and subtracted from the 1.3 mm map. The scaling factor was varied until the the residual image of M 87 showed smooth variations in the western lobe (for details see Meisenheimer *et al.* 1993). Taking into account the sidelobes of the 30 m telescope, we find a total jet flux of 495 mJy at 1.3 mm (230 Ghz).

4.5 Future Prospects

In the next two or three years significant progress will be made in studying the jet at wavelengths outside the optical and radio window: observations with the ROSAT High Resolution Imager (HRI) will yield X–ray images with about 3″ resolution (FWHM). If sufficient S/N ratio is accumulated, modern deconvolution techniques may reach almost the optical resolution of groundbased telescopes. Together with the modest energy resolution of the HRI, it should be possible to decide whether the X–ray emission represents the high frequency tail of the synchrotron emission, or whether it comes from a different emission process which might also be separated spatially from the synchrotron jet.

Calibrated HST images in the near UV will improve our knowledge of the UV flux at every position in the jet. The large near-IR arrays now available will produce images with subarcsecond resolution in a field which is large enough to use the standard techniques of optical astronomers to subtract the stellar background. Thus, the shape and curvature of the spectrum around optical frequencies will soon be determined over a much wider range ($10^{14} - 10^{15.5}$ Hz). Interferometric observations in the millimetre and sub-millimetre bands should produce maps of the jet with arcsecond resolution. These exciting new instruments will together produce overall spectra of every feature in the jet with sufficient frequency coverage that a point-wise reconstruction of the underlying electron energy distribution (cf. section 5.1) should be feasible.

5. The Spectrum of the Jet Between 10^9 and 10^{17} Hz

(Klaus Meisenheimer)

Due to the huge differences in resolution presently reached in various parts of the electromagnetic spectrum, the reconstruction of the overall spectrum at every position along the jet requires a two-step approach. First, one determines the overall shape of the spectrum for the brightest part of the jet – the complex consisting of knots A+B+C which can easily be distinguished even with resolutions of $\gtrsim 5''$. Assuming that the general shape of this spectrum does not change if one goes to a much finer scale ($\leq 1''$), one may then, in a second step, use the well resolved radio and optical maps alone to determine the spectral behaviour on subarcsecond scale by means of a differential method.

5.1 Overall Spectrum of the Complex A+B+C

After correcting the observed flux values for the various instrumental resolutions used at different wavelengths, we are able to determine internally consistent flux values for the complex A+B+C ($10\overset{''}{.}8$ to $19\overset{''}{.}3$ from the core). Details of these corrections are given in Meisenheimer *et al.* (1993).

The resulting flux values between the radio and UV are summarized in Fig. 17a, together with two model fits by theoretical synchrotron spectra:

(i) assumes that the relativistic electrons are accelerated by diffusive shock acceleration (1st–order Fermi) or a similar acceleration process (acceleration time scale independent of electron energy, *e.g.* Heavens & Meisenheimer 1987). This produces a power-law $N(E) \propto E^{-q}$. The balance of acceleration gain and synchrotron losses leads to an exponential high energy cutoff at some maximum energy E_c. In the following, we characterize electron energies by the – dimensionless – Lorentzfactor, *i.e.* $\gamma_c = E_c/m_e c^2$. Analogous to the model fits of the synchrotron spectra of radio hot spots (Meisenheimer *et al.* 1989), we allow for a flattening of the power-law below some break energy γ_b in order to reproduce the slight flattening of the observed spectrum towards radio frequencies (shown as a thin line in Figs. 17a,b).

(ii) assumes a power-law electron energy distribution $N(\gamma) \propto \gamma^{-q}$ with a very sharp cutoff at γ_c as it will occur if the acceleration time scale is a strong function of the electron energy (*e.g.* Kirk & Schneider 1987, shown as a thick line in Figs. 17a,b).

Both models are characterized by two essential parameters, the *power-law index* α_{PL} at intermediate frequencies ($10^{10} - 10^{14}$ Hz), and the *cutoff frequency* ν_c. As obvious from Fig. 17a, both models provide a sactisfactory fit to the majority of the observed flux data between 10^9 and 10^{16} Hz. A closer look at the near-infrared to optical regime ($10^{14} - 10^{15}$ Hz), however, reveals two discrepant data sets: the JHK photometry of the Australian group ($\triangle$) and the RI photometry of Biretta *et al.* (1991, $\diamond$). Since there is no possible synchrotron spectrum smoothly connecting the discrepant near-infrared values with the optical photometry, and since three independent optical measurements exclude an optical spectral index as steep as $\alpha_{opt} = -1.25$, claimed by Biretta *et al.*, we omit the discrepant values from the fit. Then the fits formally favour model *(ii)* ($\chi^2 = 12.8$ for $N_{free} = 14$, confidence level = 53% to be compared with $\chi^2 = 18.0$, $CL = 20\%$ for

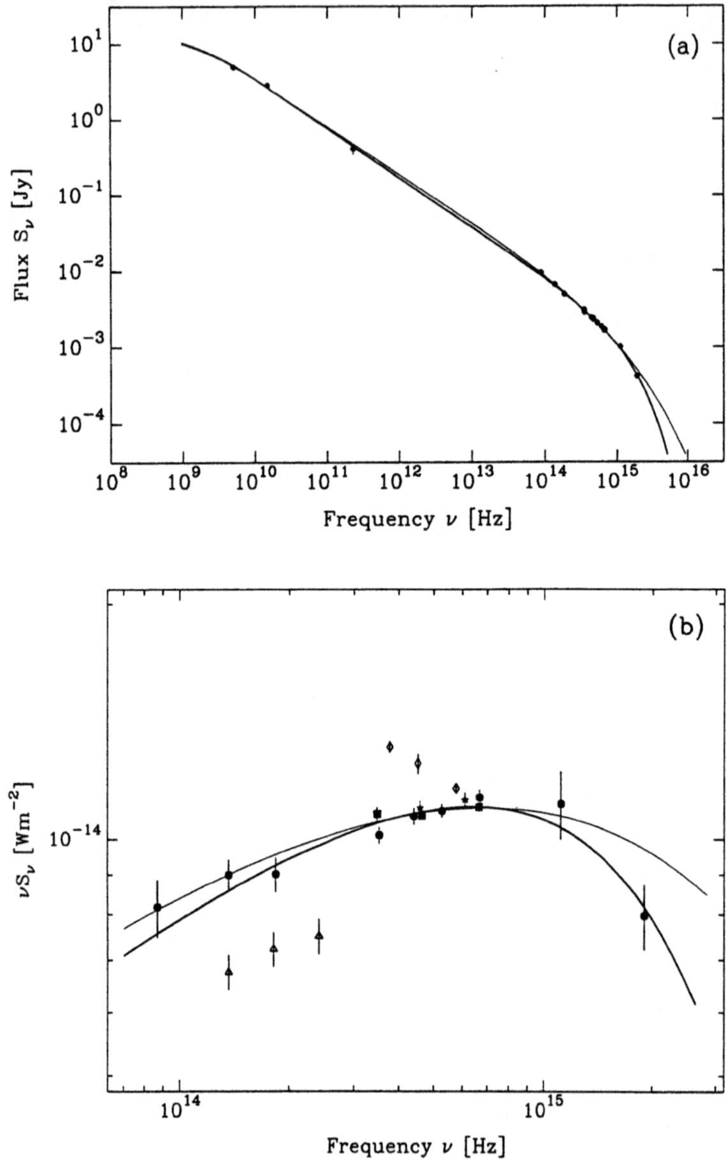

Fig. 17. **(a)** Spectrum of the brightest part of the jet (knots A+B+C) between 10^9 and 10^{16} Hz. Radio data are from Biretta *et al.* (1991), the millimetre flux is from Meisenheimer *et al.* (1993). See Fig. 17b for details of the near-IR, optical and UV part.
(b) Spectrum of the brightest part of the M 87 jet (knots A+B+C) between 10^{13} and 10^{16} Hz ($\log \nu S_\nu$ *versus* $\log \nu$). Infrared data are from Stocke *et al.* 1981 (•) and Killeen *et al.* 1984 (△), optical data are from Perez-Fournon *et al.* 1988 (⋆), Keel 1988 (•), Meisenheimer *et al.* 1993 (▪), and Biretta *et al.* 1991 (◇). UV flux is from Perola & Tarenghi (1980). The open symbols were omitted in (a).

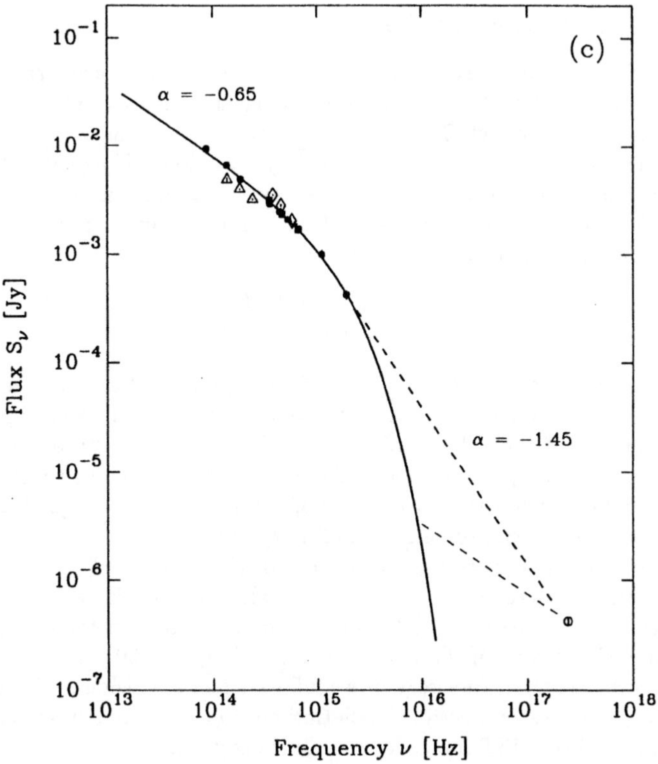

Fig. 17c. Spectrum of knots A+B+C between 10^{13} and 10^{18} Hz. The X-ray flux is from Biretta *et al.* (1991); compare Fig. 17b for the low frequency part. Two alternatives for connecting the X–ray flux to the radio-optical spectrum are indicated by dashed lines.

model *(i)*). It should be noted, however, that the preference of model *(ii)* relies mainly on the UV data which are rather uncertain.

How do the X–ray data fit into the overall spectrum? Figure 17c demonstrates that both model fits fail by a large amount to predict the observed X–ray flux.[13] One can think of two solutions for this dilemma: either the perfect radio-to-optical fit by a cutoff spectrum is purely due to coincidence – instead there is a spectral *break* by $\Delta\alpha = -0.7$ around $\nu = 10^{15}$ Hz, or the X–ray emission is due to a second component (synchrotron emission from a different place, inverse Compton radiation or thermal bremsstrahlung).

Although many authors like one-component models and thus favour the former alternative which requires extremely high energies of the relativistic electrons, we would prefer to wait for the ROSAT results before considering this exciting possibility. Therefore, in the following discussion, the X–ray data are ignored and the steep cutoff spectrum (model *(ii)*) is assumed.

[13]Some blazars with steep optical spectra show similar optical-to-X–ray spectra (*e.g.* PKS 1413+135, see Bregman *et al.* 1981).

5.2 Spectral variations along the jet

The analysis of the two-point spectral index α_{RO} between the shortest radio wavelength (2 cm, from OHC89) and the optical I band (Fig. 18a) shows only small variations ($\Delta\alpha_{RO} = \pm 0.03$) along the jet. Thus, the much larger oscillations of the optical spectral index α_{opt} (Fig. 14) seem to be mainly caused by variations of the cutoff frequency ν_c. Assuming that the shape of the spectrum remains constant, one is able to derive an unambigous function between the observed α_{opt} and the cutoff frequency ν_c. For the model spectrum *(ii)* we find heuristically:

$$\nu_c = \frac{2.3 \times 10^{15}\,\mathrm{Hz}}{-0.53 - \alpha_{opt}} \tag{1}$$

which is valid over the range $-2 < \alpha_{opt} < -0.7$. Applying this formula to α_{opt} measured along the jet yields the variation of the cutoff frequency ν_c, as displayed in Fig. 18b. As an independent test of the validity of our formula we performed the spectral fit with model *(ii)* at 14 selected positions along the jet (result is shown as thick cross-bars in Fig. 18b). Obviously within the errors, the approximation for ν_c is good and allows us to derive a continuous variation of the cutoff frequency.[14]

In order to fully appreciate the significance of these extremely small variations in ν_c (range less than a factor of 4), one has to recall that even for a constant magnetic field,[15] this corresponds to $\gamma_c \propto \nu_c^{0.5}$ being constant within a factor of 2 and that the typical lifetime of electrons of the inferred energies is $\tau_{syn} < 30$ yrs (*i.e.* $c\tau_{syn} < 10$ pc, i.e. 0.″12). Even in the inter-knot regions we fail to detect any evidence for the expected rapid synchrotron losses of electrons radiating near ν_c. This partly explains why the 2 cm radio map and the HST image look so amazingly similar.

6. Interpretation: Physics of the Jet

(Klaus Meisenheimer)

Any physical model of the jet in M 87 must at least explain the gross observational features which are:

- The morphology with its knotty (at high resolution: filamentary) appearance, including the orientation of the magnetic field.
- The occurence of (at least) mildly relativistic proper motions of several features in the jet.
- The astonishing lack of variations in the jet's spectrum both globally (along the entire jet) and locally (between knots and inter-knot regions).

At present, such a general physical explanation is not possible because it would require a full 3-dimensional MHD code which includes also the dissipation of bulk energy by

[14] It should be noted that the point-wise fits yield a small but significant oscillation of the power-law index α_{PL} (see Fig. 18a). Therefore, the assumption of a constant spectral shape ($\alpha_{PL} = -0.652$) is not strictly valid.

[15] This is a conservative assumption; it seems more likely that the field is enhanced in the knots, see below.

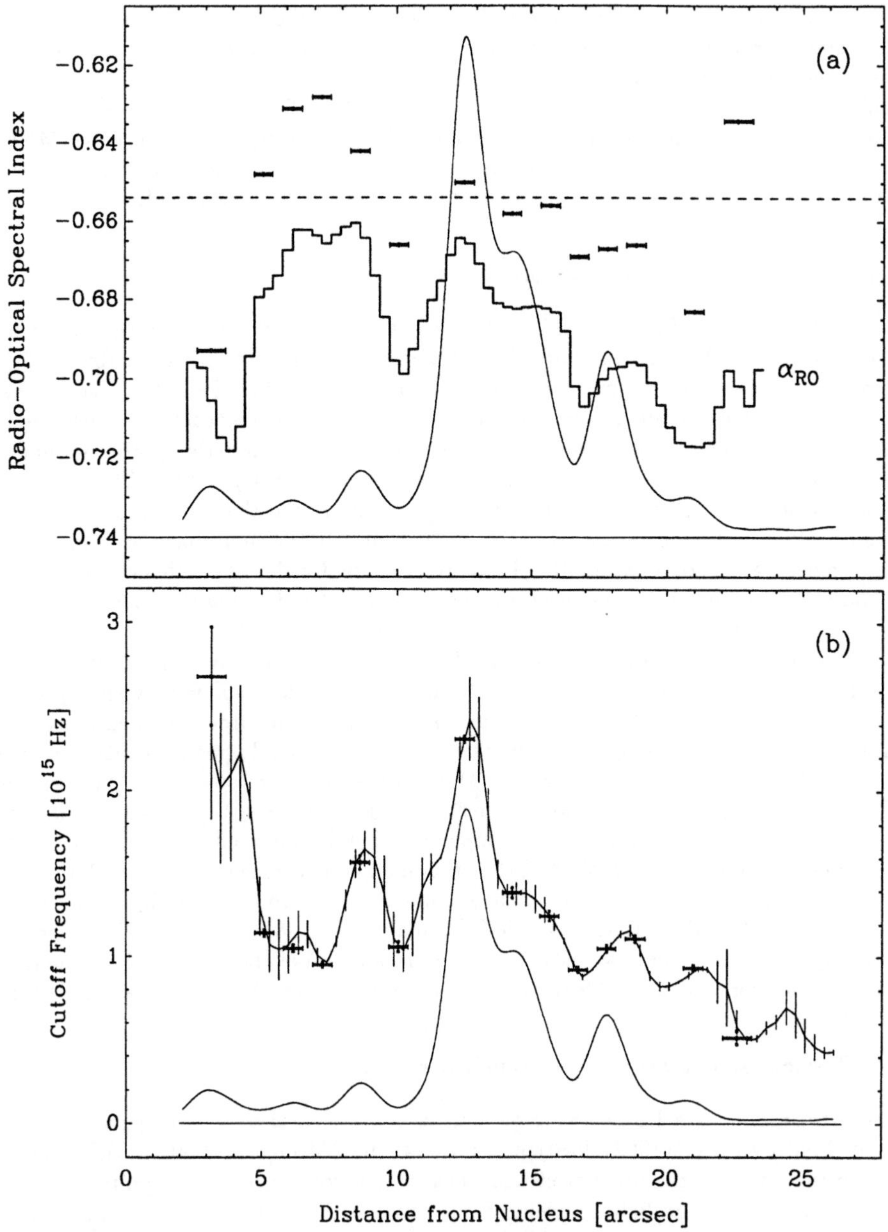

Fig. 18. (a) Radio-optical spectral index α_{RO} (two-point index between 2 cm and the I-band) and power-law index α_{PL} (⊢⊣) along the jet (from the spectral fit at individual positions along the jet). For positional reference the optical brightness profile is given. The length of the horizontal bars indicate which region was included in each fit.
(b) Run of the cutoff frequency ν_c along the jet. The thin line (with error bars) give ν_c as derived from α_{opt}. The cross bars show ν_c as derived from the spectral fit.

particle acceleration, a full treatment of the diffusion of the relativistic particles as well as a simulation of the synchrotron emission process. We have to consider, therefore, micro-models which are able to explain some selected properties in order to obtain the building blocks of a more complete theory. Note that, unlike the situation envisaged in most more distant radio galaxies, the problem for any theory about the jet of M 87 is not the lack of detailed observational input, but rather the morphological fine structure and the wide range of spectral information which has to be accounted for.

6.1 The Hydrodynamical Approach

When the first two-dimensional hydrodynamical simulations of jets were carried out (Norman *et al.* 1982), it was immediately pointed out that the knots in the M 87 jet may mark the crossed shocks commonly seen in these models. Falle & Wilson (1985) presented a more detailed simulation of the standing shock structure which should develop in the density gradient around M 87. Following an analytic approach, Sanders (1983) pointed out that recollimation of the jet and the development of strong internal shocks depend on the details of the density gradient around the radio source. A recent numerical simulation of a supersonic jet propagating into the enviroment around M 87 as derived from X–ray observations has been presented by Hardee *et al.* (these proceedings). A common feature of these models is the development of a *standing shock structure*. This is in conflict with the large proper motions observed (section 2.4).

In addition, both the overpressure in the synchrotron emitting regions (*cf.* OHC89) and the highly polarized filaments indicate that magnetic forces may play an important role in shaping the jet's morphology. Current attempts to incorporate magnetic effects into hydrodynamic simulations are limited to the case of purely passive fields (*i.e.* magnetic field pressure $\ll$ gas pressure, see Matthews & Scheuer 1990a,b) or the inclusion of the magnetic pressure in a two-dimensional simulation (*e.g.* Kössl *et al.* 1990a,b,c). Therefore, the breaking of the axial symmetry which is commonly observed in magnetized plasmas is suppressed. We feel that the main conclusions of these simulations for M 87 are that (1) highly polarized filaments may well be produced by shear layers even for passive fields and (2) that magnetic fields may help to stabilize and collimate the jet flow. But we are still some way short of reaching the goal of these simulations – namely the identification of observed features with specific components seen in the simulations.

6.2 Particle Acceleration and Synchrotron Losses

One of the main obstacles to reaching a conclusive physical theory of the jet was already pointed out by Felten (1968): for typical magnetic field strengths of 30 nT, the loss time τ_{syn} of the optically radiating electrons – near ν_c – is some orders of magnitude smaller ($\lesssim$ 30 years) than the light traveling time between the core and the brightest parts of the jet (> 3000 years at knot A). Both the detailed analysis of the optical spectral index and the high resolution optical data obtained with the HST leave little space for synchrotron losses and thus turn Feltens global argument into a local one: even the assumption of permanent re-acceleration in the knots (which the hydrodynamical models identify with strong shocks in which efficient particle acceleration could occur) is not sufficient to explain the extremely smooth ν_c distribution along the jet, from which we inferred a maximum range of a factor 2 in the maximum energy γ_c. Three different ways around this dilemma have been proposed:

(1) Owen *et al.* (1989a,b) point out that the short synchrotron loss time $\tau_{syn} \lesssim 30\,\mathrm{yrs}$ applies only to the synchrotron *emitting regions* for which the estimated equipartition magnetic field should be a resonable estimate. However, if a field-free channel ($|B| \leq 0.3\,\mathrm{nT}$) should exist around the axis of the jet, electrons (positrons), which are accelerated in the core, could easily travel several kpc without losing much energy. The synchrotron emitting regions forming a hollow cone or cylinder would then be fed by a fraction of the ultra-relativistic particle beam diffusing into the regions of high magnetic field. In fact, the brightness profiles across the jet (Figs. 4,6) and the polarization structure (Figs. 10,15) seem to support such a hollow cylinder for the synchrotron emitting regions (see also Meisenheimer 1992). Although this scenario could explain the observed constancy of the spectrum in a straightforward way, it has one major flaw already pointed out by Felten (1968): the strong particle–field interaction in the boundary layer between the field-free channel and the emission regions would produce violent plasma waves and MHD turbulence which should break up the channel within a few 100 years. Furthermore, very rapid energy losses cannot be avoided in this model beyond knot C where the ordered jet structure is obviously destroyed – nevertheless, we trace the synchrotron light emission several 100 pc further out.

(2) Begelman and Kirk (*priv.comm.*) pointed out that in a highly relativistic jet flow ($\Gamma_{jet} > 10$) towards the observer, time dilatation and superluminal motion would lead to an underestimation of τ_{syn} in the order of $\Gamma_{jet}^{-3} \lesssim 0.001$. The true value would then be $\tau_{syn} \gtrsim 20000\,\mathrm{yrs}$, just consistent with the (deprojected) length of the jet. However, this scenario demands that the relativistic flow is maintained well beyond knot C. But then – where is the hot spot in which the jet is decelarated from $\Gamma \gtrsim 10$ to $\Gamma = 0$?

(3) Felten (1968) and – enforced by recent observational evidence – Meisenheimer (1989,1992) argue that the observational facts leave little space to circumvent the demand for in situ particle acceleration *all along the jet.* The lack of significant inter-knot energy losses indicates that this (re-)acceleration cannot be confined to a dozen or so localized accelerators (*i.e. the knots*) but must persist as an almost permanent process throughout the entire length of the jet. The main challenge to this interpretation in terms of a *local* acceleration process is certainly the perfect *global* constancy of the resulting electron spectrum (details in section 6.3). This can only be understood if some sort of self-regulation process controls the production of a common powerlaw spectrum $N(\gamma) \propto \gamma^{-2.3}$ up to an almost constant maximum energy γ_c. We believe that only electromagnetic processes which convert field energy into relativistic particles are able to do this. The small variations in the spectrum would then argue for a dominance of the electromagnetic field pressure in the entire jet. Only a small fraction of this energy is dissipated out to knot G. Indeed, the jet's emission ($P_{5\mathrm{GHz}} = 1.5\ 10^{22}\mathrm{W\,Hz^{-1}}$) is only a tiny fraction of the total power emitted by the radio source Virgo A ($P_{5\mathrm{GHz}} = 2\ 10^{23}\mathrm{W\,Hz^{-1}}$) and much more energy seems to be stored in the lobes.

6.3 A Heuristic Model for the Jet's Synchrotron Emission

As pointed out in section 5.2, the observed cutoff frequency and accordingly the maximum energy γ_c of the underlying electron distribution remain rather constant along the jet. So it is natural to assume that γ_c is indeed a constant. Then, from standard synchrotron theory, the observed variations of the cutoff frequency

$$\nu_c = 42\ |\mathbf{B}(\mathrm{nT})| \sin\theta\ \ \gamma_c^2\ \ [\mathrm{Hz}] \tag{2}$$

have to be attributed to variations in the projected field strength $|\mathbf{B}|\sin\theta$. Since the emitted synchrotron power per unit volumn, ϵ_ν, also depends strongly on $|\mathbf{B}|\sin\theta$, one expects that the variations of $|\mathbf{B}|\sin\theta$ (as inferred from ν_c) also determine the brightness variations $B_\nu(z)$ along the jet. Indeed we will demonstrate in the following that variations of the field strength alone are sufficient to explain the local variations in both $\nu_c(z)$ and $B_\nu(z)$.

In order to account for the global decrease of ν_c along the jet (Fig. 18b), we allow for a gradual decrease of $\gamma_c(z)$. We parameterize γ_c as a function of the jet radius R (from the 20% contour of the radio map, Fig. 10): $\gamma_c(z) \propto R^{-s}(z)$. In Fig. 19 we show $\gamma_c(z)$ for $s = 1/3$ which yields the best description of $\nu_c(z)$ in the inter-knot regions of Fig. 18b.[16]

If one ignores the (unknown) geometrical factor $\sin\theta$, the observed $\nu_c(z)$ can be directly converted into the field strength

$$|\mathbf{B}(z)| = 30 \frac{\nu_c(z)}{10^{15}\mathrm{Hz}} \left(\frac{\gamma_c(z)}{8.8\ 10^5}\right)^{-2} \quad [\mathrm{nT}] \tag{3}$$

where the conversion factor is determined by an equipartition analysis of knot A (B_A = 70 nT, see Meisenheimer *et al.* 1993). Assuming a helical field $|\mathbf{B}(z)| = \sqrt{\mathrm{B}_z^2 + \mathrm{B}_\phi^2}$ with a constant poloidal component B_z and a toroidal component B_ϕ, which is locally enhanced by axial compressions (compression ratio: $r(z)$), one finds:

$$|\mathbf{B}(z)| = \mathrm{B}_z \sqrt{1 + b_A^2 \cdot \left(\frac{r(z)}{r_A}\right)^2} \tag{4}$$

where b_A and r_A are the ratio of the toroidal field to poloidal field component and the compression ratio in knot A, respectively. From the observed polarization vector in knot A (Fig. 15), we conclude that b_A is significantly larger than 1 (*cf.* Meisenheimer 1992). With these approximations we are able to predict the synchrotron emissivity (per unit length) along the jet:

$$\epsilon_\nu R^2 \propto n_0(z) R^2 |\mathbf{B}(z)|^{1-\alpha}$$

with n_0 being the density of relativistic electrons with energies $\gamma = \gamma_c$. For a common acceleration process this density is determined by the axial compressions alone, *i.e.* $n_0(z) = n_0 \frac{r(z)}{r_A}$. The predicted optical brightness distribution (per unit length) is then:

$$B_{opt} = B_{opt}(A) \cdot \frac{r(z)}{r_A} |\mathbf{B}(z)|^{1-\alpha_{opt}} \tag{5}$$

[16] Due to a misprint $\gamma_c \sim R_{jet}^{-1}$ instead of $\gamma_c \sim R_{jet}^{-1/3}$ was given in Meisenheimer (1992).

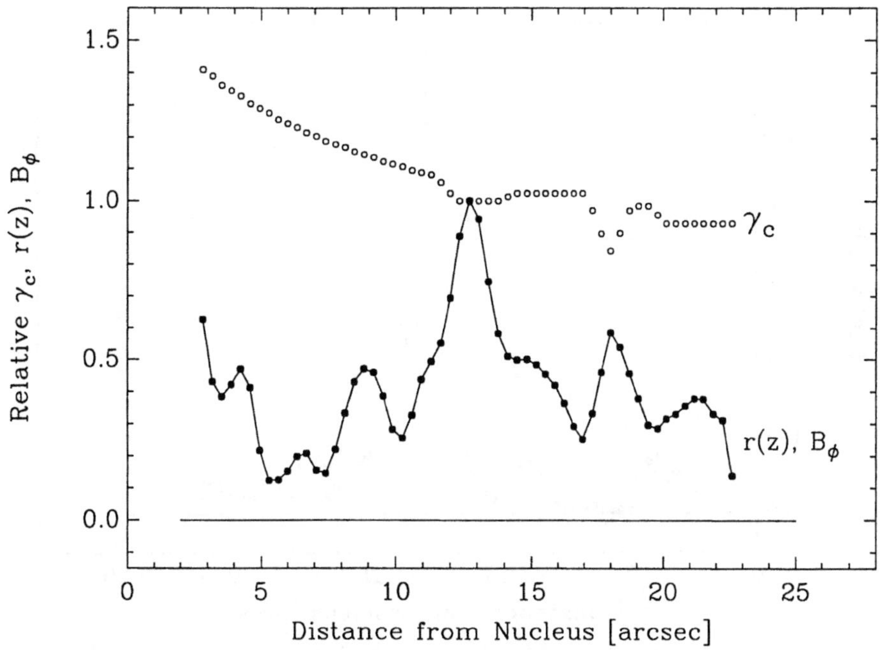

Fig. 19. Parameters of the model for the jet's synchrotron emission. It is *assumed* that the cutoff energy γ_c depends on the jet radius: $\gamma_c \propto R_{jet}^{-1/3}$ (o o o). The model *predicts* the run of the compression ratio $r(z)$ along the jet which directly determins the toroidal fiels component $B_\phi \propto r(z)$ (•—•—•).

If we take $|\mathbf{B}(z)|$ from (3), the predicted brightness B_{opt} depends on 2 free parameters only, s and b_A. For $s = 1/3$ and $b_A = 4.2$ the prediction (5) reproduces the observed brightness distribution rather well (Fig. 20). This choice also determines the run of $B_\phi \propto r(z)$ along the jet (Fig. 19).

In fact, we see only two main discrepancies between the predicted and the observed brightness distribution:

- Upstream of the knots, the observed B_{opt} starts its rise in advance of the prediction (see knots F,A,C). Within our model, this would indicate that the jet is inclined by some angle $\gtrsim 30°$ out of the plane of sky.
- The predicted brightness is definitely too small around knot B. A natural explanation for this would be that here γ_c is smaller than the global run $\gamma_c \propto R^{-1/3}$. The small but significant deviations of the radio and optical polarization vectors (Fig. 15) also indicate that synchrotron losses play an important role immediately downstream of knot A.

In spite of these details we conclude from the close match between predicted and observed brightness distribution and from the observed 1:1 correspondence between radio and optical polarization that:

(i) local variations of γ_c (on scales of 100 pc or less) are extremely small. This demands a continuous re-acceleration which leads to a common γ_c even if the mean magnetic field $|\mathbf{B}(z)|$ varies by a factor of > 2. This is only possible if the accelera-

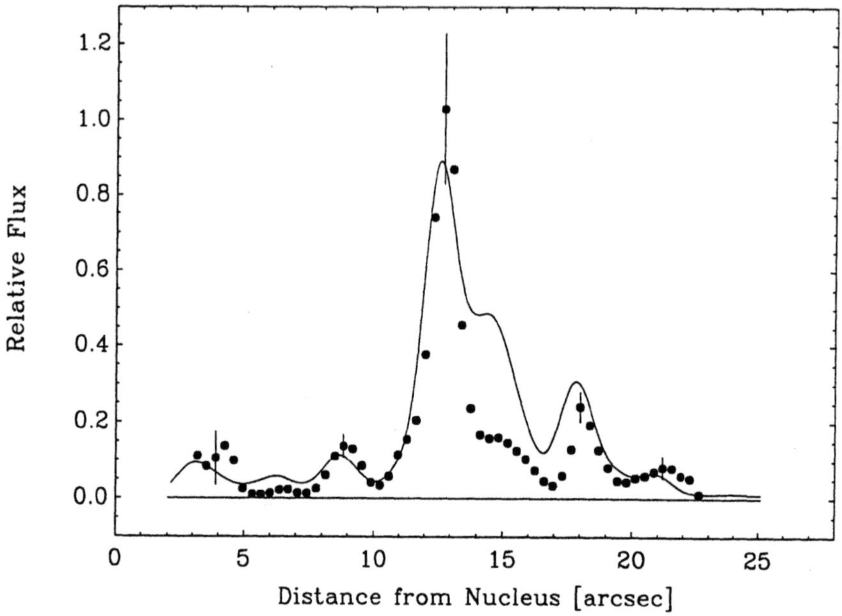

Fig. 20. Comparison between optical brightness distribution B_{opt} as predicted by the model (•) and the observations (——). The errors of the observed cutoff frequency ν_c limit the accuracy of the prediction. This is indicated by error bars at several positions along the jet.

tion time scale closely matches the synchrotron loss time $\tau_{syn} \sim |B|^{-2}$. This favours electromagnetic processes determining *both* the acceleration and loss processes.

(ii) The pattern of the magnetic field structure is the major factor in determining the morphology. Observed proper motions thus refer to the pattern speed of the field structure.

(iii) Besides the in situ acceleration process, the synchrotron losses also seem to happen more or less "on the spot" since there is little evidence for the expected offsets between radio and optical emission. Obviously, the ultrarelativistic electrons are trapped near their place of acceleration. This implies that the jet is not able to feed relativistic electrons into the lobes and additional acceleration processes have to be responsible for the lobe emission.

Unfortunately, we are not in the position to propose any refined theory of an acceleration mechanism which would naturally produce a common electron spectrum along the entire jet. Thus our *ad hoc* model should be regarded as a more stringent formulation of Felten's dilemma, rather than a solution of the particle acceleration problem in the M 87 jet.

7. Conclusions and Future Prospects

At the present state we believe that it would be most useful to summarize our knowledge of the physics in the prototype extragalactic radio jet of M 87 in order of increasing uncertainty.

Without much interpretation, the following **facts** about the jet of M 87 are obvious from the observations:
The jet's emission is entirely due to synchrotron radiation, at least in the frequency range 10^8 to 10^{15} Hz. The jet is well collimated on all (projected) scales between $0\overset{''}{.}01$ (0.8 pc) and 24″ (2 kpc), with some recollimation ocurring between $0\overset{''}{.}2$ and 1″ and at 12″ from the core (at knot A). While the first recollimation is not clearly associated with a morphological feature, the one near knot A is marked by a sharp increase in surface brightness by a factor of 8 or more.
All along the collimated jet, the radio-optical synchrotron spectrum is a power-law $S_\nu \propto \nu^{-0.65}$ with a sharp "cutoff" or "break" at frequencies $\nu_c \gtrsim 10^{14}$ Hz. Towards radio frequencies the spectrum flattens slightly to $S_\nu \propto \nu^{-0.5}$. The spectrum is extremely constant with only minor changes between knot and inter-knot regions. All variations can be ascribed to a shift in the cutoff frequency ν_c by a factor of less than 4. Somewhat downstream of knot C, the radio jet loses its collimation and gradually merges into the surrounding radio lobe. At the same place the optical spectrum steepens until the optical jet fades away. There is no indication of a "hot spot" near the jet's terminus which could be identified with the working surface expected for a highly supersonic jet.
The jet's brightness (and polarization) pattern, which seems to be dominated by highly magnetized filaments, shows proper motions with an apparent velocity of 0.3 to 0.8c on scales between several 10 mas and 20″. Although jet features move predominantly outwards there is also evidence of proper motion components perpendicular to the jet axis in the outer regions of the jet.

Although various exotic interpretations have been proposed, we think the **common sense** about the jet and radio source of M 87 can be summarized as follows:
The prominence of bright highly polarized filaments on the 2 cm maps and the success of the heuristic model (6.3) indicate that a helical magnetical field dominates the jet pressure. Further support for this hypothesis comes from the constancy of the synchrotron spectrum since plasma compressions (*e.g* in Knot A) seem to have only minor effects on the efficiency of particle acceleration. As a consequence, the magnetic field can also play an important role in collimating the jet flow (the thermal pressure of the X–ray gas around M 87 is insufficient to balance the relativistic pressure in the jet, see OHC89).
In a magnetically dominated jet, the Alvén speed which could reach a large fraction of the speed of light is the characteristic velocity. With respect to this Alvén speed the jet seems to be subsonic, or at most just transsonic, since there is no "hot spot", but the jet merges gradually into the radio lobe. The constancy of the synchrotron spectrum and the lack of significant radio-optical displacements can only be understood if magnetic field energy is continuously fed into the acceleration of relativistic electrons (positrons). However, the conversion rate of the field energy has to be small in order to avoid significant changes in the physical state of the jet which would manifest themselves in more pronounced spectral changes than observed. Both the collimating force and the acceleration efficiency for ultrarelativistic electrons seem to vanish around knot G. Here

we also observe the most pronounced difference between radio and optical polarization. The observed X–ray emission of the jet could either represent the high frequency tail of the synchrotron emission (this would require very high particle energies: $\gamma_x \simeq 10^7$), but may be just as well be generated in a ultrahot boundary layer caused by magnetic waves from the jet interacting with the surrounding material.
Since the observational evidence points to synchrotron losses happening "on the spot", the synchrotron emission of the western lobe is not likely to be caused by electrons which have been accelerated in the jet. This requires some additional *in situ* acceleration. Thus, it is the kinetic energy (pointing flux) rather than the relativistic particles from the jet that powers the radio lobe.

Finally, if we try to address the important questions of the jet's one-sidedness and the speed of its flow, we obviously reach a point at which **speculations** come into play:
If one regards the observed pattern speed from the proper motion studies as a fair estimate of the Alvén speed, and if one also takes into account the morphological indications that the jet flow is just transsonic, one deduces a mildly relativistic jet speed ($v_{jet} = 0.3 \ldots 0.8\,c$). Such speeds are certainly not sufficient to boost the jet's brightness by factors > 100 which would be required to make a symmetrical counter-jet invisible. Thus, there seems to exist an asymmetry in the core of M 87 which leads to the formation of a highly collimated bright jet on one side only. However, the radio and millimetre maps of Virgo A show a quite symmetrical double lobe structure within the inner 35″ around the core. Obviously, an almost equal amount of energy is fed into the "unjetted" eastern lobe. Perhaps the eastward outflow is not sufficiently collimated or magnetized to produce detectable radio emission. In this scenario, the bow-shaped patch of enhanced radio brightness could indeed be identified with a "hot spot" in which the eastward outflow is rapidly decelerated giving rise to efficient particle acceleration. The detection of an optical counterpart to the patch supports this interpretation. The formation of a hot spot would argue for a supersonic eastward outflow. The high Mach number may be explained by a lower sound speed than that inferred for the radio jet. Thus, M 87 might be the first example showing evidence for an intrinsically asymmetric outflow which, nevertheless, is able to produce an almost symmetrical double lobe structure. This would question the generally accepted paradigm that Doppler boosting is the sole reason for jet asymetries in powerful radio galaxies and quasars.

In view of this exciting possibility the aims for future investigations into the kiloparsec radio source in Virgo A seem obvious:

- *Understand what makes the radio jet in M 87 shine so brightly* in the entire frequency regime between 10^9 and $\gtrsim 10^{15}$ Hz. On the observational side, this requires a detailed investigation of the jet spectrum with subarcsecond resolution. The present developments of millimetre arrays and seeing compensating techniques at near infrared and optical wavelengths should make this goal reachable within the next years. On the theoretical side, we need to improve our understanding of how magnetized plasma flows do accelerate the synchrotron emitting ultra-relativistic particles which make the radio jets visible.
- *Establish how asymmetric the central source of Virgo A really is.* This requires the determination of the synchrotron spectra of the western and eastern lobe, respectively. The confirmation that the eastern hot spot does indeed emit synchrotron

radiation up to 10^{15} Hz would be of great significance in this context. A proper motion measurement could provide additional insight into the nature of this feature.
- *Search for an outflow or jet towards the eastern lobe.* We propose two complementary observing strategies: a highly collimated Doppler-dimmed outflow would show up best on radio maps with high resolution (VLA, VLBA, MERLIN at $\lambda \lesssim 6$ cm), while a less collimated, low brightness outflow would most likely be detected on very high signal-to-noise low frequency maps with arcsecond resolution, or perhaps on very deep X–ray images.

Even if these questions, which are based on our present preoccupations, were to prove as inappropriate for a deep understanding of Virgo A and its jet – the proposed observational and theoretical investigations would significantly improve our knowledge about the M 87 jet and extragalactic radio source in general.

References

Arp, H.: 1967, *Astrophys. Letters* **1**, 1.
Baade, W.: 1956, *Astrophys. J.* **123**, 550.
Baade, W. and Minkowski, R.: 1954, *Astrophys. J.* **119**, 221.
Baum, S.A., Heckman, T.M. and van Breugel W.: 1990, *Astrophys. J. Suppl.* **74**, 389.
Baum, S.A., Heckman, T.M. and van Breugel W.: 1992, *Astrophys. J.* **389**, 208.
Biretta, J. A., Owen, F. N., and Cornwell, T. J.: 1989, *Astrophys. J.* **342**, 128.
Biretta, J. A., Owen, F. N.: 1990, in *Parsec-scale Jets*, eds. J. A. Zensus and T. J. Pearson, (Cambridge Univ. Press: Cambridge), p. 125.
Biretta, J.A., Stern, C.P. and Harris, D.E.: 1991, *Astron. J.* **101**, 1632 (BSH91).
Biretta, J. A. and Reid, M. J.: 1993, in preparation.
Biretta, J. A. and Owen, F. N.: 1993, in preparation.
Blandford, R.D. and Rees, M.J.: 1974, *Mon. Not. R. astr. Soc.* **169**, 395.
Boksenberg, A. *et al.*: 1992, *Astron. Astrophys.* **261**, 393.
Bolton, J.G., Stanley, G.J., Slee, O.B.: 1949, *Nature* **164**, 101.
Bregman, J.N., Lebofsky, M.J., Aller, M.F., Rieke, G.H., Aller, H.D., Hodge, P.E., Glassgold, A.E. and Huggins, P.J.: 1981, *Nature* **293**, 714.
Burbidge, G.R.: 1956, *Astrophys. J.* **124**, 416.
Curtis, H.D.: 1918, *Lick Obs. Publ.* **13**, 11.
de Vaucouleurs, G. and Nieto, J-L.: 1979, *Astrophys. J.* **231**, 364.
Falle, S.A.E.G. and Wilson, M.J.: 1985, *Mon. Not. R. astr. Soc.* **216**, 79.
Fanaroff, B.L. and Riley, J.M.: 1974, *Mon. Not. R. astr. Soc.* **167**, 31p.
Felten, J.E.: 1968, *Astrophys. J.* **151**, 861.
Ford, H.C. and Butcher, H.: 1979, *Astrophys. J. Suppl.* **41**, 147.
Forster, J. R.: 1980, *Astrophys. J.* **238**, 54.
Fraix-Burnet and J., Nieto, J.-L.: 1988, *Astron. Astrophys.* **198**, 87.
Hardee, P.E., White, R.E. and Norman, M.L.: 1993, *these proceedings, p. 193.*
Hargrave, P.J. and Ryle, M.: 1974, *Mon. Not. R. astr. Soc.* **166**, 305.
Heavens, A.F., Meisenheimer, K.: 1987, *Mon. Not. R. astr. Soc.* **225**, 335.
Hiltner, W.A.: 1959 *Astrophys. J.* **130**, 340.
Jarvis, B.J.: 1990, *Astron. Astrophys. (Lett.)* **240**, L8.
Kaneko, N., Mishimura, M. and Toyama, K.: 1973, *Publ. Astron. Soc. Japan* **25**, 175.
Keel, W.C.: 1988, *Astrophys. J.* **329**, 532.
Killeen, N.E.B., Bicknell, G.V., Hyland, A.R. and Jones, T.J.: 1984, *Astrophys. J.* **280**, 126.

Kirk, J.G. and Schneider, P.: 1987, *Astrophys. J.* **322**, 256.
Kössl, D., Müller, E. and Hillebrandt, W.: 1990a, *Astron. Astrophys.* **229**, 378.
——: 1990b, *Astron. Astrophys.* **229**, 397.
——: 1990c, *Astron. Astrophys.* **229**, 401.
Laing, R.A.: 1980, *Mon. Not. R. astr. Soc.* **193**, 427.
Matthews, A.P. and Scheuer P.A.G.: 1990a, *Mon. Not. R. astr. Soc.* **242**, 616.
——: 1990b, *Mon. Not. R. astr. Soc.* **242**, 623.
Meisenheimer, K.: 1989, *Rev. Mod. Astronomy* **2**, 129.
Meisenheimer, K.: 1992, in W. Duschl & S. Wagner (eds.): *Physics of Active Galactic Nuclei*, Springer Berlin, Heidelberg etc., p. 512.
Meisenheimer, K. and Hippelein, H.: 1989, *Sterne und Weltraum* **28**, 292.
Meisenheimer, K., Röser, H.-J., Hiltner, P., Yates, M.G., Longair, M.S., Chini, R. and Perley, R.A.: 1989, *Astron. Astrophys.* **219**, 63.
Meisenheimer, K., Schlötelburg, M. and Röser, H.-J.: 1993, in preperation.
Mould, J., Aaronson, M. and Huchra J.: 1980, *Astrophys. J.* **238**, 458.
Nieto, J.-L. and Lelièvre, G.: 1982, *Astron. Astrophys.* **109**, 95.
Norman, M.L., Smarr, L.L., Winkler,K.-H.A. and Smith, M.D.: 1982, *Astron. Astrophys.* **113**, 285.
Owen, F. N.: 1992, private commumication.
Owen, F.N., Hardee, P.E. and Bignell, R.C.: 1980, *Astrophys. J. (Lett.)* **239**, L11.
Owen, F.N., Hardee, P.E., Cornwell, T.J., Hines, D.C. and Eilek, J.A.: 1989a, in K. Meisenheimer and H.-J. Röser (eds.): *Hot Spots in Extragalactic Radio Sources*, Lecture Notes in Physics Vol. **327**, Springer Heidelberg, p. 77.
Owen, F.N., Hardee, P.E. and Cornwell, T.J.: 1989b, *Astrophys. J.* **340**, 698 (OHC89).
Owen, F. N., Eilek, J.A. and Keel, W. C.: 1990, *Astrophys. J.* **362**, 449.
Oort, J.H. and Walraven, T.: 1956, *Bull. Astron. Inst. Netherlands* **12**, 285.
Perez-Fournon,I., Colina, L., Gonzalez-Serrano, J.I. and Biermann, P.L.: 1988, *Astrophys. J. (Lett.)* **329**, L81.
Perola, G.C. and Tarenghi, M.: 1980, *Astrophys. J.* **240**, 447.
Reid, M.J., Biretta, J.A., Junor, W., Muxlow, T.W.B and Spencer, R.E.: 1989, *Astrophys. J.* **336**, 112.
Röser, H.-J. and Meisenheimer, K.: 1991, *Astron. Astrophys.* **252**, 458.
Sanders, R.H.: 1983, *Astrophys. J.* **266**, 73.
Salter, C.J., Chini, R., Haslam, C.G.T., Junor, W., Kreysa, E., Mezger, P.G., Spencer, R.E., Wink, J.E. and Zylka, R.: 1989, *Astron. Astrophys.* **220**, 42.
Scheuer, P.A.G.: 1974, *Mon. Not. R. astr. Soc.* **166**, 513.
Schlötelburg, M., Meisenheimer, K. and Röser, H.-J.: 1988, *Astron. Astrophys. (Lett.)* **202**, L23.
Schreier,E.J., Gorenstein, P. and Feigelson, E.D.: 1982, *Astrophys. J.* **261**, 42.
Shklovsky, I.S.: 1953, *Doklady Akad. Nauk. U.S.S.R.* **90**, 983.
Smith, R. M., Bicknell, G.V., Hyland, A.R. and Jones, T.J.: 1983, *Astrophys. J.* **266**, 69.
Sparks, W.B., Fraix-Burnet, D., Machetto, F. and Owen, F.N.: 1992, *Nature* **355**, 804.
Spencer, R. E. and Junor, W.: 1986, *Nature* **321**, 753.
Stiavelli, M., Biretta, J., Møller, P. and Zeilinger, W.W.: 1992, *Nature* **355**, 802.
Stocke, J.T., Rieke, G.H. and Lebofsky, M.J.: 1981, *Nature* **294**, 319.
Turland, B.D.: 1975, *Mon. Not. R. astr. Soc.* **170**, 281.
Vashakidze, M.A.: 1954, *Astr. Circ.* **147**, 11.
Warren-Smith, R.F., King, D.J. and Scarrott, S.M.: 1984, *Mon. Not. R. astr. Soc.* **210**, 415.
White, R.E. and Sarazin, C.L.: 1988, *Astrophys. J.* **335**, 688.
Wisniewski, W.Z. and Kleinmann, D.E.: 1968, *Astron. J.* **73**, 866.

Modelling the Jet and Cooling Flow in M 87

Philip E. Hardee [1], Raymond E. White III [1],

Michael L. Norman [2]

[1]Department of Physics & Astronomy
University of Alabama
Tuscaloosa, AL 35487, USA
[2]Beckman Institute & Department of Astronomy
University of Illinois
Urbana, IL 61820, USA

1 Introduction

The jet in M 87 provides a unique opportunity for numerical modelling. Detailed observations of the jet at VLBI scales (Reid *et al.* 1989) and VLA scales (Owen, Hardee, & Cornwell 1989; Biretta, Owen, & Cornwell 1989, these proceedings) have been performed. The jet is observed to be highly collimated with an opening angle that may be as large as 10 degrees within 0.05 arcsec of the center. Feature N2 at $\sim$ 0.02 arcsec (or 2 pc) from the center has a proper motion equivalent to $0.28c$ (1 arcsec $\sim$ 78 pc for a distance of 16 Mpc to M 87). At VLA scales from feature D at $\sim$ 3 arcsec from the center out to feature A at 12 arcsec the opening angle is nearly constant and is about 6.5 degrees. At feature A jet expansion ceases, the jet is observed to oscillate sideways, and loses its coherence at feature C, 18 arcsec from the center. Proper motions have been found from feature D out to feature C equivalent to speeds higher than c outwards to slowly backwards (see Biretta & Meisenheimer, these proceedings). On average, proper motions are consistent with speeds of at least $0.3c$, and theoretical mechanisms capable of producing the moving features require that the jet flow speed be at least this large (see Owen, Hardee, & Cornwell 1989).

X-ray observations of the Virgo cluster indicate that the jet and radio source are buried at the center of a cooling accretion flow. Extensive modelling of this accretion flow has been performed by White and Sarazin (1988). We used similar techniques to explore various cooling flow models in order to match the X-ray surface brightness profile observed by the *Einstein* Observatory IPC (Fabricant & Gorenstein 1983) and HRI (Schreier, Gorenstein, & Feigelson 1982). The nature of the cooling flow is determined by the initial temperature of gas just outside the flow, the total accretion rate, the age of the flow, and the gravitational potential. The Virgo cluster is observed to be isothermal outside the inner cooling flow region, with a temperature of $\sim$ 3 keV (Hutsukade 1989;

Edge 1989). The gravitational potential is estimated from the observed dynamics of Virgo cluster galaxies and the light and dynamical distribution of stars in M 87 itself. We adopt a two-component mass model, one component due to the luminous parts of the galaxy, the other to a dark matter component. The mass profile of the luminous galaxy is derived from the photometry of Young *et al.* (1978), and the spectroscopy of Sargent *et al.* (1978). The velocity dispersion characterizing the asymptotically isothermal dark-matter distribution of the inner regions of the cluster is taken to be 580 $km\ s^{-1}$ (Huchra 1985). The core radius of the dark matter distribution is determined by fits to the X-ray surface brightness profile and is approximately 15 kpc. Simulations which lose gas density at a rate proportional to the local gas density divided by the local cooling time are successful in matching the observed IPC and HRI X-ray surface brightness profiles. Cooling flows which do not lose gas density out of the flow lead to surface brightness profiles which are too centrally peaked.

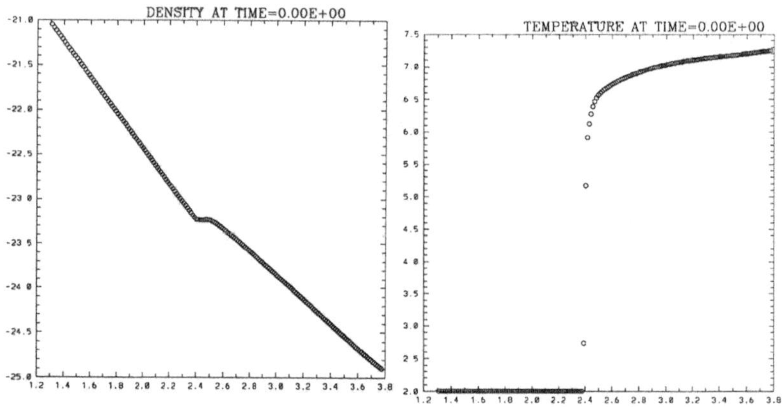

Fig. 1. Run of density (gcm^{-3}) and temperature in the cooling flow atmosphere between 20 and 6000 pc. Scales are logarithmic.

2 Model Constraints

Jet velocity must be at least 0.3c and the emission associated with the jet must expand approximately constantly out to feature A with opening angle of about 6.5 degrees. At feature A the jet must collimate and be capable of side to side oscillation and disruption by about the position of feature C. We do not need to model the fine structure on the jet as it is not relevant to the gross dynamics and morphology. Thus, out to feature A the jet can be modelled numerically as axisymmetric. Much beyond feature A the jet is clearly not axisymmetric. However, an axisymmetric simulation will provide accurate internal conditions in the jet at collimation. From these conditions we can calculate the expected subsequent asymmetric behavior from stability theory (see Hardee & Norman this proceedings).

The cooling flow atmosphere does not provide a pressure gradient consistent with constant expansion of the jet, and this requires that the jet be overpressured - as is observed to be the case. In modelling the jet using a non-relativistic hydrodynamics code we set the jet velocity equal to $0.3c$. We set the temperature at the inlet $T = 8 \times 10^8\ K$ to give a sound speed consistent with the observed opening angle of the jet. Since the inlet in the numerical simulation is not at the center and the jet is not necessarily in the plane of the sky, this temperature is an upper limit. A lower limit to the jet's density at the inlet is set by the observed source luminosity, and by the neccessity for a large overpressure at the inlet relative to the surrounding atmosphere.

3 Numerical Model

The simulation was performed using the two dimensional MHD code ZEUS-2D (Stone and Norman 1992a;b). The cooling flow atmosphere was initialized on a spherical polar grid with 184 geometrically stretched zones in the radial direction spanning a distance from 20 pc to 6000 pc. The grid resolution is 0.8 pc in the radial direction at 20 pc, and

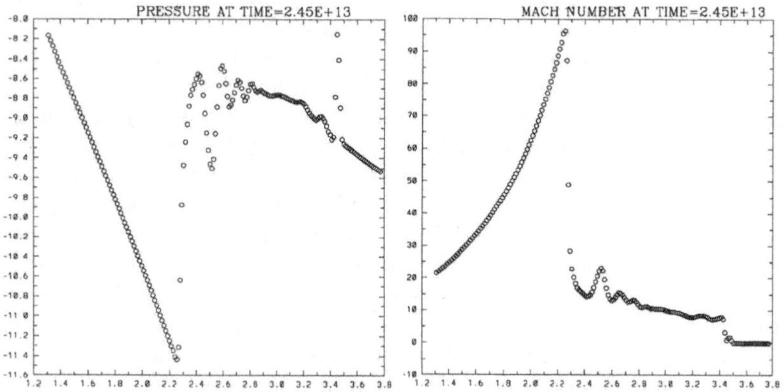

Fig. 2. Pressure ($dyn\ cm^{-2}$) and Mach number along an axial slice from 20 to 6000 pc when the jet front is at 2500 pc. Time is in seconds.

175 pc at 6000 pc. There are 60 zones in the transverse direction, the outer 45 zones stretched geometrically. The inner 15 uniform angular zones with angular size of 3.8×10^{-3} *radian* span the jet's radius at the inlet. At a radial distance of 20 pc the angular resolution on the jet corresponds to a transverse resolution of 0.08 pc. Density and temperature in a cooling inflow atmosphere which has a constant mass flux, *i.e.*, no mass condenses out of the flow, is shown in Fig. 1 between 20 and 6000 pc. The mass flux is about 30 solar masses/year.

In this model there is a sonic radius at 250 pc from the center where the temperature plummets to an imposed 100 K minimum yielding a pressure change by four and one-half orders of magnitude. The inwards flow velocity is approximately 400 $km\, s^{-1}$ within the sonic radius, and the atmospheric density varies as r^{-2}. Outside the sonic radius the temperature slowly rises, the density falls as $r^{-5/4}$, and the thermal pressure falls as r^{-1}. While this atmosphere does not sucessfully model the X-ray surface brightness inside 10 kpc, it provides a limiting case in which the sonic radius is at its maximum extent from the center and the jet front experiences the maximum possible external density. Thus, we can study the effect of the sonic radius on jet dynamics and features when the sonic radius is far from the inlet.

At the inlet the jet is initialized with a density of 6.75×10^{-26} gcm^{-3} for a kinetic luminosity of 2×10^{42} $erg\ s^{-1}$, and with a constant opening angle of 6°.5 This density provides an overpressure at the inlet of about three orders of magnitude, and a kinetic luminosity about five times the source luminosity. This density is chosen because a constantly expanding jet with these initial conditions will stall out at about 1 kpc from the center, *i.e.*, the jet's ram pressure will become comparable to the thermal pressure in the external medium. However, with this choice the constantly expanding jet will be underpressured at the pressure wall associated with the sonic radius. The consequences of this underpressure at the pressure wall are shown in Fig. 2. Beyond the bow shock at 2800 pc, properties are those of the initial cooling flow atmosphere.

The initial overpressure has caused the jet to expand constantly out to the sonic radius. Adiabatic expansion has cooled the jet by one and one-half orders of magnitude, and the Mach number has increased correspondingly. The high Mach number guarantees stability of the jet to non-axisymmetric modes of disruption. At the sonic radius the jet is reconfined, and at larger distance the jet expands slowly following the pressure gradient in the external medium. As a result the jet front does not stall out at 1 kpc. Internal shocks at the sonic radius produce a rise in jet density by about an order of magnitude, and result in a rise in jet temperature by about two orders of magnitude. Outside the sonic radius, the Mach number decreases to about ten. With this Mach number a jet will become unstable to disruption by helical and filamentary modes with a disruption length of about 5 M_jR (see Hardee & Norman this proceedings). Since the jet's radius is about 14 pc at the sonic radius, and the jet continues to expand slowly, the corresponding disruption distance of a fully three dimensional jet is between 1 and 2 kpc from the center and a lobe would form at about this distance.

While the jet does not continue its initial constant expansion the gray-scale temperature image shown in Fig. 3 reveals that the cocoon around the jet does expand approximately constantly out to 1 kpc and the cocoon is hotter than the jet.

4 Implications for Further Modelling

The present model is successful in the sense that the jet would disrupt at about the observed distance if axisymmetry were relaxed, and it is tempting to associate the observed brightening at feature D (about 235 pc from the center) with reconfinement of the jet at the sonic radius. However, the observed constant expansion out to knot A is reproduced only if the observed emission comes from the hot nearly constantly

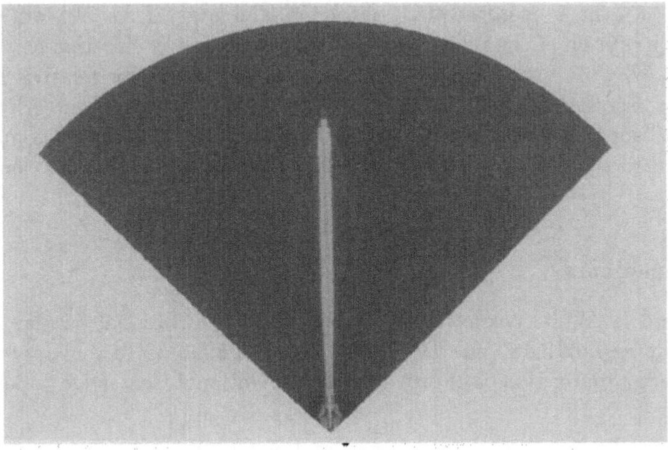

Fig. 3. Grey scale image of the temperature. Darker shades correspond to colder temperature.

expanding cocoon surrounding the colder and narrower jet. This cannot be ruled out on dynamical grounds because the pressure gradient in the ambient medium is such that cocoon material is expected to flow outwards far behind the jet front (see Hardee *et al.* 1992), and this cannot be immediately ruled out by the observations because the emission associated with the jet is observed to be center darkened (Owen *et al.* 1989).

The atmosphere model that best fits the X-ray emission requires that significant mass drop out of the cooling accretion flow inside 10 kpc from the center. This has the effect of moving the sonic radius inwards towards the center. This effect combined with the fact that any orientation of the observed jet out of the plane of the sky moves the true position of feature D outwards makes the association of this feature with the sonic radius unlikely. For the jet to show no observable feature at the sonic radius implies that there be no sonic radius or that the jet be overpressured at the sonic radius. This can be explained most simply by jet overpressure out to feature A. Within the present context this will require an increase in the inlet density by many orders of magnitude if jet expansion to A is adiabatic. In this case the jet will be much more dense than the surrounding atmosphere, and have a kinetic luminosity many orders of magnitude greater than the observed luminosity.

If constant expansion requires that the jet be overpressured out to feature A, collimation at feature A will be relatively gentle. Heating at collimation may be insufficient to reduce the adiabatically cooled jet's Mach number to the small number required for the jet disruption and lobe formation that is observed. If the jet is significantly out of the plane of the sky a higher initial Mach number is required to be consistent with a smaller true opening angle. A partial solution to the problem posed by the stability of a high Mach number jet beyond feature A may be found if the jet is relativistic in the rest frame of the fluid. In this case the sound speed can be $c_s = c/\sqrt{3}$ and remain independent of adiabatic cooling during jet expansion. The opening angle of the jet then requires that the jet flow be highly relativistic with a relativistic Mach number $\gamma_j v_j / \gamma_s c_s \sim 17$ if the jet is in the plane of the sky, and larger if the jet is out of the plane of the sky. Disruption of the relativistic jet will occur after collimation at a distance of about $5(\gamma_j v_j / \gamma_s c_s)$ R - see Königl (1980), and Hardee (1987) for relativistic dynamics and stability. Another potential solution occurs if the jet remains sub-Alfvenic out to feature A and super-Alfvenic beyond feature A. The super-Alfvenic trans-magnetosonic jet is doomed to rapid disruption by instability (see Hardee & Norman these proceedings).

Acknowledgements

P. Hardee and R. White acknowledge support from the National Science Foundation through grant AST-8919180 and EPSCoR grant RII-8996152 to the University of Alabama. Supercomputer time was provided by the National Center for Supercomputing Applications.

References

Biretta, J.A., Owen, F.N., Cornwell, T.J.: 1989, *Astrophys. J.* **342**, 128.
Edge, A.C.: 1989, Ph.D. thesis, Leicester University.
Fabricant, D. and Gorenstein, P.: 1983, *Astrophys. J.* **267**, 535.
Hardee, P.E.: 1987, *Astrophys. J.* **318**, 78.
Hardee, P.E., White, R.E. III, Norman, M.L., Cooper, M.A., Clarke, D.A.: 1992, *Astrophys. J.* **387**, 460.
Hatsukade, I. 1989, Ph.D. Thesis, Osaka University.
Huchra, J.P.: 1985, in *ESO Workshop on the Virgo Cluster*, eds. O.-G. Richter, B. Binggelli, Garching: ESO, p.181.
Königl, A.: 1980, *Phys.Fluids* **23**, 1083.
Owen, F.N., Hardee, P.E., Cornwell, T.J.: 1989, *Astrophys. J.* **340**, 698.
Reid, M.J., Biretta, J.A., Junor, W., Muxlow, T.W.B., Spencer, R.E.: 1989, *Astrophys. J.* **336**, 112.
Sargent, W.L.W., Young, P.J., Boksenberg, A., Shortridge, K., Lynds, C.R., Hartwick, F.D.A.: 1978, *Astrophys. J.* **221**, 731.
Schreier, E.J., Gorenstein, P., Feigelson, E.D.: 1982, *Astrophys. J.* **261**, 42.
Stone, J.M., Norman, M.L.: 1992a, *Astrophys. J. Suppl.* **80**, 753.
——: 1992b, *Astrophys. J. Suppl.* **80**, 791.
White, R.E. III, Sarazin, C.L.: 1988, *Astrophys. J.* **335**, 688.
Young, P.J., Westphal, J.A., Kristian, J., Wilson, C.P., Landauer, F.P.: 1978, *Astrophys. J.* **221**, 721.

The Jet of the Quasar 3C 273

Robin G. Conway [1] and Hermann-Josef Röser [2]

[1]Nuffield Radio Astronomy Laboratories
Jodrell Bank — University of Manchester
Macclesfield, Cheshire SK11 9DL, U.K.

[2]Max-Planck-Institut für Astronomie
Königstuhl 17
D 6900 Heidelberg 1
Germany

Abstract: A review is given of observations of the jet of 3C 273 in all accessible wavebands. In addition we describe some recent results obtained by our study of this object at radio and optical wavelengths. We conclude with some comments about the nature of the jet as deduced from the observations described here.

1. Identification and early observations

The jet of 3C 273 is one of the four optically detectable extragalactic radio jets and the only one from a quasar. It was discovered by Schmidt (1963) during identifications of 3C radio sources as the counterpart of 3C 273A, the stellar quasar core being the counterpart to 3C 273B (Hazard *et al.* 1963). As Schmidt reported, the optical jet coincides in position angle with the line connecting radio components A and B, but is not visible within 11″ of the quasar core and terminates 1″ east of radio component A. With the quasar's redshift of 0.158 the (projected) linear scale is then approximately 40 kpc[2]. In their detailed study of the first identified quasars 3C 273 and 3C 48 Greenstein and Schmidt (1964) also briefly discuss this jet. Their spectrum of its outer end showed a featureless, bluish continuum and they assumed the optical to be of the same origin as the radio emission namely synchrotron radiation from relativistic particles. The study of the jet in various wavebands, particularly to establish the synchrotron nature of the optical emission, has remained a major issue ever since.

Before we describe results in other wavebands we will first summarize the radio data. Here we have — due to modern synthesis instruments — the most detailed information

[2] For $q_0 = 0.05$ and $H_0 = 100\,h\,\mathrm{km\,sec^{-1}\,Mpc^{-1}}$ one arcsecond equals $1.84\,h^{-1}$ kpc.

available. We conclude with a brief account of current ideas about the physics of the jet.

Table 1. Total flux of the jet of 3C 273 at various frequencies. The panels refer to lunar occultation measurements, early interferometric data, VLA- and MERLIN observations and IR, optical and X-ray fluxes.

frequency [MHz]	flux [Jy]	ref.
136.	89.0	1
195.	68.0	1
213.	64.0	1
240.	54.7	3
245.	57.8	4
405.	38.7	4
408.	37.0	2
410.	43.1	1
430.	37.5	1
612.	34.2	3
735.	29.0	4
950.	23.7	11
1420.	16.0	1
2695.	10.7	4
408.	44.0	5, 6
610.	30.3	7
962.	24.4	8
2695.	9.5	5,9
4995.	6.5	10
151.	80.6	12
408.	39.8	12
1465.	16.0	12*
1666.	13.0	12
4385.	5.2	12
14435.	2.3	12
136.3×10^6	$(257 \pm 19) \times 10^{-6}$	13
239.8×10^6	$(149 \pm 15) \times 10^{-6}$	13
347.5×10^6	$(52.2 \pm 2.8) \times 10^{-6}$	14
462.4×10^6	$(35.7 \pm 1.4) \times 10^{-6}$	
663.7×10^6	$(22.1 \pm 0.5) \times 10^{-6}$	
4.84×10^{11}	90×10^{-9}	15

References:[1] Hazard *et al.* (1966), [2] Bailey *et al.* (1964), [3] Conway *et al.* (1964), [4] von Hoerner (1965), [5] Conway, Stannard (1972), [6] Stannard (1967), [7] Wardle (1968), [8] Davis (1974), [9] Conway (1979), [10] Ryle *et al.* (1975), [11] Hughes (1965), [12] Conway *et al.* (1992), [12*] Holmes, Saika, unpublished (quoted in ref. 12), [13] Henry *et al.* (1984), [14] Röser, Meisenheimer (1991), [15] Harris, Stern (1987).

2. Observations

2.1 Radio

Integrated spectrum: Early radio measurements were of relatively low angular resolution. Their value lies in providing accurate flux density measurements at a large number of wavelengths to study the run of the jet's flux with frequency. In contrast to 3C 273B, simultaneous observations are not required here as variability has not been reported (and is also not expected). We plot in Fig. 1 the radio data from Tab. 1 to study the integrated spectrum of the jet. These data are taken mainly from the compilation by Conway *et al.* (1992).

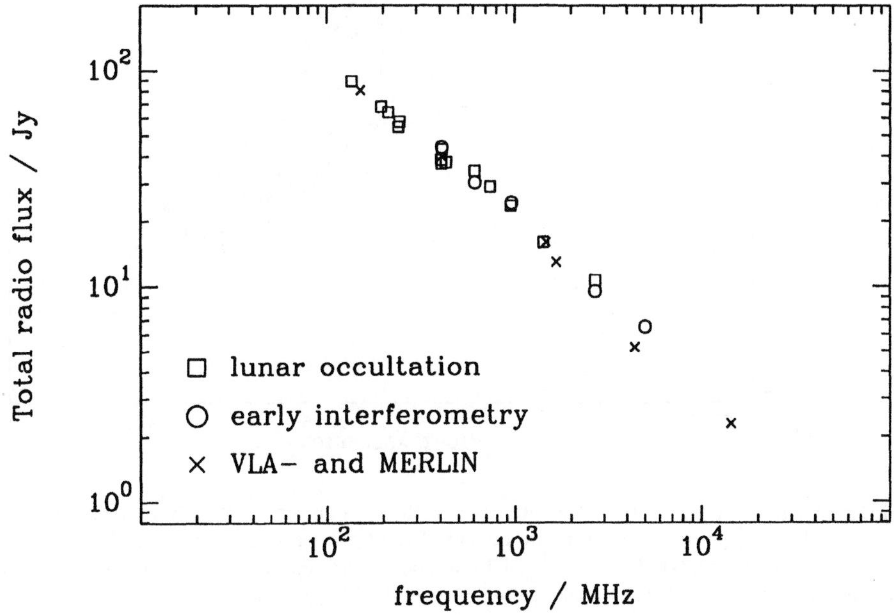

Fig. 1. Integrated radio spectrum of the jet of 3C 273 (see Tab. 1).

The mean spectral index[18] over the range 100 MHz to 5 GHz is $\alpha_r = -0.78$. There is an indication for a slight steepening of the continuum around 1 GHz.

Early polarization measurements:

The first polarization measurements for the jet were performed by Conway and Stannard (1972) at 11 and 73 cm at Greenbank and Jodrell Bank respectively. They found the magnetic field to run parallel to the jet's body in the inner part but turning by 90° at the outer tip. The rotation measure *RM* they determined amounts to 1.5

[18]We define the spectral index α via $f_\nu \propto \nu^\alpha$.

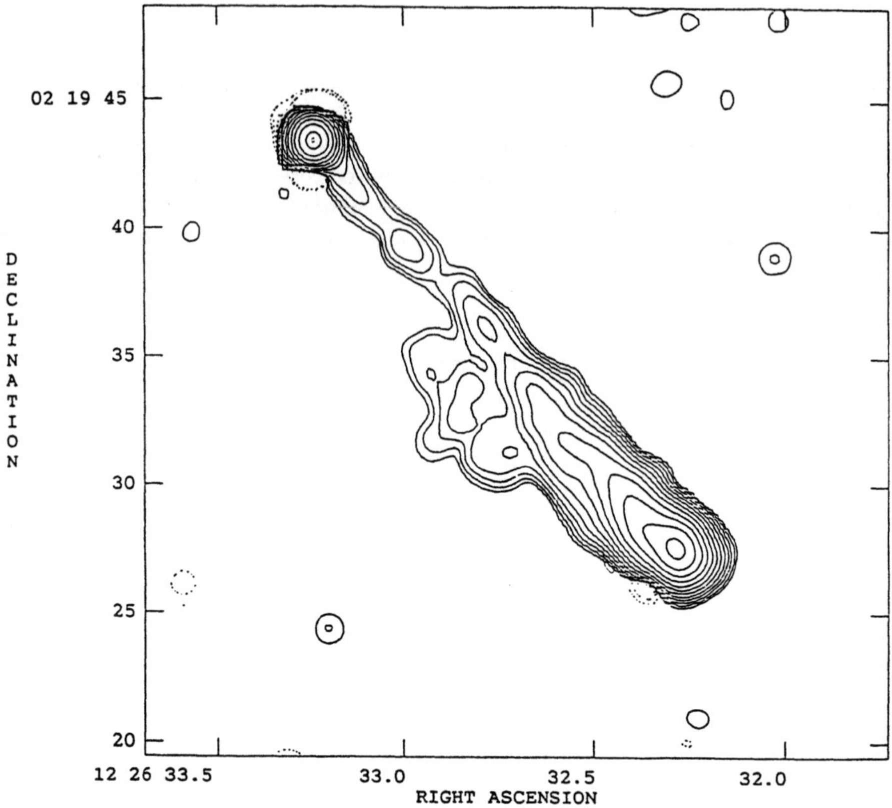

Fig. 2. MERLIN map of 3C 273 at 408 MHz with a resolution of 1″. Component A is the bright end of the jet at lower right, component B the QSO's core at upper left. The "lobe" is conspicuous below center. Peak flux is 13.325 Jy/beam with contour levels at (−10, −5, 5, 10, 20, 40, 80, 160, 320, 640, 1280, 2560, 5120, 10240)× 0.001 Jy/beam).

rad m^{-2}, consistent with the result by Gardner *et al.* (1969) derived from the integral polarization.

Maps: The first attempt to map the intensity distribution of the whole source was by Hughes (1965), who used lunar occultation measurements at different position angles. He deduced a flux ratio between components A and B of 1:0.92 at 950 MHz. Hazard *et al.* (1966) also presented high-resolution lunar occultation studies of this source. These measurements confirmed the coincidence of component A with the faint optical jet detected by Schmidt (1963). The first synthesis observations of the jet were conducted by Conway and Stannard (1975) at various wavelengths between 18 and 74 cm. These measurements showed the now well known steep rise of the intensity distribution towards the outer end of the radio jet.

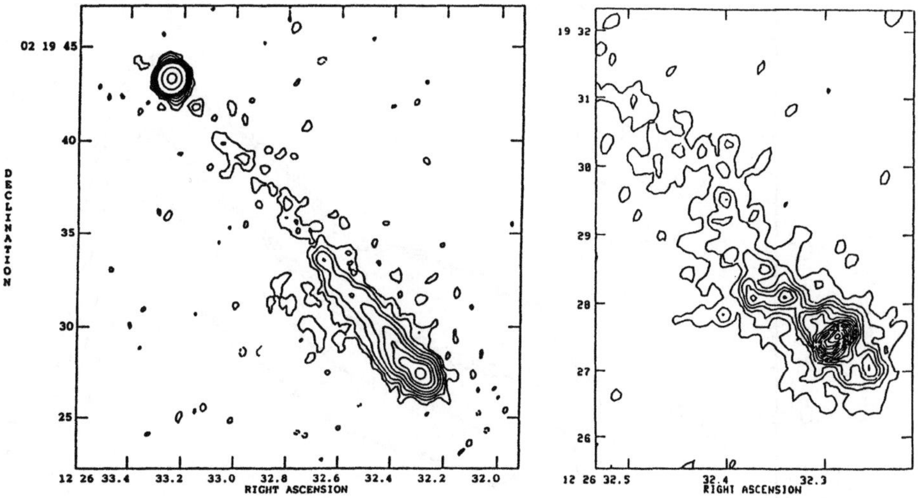

Fig. 3. Map of 3C 273 at 2 cm wavelength. **(left)** VLA B- and C-Arrays. (Resolution $0\farcs6$, peak flux 29.612 Jy/beam, contours at (−1, 1, 2, 4, 8, 15, 30, 50, 100, 1000, 6000) × 0.003 Jy/beam.) **(right)** VLA A-Array. (Resolution $0\farcs22$, peak flux 25.35 Jy/beam, contours at (−1, 1, 2, 3, 4, 5, 6, 7, 8, 9, 10, 11, 12, 13, 14, 15) × 0.006338 Jy/beam.)

With the advent of the VLA and MERLIN high resolution maps could be constructed for the jet, being published *e.g.* by Perley (1981, 1984) and by Conway *et al.* (1981), Foley and Davis (1985), Flatters and Conway (1985), Davis *et al.* (1985). More recent data have been collected by Conway *et al.* (1992) from whom we summarize the radio properties in the following.

2.2.1 Radio properties

The radio maps of 3C 273 show a compact object (the QSO), accompanied by a single narrow, structured jet in $PA\,222°$. We choose as an example the 408 MHz map taken with MERLIN (Fig. 2). There is no sign of a counter-jet, to a ratio of more than 10000:1 (Davis *et al.* 1985). The radio image extends more than $22''$ and terminates with a bright head that at the highest resolution has at least three sub-components (Fig. 3). At each wavelength within the radio regime (2 to 73 cm) the profile of intensity along the ridgeline of the jet rises continuously (Fig. 4), and closely follows an exponential law $\exp(r/1\farcs7)$). The profile is almost independent of wavelength, and the radio spectral index $\alpha_{\rm r}$ lies between −0.6 and −1.0 (Fig. 4). On the southern side of the jet a "lobe" is most prominent at low frequencies due to its steep spectral index ($\alpha \approx -1.5$).

The polarization images (Figs. 5,7) show a relatively simple pattern of magnetic field, parallel to the jet axis up to a distance $r = 20\farcs5$. Thereafter, in the head, the field suddenly "flips" and becomes transverse to the axis. The polarization in the jet itself

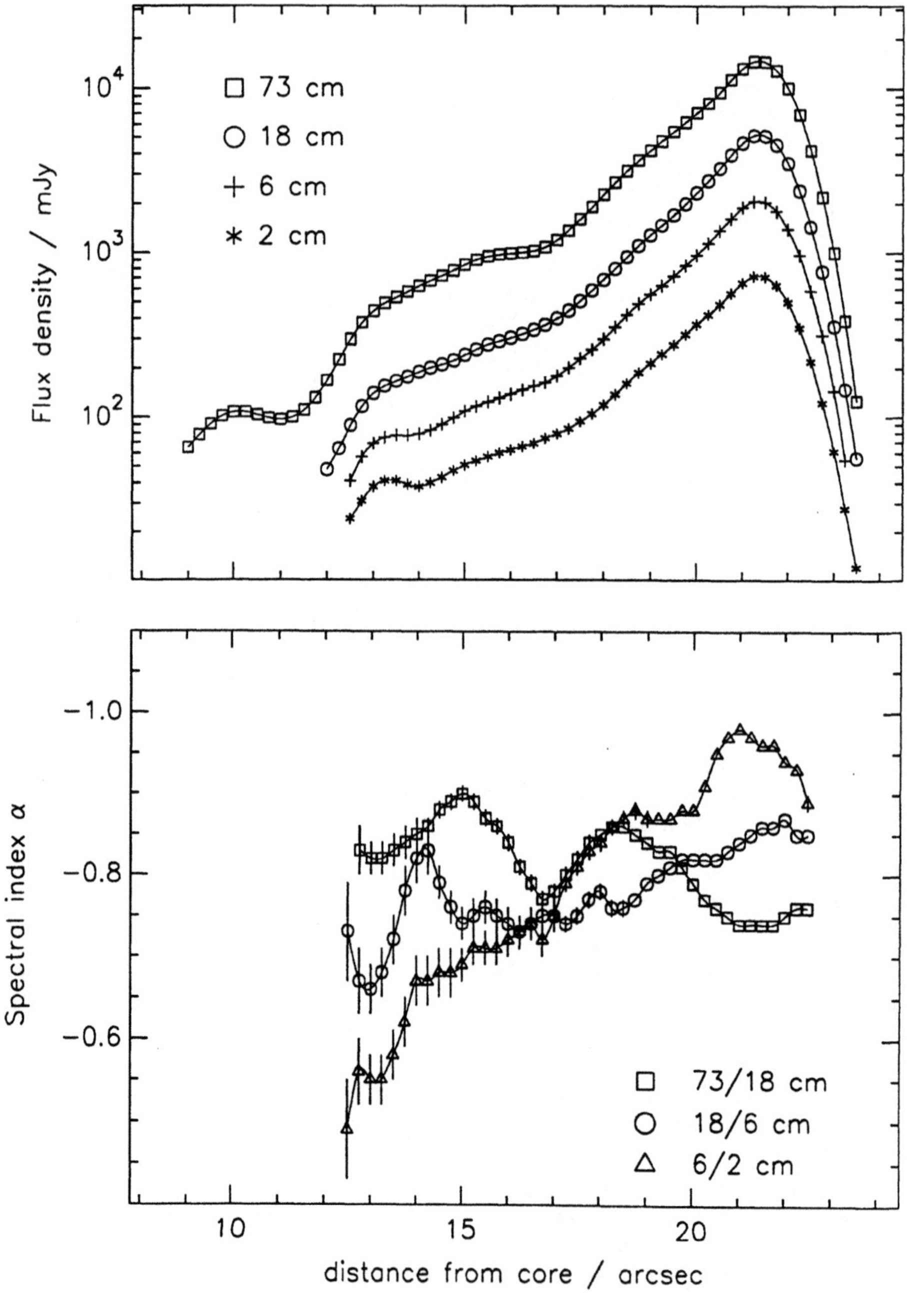

Fig. 4. (top) Intensity profiles along the ridgeline at four radio wavelengths (resolution 1.″20). **(below)** Spectral index variation along the ridgeline.

is of order 10–15% (Fig. 6). The polarization in the head, beyond the "B-flip" is about 25%[19].

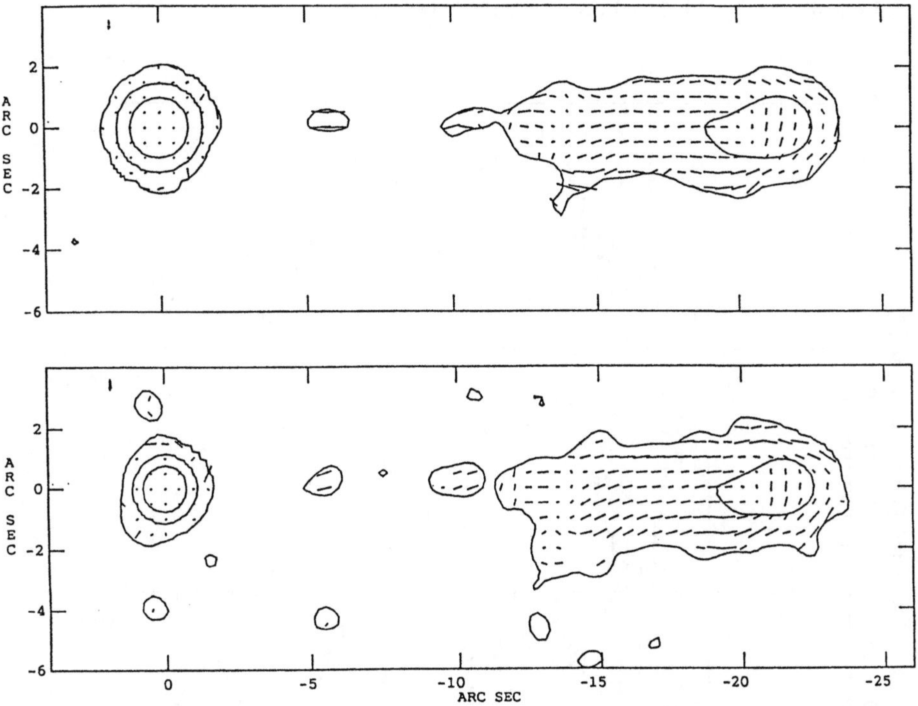

Fig. 5. Polarization images of 3C 273 at 1″ resolution rotated by 47.8° against north. **(top)** 6 cm wavelength **(bottom)** 18 cm wavelength Contours: Intensity (Stokes I); bars show fractional polarization, oriented along the local B-vector (vector of length 1″ corresponds to 50% polarization).

The rotation measure (RM) over the whole source is low, no more than $2 - 4\,\mathrm{rad/m^2}$, in marked contrast to the situation in M 87 (Owen *et al.* 1990), even though the sources are only 10° apart in direction on the sky. In addition to the data shown here, observations at wavelengths of 11 and 31 cm indicate that there are no ambiguities of PA that might falsify the RM. The RM in the head is $1.8\,\mathrm{rad/m^2}$, while that in the jet (at $r = 19''$) is $1.3\,\mathrm{rad/m^2}$, the difference being certainly no more than $0.5\,\mathrm{rad/m^2}$. The lack of any appreciable change in RM at the B-flip, (where the line-of-sight component of B must change substantially) argues that the Faraday rotation is external.

The RM-map (Fig. 7) suggests that there is a small but systematic increase in RM across the jet from south to north, consistent with a toroidal component of field. If

[19] A circular pattern of polarization in the outer parts makes the observed degree of polarization somewhat resolution-dependent, but the true polarization at the head is at least 20%.

the jet carries a longitudinal current $I_{\rm jet}$, Faraday rotation in an ambient medium of density n will give rise to such a pattern, provided that the return current flows outside, rather than inside, the jet boundary. Assuming that n is of order $100\,{\rm m}^{-3}$ (the value expected for a poor cluster such as that surrounding 3C 273), the required current $I_{\rm jet}$ is about 4×10^{15} A, directed positive outwards.

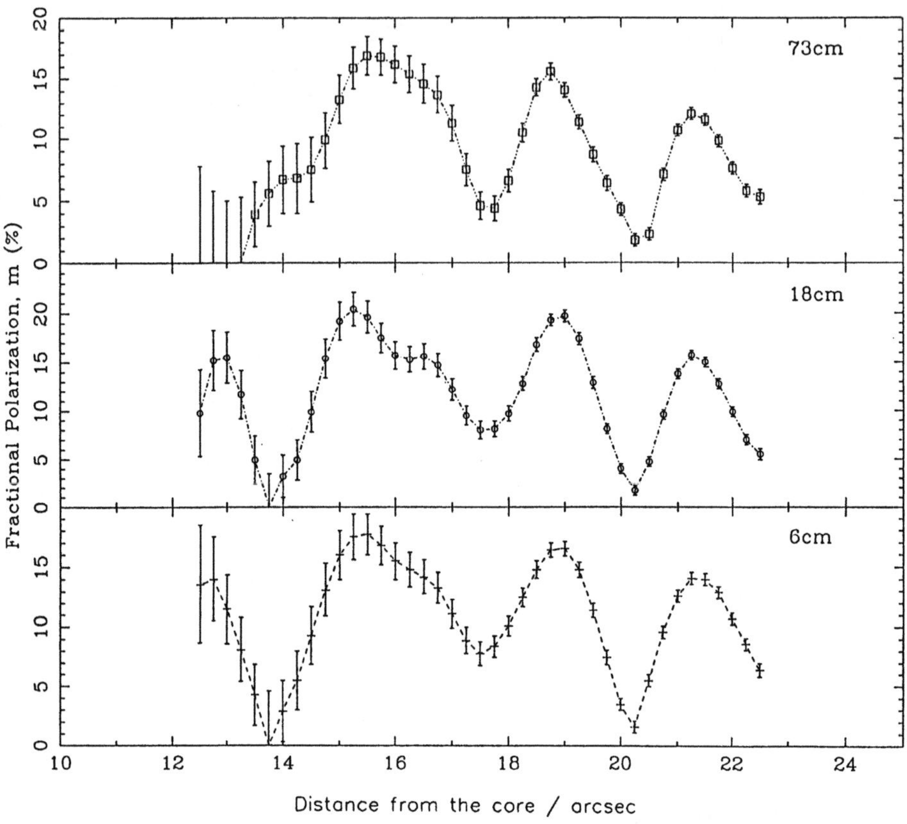

Fig. 6. Fractional polarization along the jet at three radio wavelengths.

2.2.2 VLBI

Results from recent VLBI-experiments for 3C 273 are described in the contributions by Krichbaum and by Zensus in this volume. Other reviews include the contributions by Unwin, Wardle *et al.*, Krichbaum, and by Bååth in the conference proceedings on *"Parsec-Scale Radio Jets"* edited by A. Zensus and T. Pearson (1990).

3C 273 was the first object for which superluminal expansion was demonstrated by a series of hybrid maps (Pearson *et al.* 1981): on the parsec-scale knots of emission seem

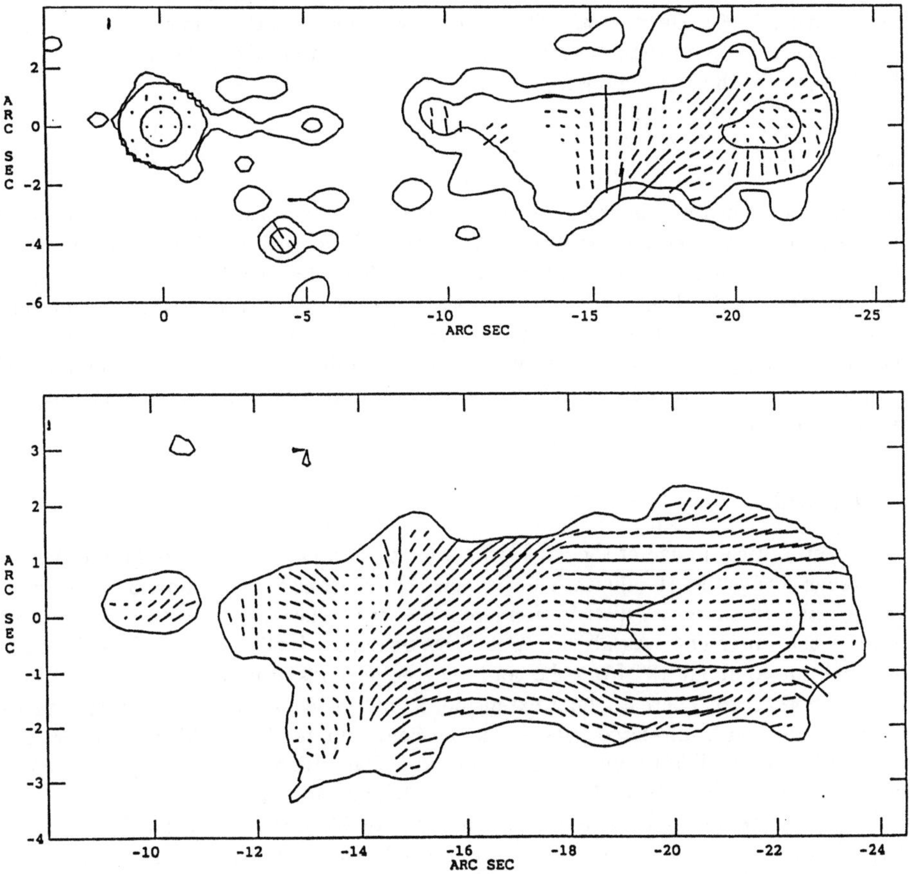

Fig. 7. (top) Polarization image at 73 cm wavelength with contours and bars as in Fig. 5. **(bottom)** Rotation Measure Image. Bars show the *PA* difference (73 cm − 18 cm) due to Faraday rotation.

to emerge from the QSO's core with speeds — as deduced from the cosmological redshift and the rate of separation from the core — of several times the speed of light. It is now generally accepted that this is an illusion explained by effects of special relativity, if relativistic motion occurs almost along the line of sight. It could be demonstrated that the emergence of new (superluminal) knots is accompanied by brightness flares at mm- and optical wavelengths (see contributions by Robson and Litchfield and by Bregman in this volume).

The orientation of the large-scale jet is still unsettled, *i.e.* whether it is directed close to the line of sight as on milliarcsec scales (in which case the overall length is very long), or whether it is bent to a larger angle somewhere beyond the range covered by present

VLBI maps. In this context it should be pointed out that the apparent position angle of the jet close to the core (−128°) is different from the angle on the kpc-scale (−138°, see Zensus *et al.* 1990). Davis *et al.* 1991 have shown, however, that superluminal motion continues out to more than 100 mas from the core and that there the orientation is in the direction of the kpc-scale jet.

2.3 Infrared

The only published measurement of the jet at infrared wavelengths we are aware of is the aperture photometry by Henry *et al.* (1984). They used a 7.″7 aperture in the H and K band to measure the flux at two locations along the jet (13.″8 and 21.″4 from the core, see Tab. 1). As their results were not corrected for contributions from the quasar's host galaxy the true value should be less than the value which they quote, at least at their inner location. A modern observation with a twodimensional array detector would be most helpful to provide reliable values. No infrared polarization measurements are available todate.

2.4 Optical

A deep optical image of the 3C 273 field (Fig. 8) shows the jet stretching out from the stellar core across the quasar's host galaxy. In contrast to what was found at radio wavelengths, the intensity distribution along the jet is relatively flat with a pronounced knot at the outer end. As already noted by Schmidt (1963) no optical radiation is detected out to about 12″ from the core and the jet ends well before the bright radio hot spot is reached. There are faint extensions at both ends. Most likely they are chance superpositions with unrelated galaxies in the field (but see Röser, Meisenheimer 1991 as regards the possible nature of the inner extension). High resolution optical images are presented by Evans *et al.* (1989) and by Hutchings *et al.* (1991).

The first attempt to derive a flux estimate was by Kron *et al.* (1972) and the colours they obtained were long taken as an argument for a synchrotron origin also for the optical radiation (but see Guthrie and Napier 1975). Their observation was, however, hampered by the fact that the telescope's diffraction spike was running along the jet spoiling the measurement. With the advent of CCD-detectors isolation of the jet's flux from its surrounding became much easier (see also Tyson *et al.* 1982). For a summary of more recent optical flux measurements see Röser, Meisenheimer (1991) from which we also quote the optical values in Tab. 1. From these data it is evident that the optical continuum varies in its spectral index along the jet (Fig. 9). In the inner part it is considerably bluer than at the outer end of the optical jet. The very outer tip at the optically faint radio hot spot becomes bluer again.

First optical polarization measurements were attempted by Schmidt *et al.* (1978). They found no significant polarization for the jet's total radiation with a limit of $(3.7 \pm 4.1)\%$. Arp (1981) could also not detect significant polarization. First CCD measurements by Röser and Meisenheimer (1986) finally demonstrated that at least the outer end of the optical jet is substantially polarized and that here the polarization is in accordance with the radio polarization. This proved that the optical radiation is also of synchrotron origin. Polarization of the inner part of the visible jet remained unclear.

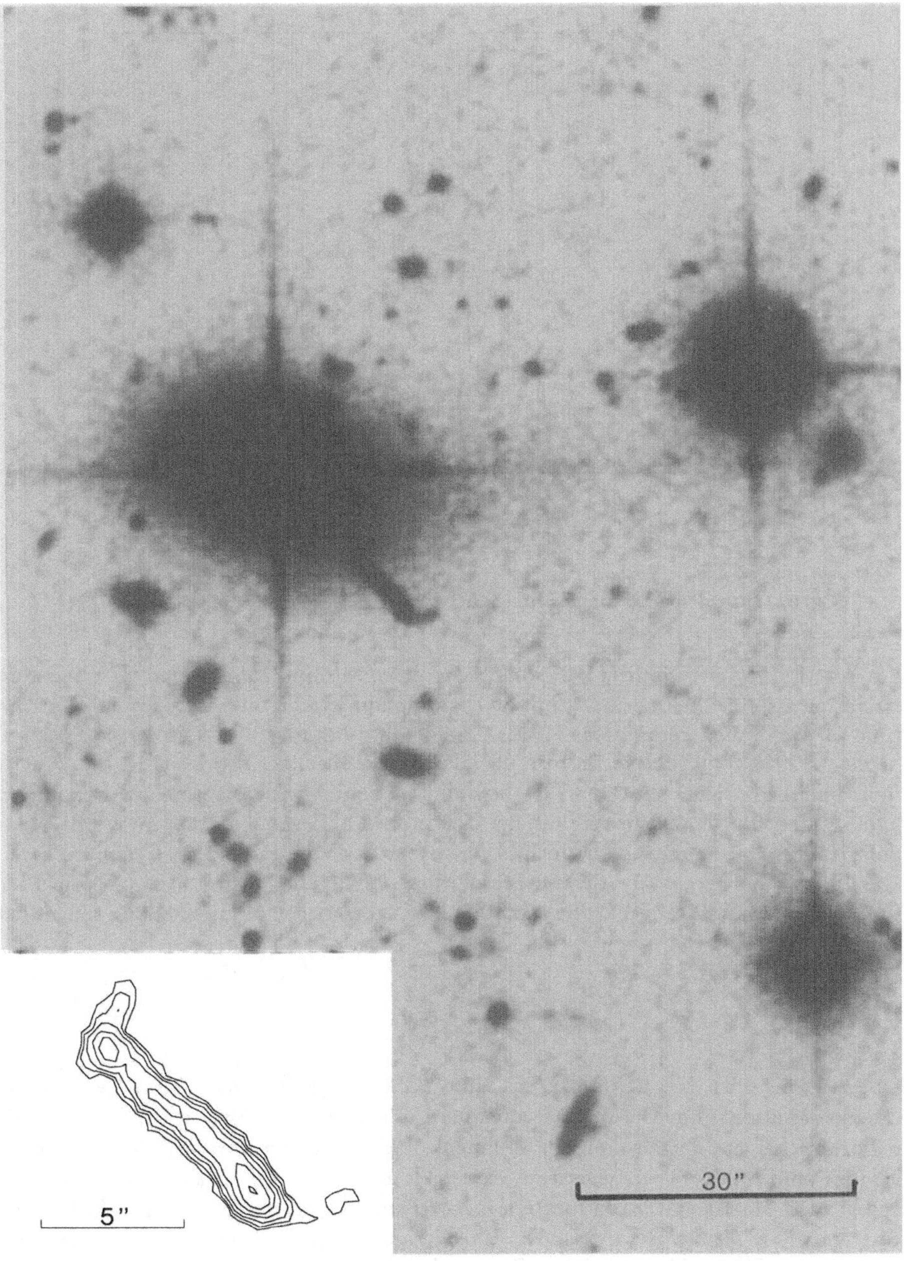

Fig. 8. The optical field of 3C 273 showing the jet and the quasar's host galaxy. Insert shows jet after subtraction of the local background (contour intervalls increase by $\sqrt{2}$). Note the two extensions at both ends, which are not present on the radio map (Fig. 2). North is up and east to the left; scales have lengths of 5″ and 30″ respectively.

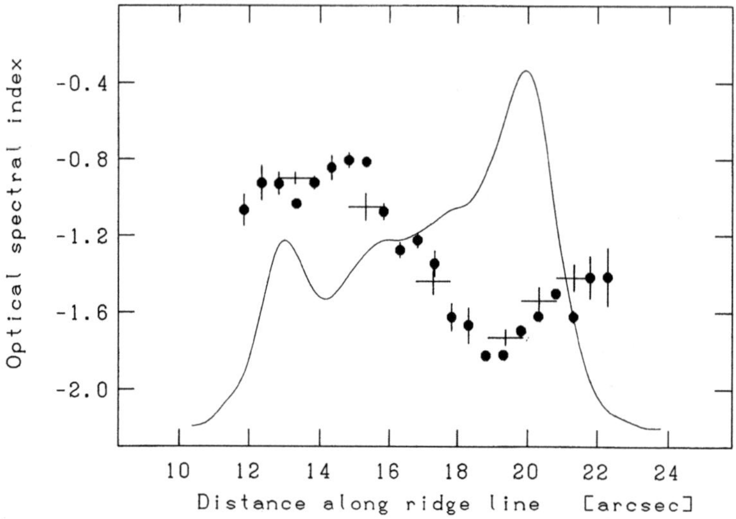

Fig. 9. Optical spectral index along the jet of 3C 273 (Röser, Meisenheimer 1991).

Scarrott and Warren-Smith (1987) and Scarrott and Rolph (1989) found high linear polarization also in the inner part, but no corresponding radio data were available at that time for a detailed comparison. With considerably improved data Röser and Meisenheimer (1991) could derive the run of polarization all along the jet (Fig. 10). For the first time the optical counterpart to the radio hot spot was clearly detected in polarized light by the flip in polarization angle. These data also show a drop in the degree of polarization at the inner end which is consistent with the radio data (compare Figs. 6 and 10 but be aware of the different resolution!). A detailed comparison of optical and radio polarization shows that the radiation is of synchrotron origin all along the jet (see below).

2.5 X-rays

Observations with the EINSTEIN satellite (Giacconi 1980) have shown that besides strong radiation from the quasar core there is also X-ray emission coming from the region of the jet (Feigelson 1980, Willingale 1981, Harris and Stern 1987). Its exact location and flux level are, however, extremely uncertain due to the proximity of the bright core. Harris and Stern locate the X-ray emission at a distance of $15''\!.8$ from the core with $f_\nu = 90\,\mathrm{nJy}$. Here we expect considerable progress once our ROSAT-HRI data are reduced (Röser, Meisenheimer, Conway in prep.).

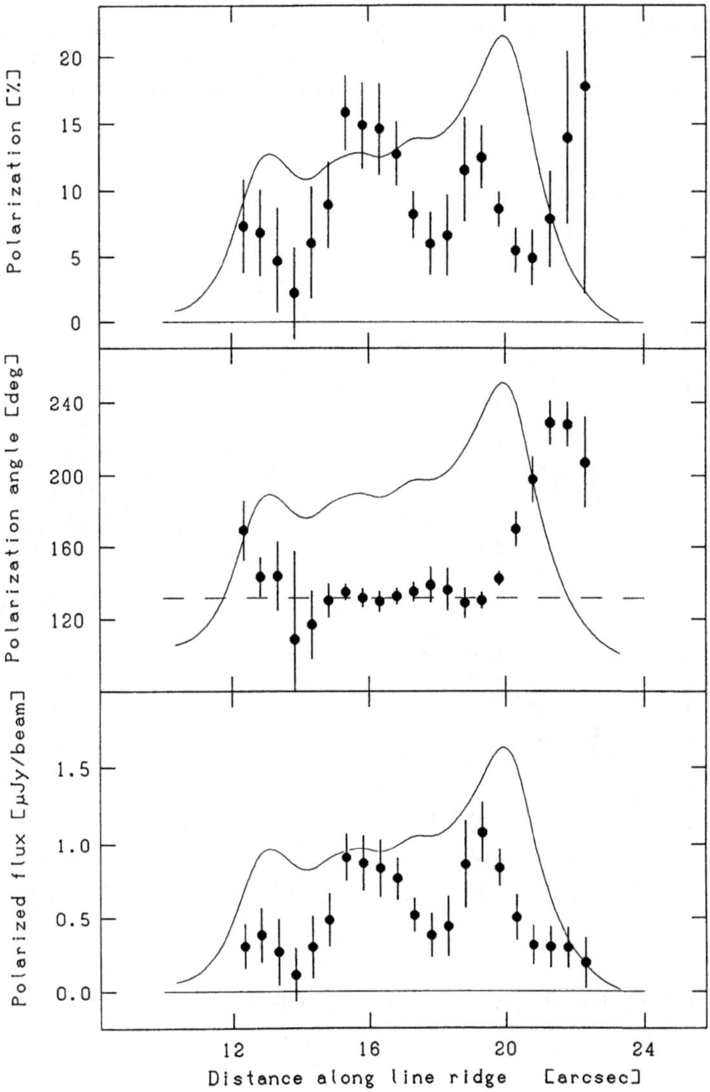

Fig. 10. Optical polarization along the jet of 3C 273 (Röser, Meisenheimer 1991).

3 Radio / optical comparison

We are currently preparing a paper on the comparison of optical data (Röser, Meisenheimer 1991) with the radio data (Conway *et al.* 1992) for the jet of 3C 273. Some preliminary results of these studies are given in the following paragraphs.

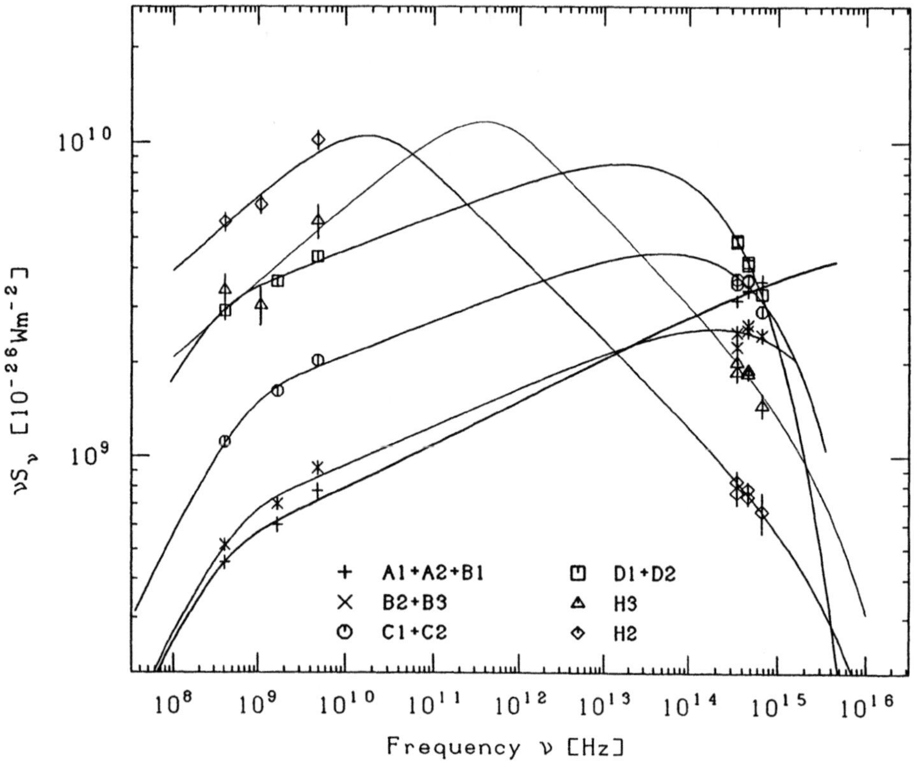

Fig. 11. Overall spectra of various components in the jet of 3C 273.

3.1 Overall Spectra of Jet

Röser and Meisenheimer (1991) have decomposed the intensity along the ridgeline of the jet into a series of Gaussians, whose positions were fixed at the locations of knots on high-resolution radio maps. This procedure was also applied to the radio data as published by Conway *et al.* (1991). Thus, we now have the possibility to use this decomposition to derive the overall spectra of the individual knots at a common resolution[20]. Synchrotron

[20]The spectra are preliminary and thus subject to changes.

cutoff spectra were fitted to these datapoints following the procedure by Meisenheimer *et al.* (1989) (Fig. 11).

Several features deserve attension. The continua of knots along the central body of the jet (B, C and D) are very similar to one another. The spectrum of the hot spot counterpart (H2) is, however, considerably different in that its radio-to-optical spectral index is much steeper, resulting in the optically weak counterpart. The spectrum of the innermost part (A) is blue with a constant spectral index from the radio to beyond the optical. Extrapolation of this continuum is consistent with the X-ray-emission originating from this region.

3.2 The lobe and the "backflow"

In general, radio and optical polarization agree very well with each other. This proves that the optical radiation is also of synchrotron origin all along the jet. However, at a distance of 17.″5 to 19.″0 from the core, the radio polarization is considerably higher than the optical polarization (comparison is based on maps brought to a common resolution of 1.″6). We attribute this to a region of increased radio polarization at the southern rim of the jet falling into the beam (for a map of radio polarization at 1″ resolution see Conway 1987). This region is obviously optically faint. Its existance is also exemplified by crosscuts through the jet at various frequencies in the radio and the optical (common resolution 1.″3) in Fig. 12: beyond the optical knot at 19.″8 from the core all crosscuts coincide within the error. Starting here a southern wing of radio emission gradually emerges from the jet's main body until the lobe is reached at 14.″8 from the core. The steepening of the lobe emission for decreasing r is also evident from these crosscuts. This highly polarized, steep-spectrum radio component may be connected with a backflow from the terminal shock found by hydrodynamic simulations (see *e.g.* Norman and Balsara, this volume and references therein).

4. Interpretation

The jet of 3C 273 shows many of the expected features of the canonical picture of a radio source (see *e.g.* Begelman, Blandford and Rees 1984), and if the effects of Doppler beaming could be taken into account, one could estimate directly from observation the values of various physical parameters and their variation along the jet.

This behaviour in 3C 273 differs from the usual scenario, whereby particles are accelerated to relativistic energies in the head, and subsequently flow into a surrounding cocoon. Rather it appears that the synchrotron-emitting plasma must be excited in situ, especially the particles which radiate at optical frequencies. Plasma emanating from the region of the optical knot forms an optically faint (or quiet) backflow, as predicted by numerical simulations of jets.

The superluminal motion in the core shows that matter enters the jet at a relativistic velocity along a direction close to the line of sight, and continues to do so out to kpc scales (Davis *et al.* 1991). If the principle deceleration occurs at a terminal jet-shock, near the head, this could be identified with the B-flip at $r = 20.''5$, since downstream from such a shock the magnetic field will become transverse from compression (Cawthorne

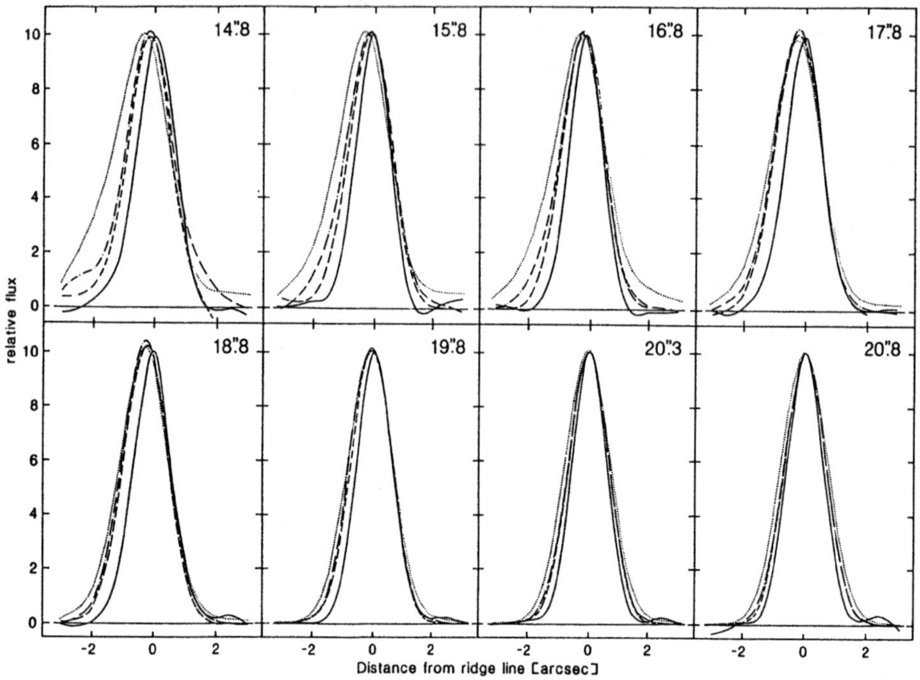

Fig. 12. Normalized crosscuts at 1″.3 resolution perpendicular to the jet of 3C 273 at various radio frequencies and in the optical (full line). Distance to the quasar core is given in the upper right corner for each panel. The southern wing we attribute to the backflow is evident for $r \leq 19''.8$.

and Wardle 1988). The pressure in the head, which can be determined from the radio brightness, is governed by the density of the ambient intergalactic medium and the velocity of the bow-shock surrounding the head. If the velocity is everywhere high (greater than 0.92 c), the absence of a counter-jet can be explained by Doppler beaming alone.

Unfortunately, the hypothesis of a high-velocity head proves to be self-contradictory. Because 3C 273 is part of a small cluster (Stockton 1980) the ambient density cannot be much below $100\,\mathrm{m}^{-3}$, and if the bowshock were faster than about $0.3\,c$ the pressure in the head, and hence the brightness, would be much greater than observed (Conway *et al.* 1981). Any model in which the jet-shock forming the head rapidly overtakes the bowshock could represent only a transitory situation, lasting much less than the lifetime of the source.

As Flatters and Conway (1985) and Meisenheimer and Heavens (1986) have discussed, the polarization of the radio head (H2) resembles that of a compression sheet, but requires that the sheet be orientated at about 45° to the line of sight (Laing 1981).

This is incompatible with the orientation deduced from the superluminal motion (see *e.g.* Zensus, this volume). Flatters and Conway tried to resolve the problem by invoking relativistic aberration in the head. Again, the velocity required is high, and the argument given above makes this unlikely. The aberration interpretation is, therefore, no longer sustained. Instead, we believe either that the jet must have been bent before reaching H2, or that H2 is not in the form of a sheet normal to the jet axis.

In either case, if the velocity of the head is not more than 0.3 c, the Doppler shift will be small, and in particular cannot explain the absence of a counter-jet. It follows that a completely symmetric source can be ruled out by the observations, since the counterpart to H2 would easily be detected. In 3C 273 the jet asymmetry must be basically intrinsic — there is indeed no counter-jet.

References:

Arp, H.: 1981, in *Optical Jets in Galaxies*, Proceedings of the second ESO/ESA workshop, Munich 1981, ESA SP-162 (March 1981), p. 53.

Begelman, M.C., Blandford, R.D., Rees, M.: 1984, *Rev. Mod. Phys.* **56**, 255.

Bignami, G.F. *et al.*: 1981, *Astron. Astrophys.* **93**, 71.

Cawthorne, T.V., Wardle, J.F.C.: 1988, *Astrophys. J.* **332**, 696.

Conway, R.G.: 1987, in *Magnetic Fields and Extragalactic Objects*, proceedings of Cargèse workshop, ed. E. Asseo, D. Grésillon, Editions de Physique.

Conway, R.G., Davis, R.J., Foley, A.R., Ray, T.P.: 1981, *Nature* **294**, 540.

Conway, R.G., Stannard, D.: 1972, *Nature Physical Science* **239**, 22.

Conway, R.G., Stannard, D.: 1975, *Nature* **255**, 310.

Conway, R.G., Garrington, S.T., Perley, R.A., Biretta, J.A.: 1993, *Astron. Astrophys.* **267**, 347.

Davis, R.J., Muxlow, T.W.B., Conway, R.G.: 1985, *Nature* **318**, 343.

Davis, R.J., Unwin, S.C., Muxlow, T.W.B.: 1991, *Nature* **354**, 374.

Evans, I.N., Ford, H.C., Hui, X.: 1989, *Astrophys. J.* **347**, 68.

Feigelson, E.D.: 1980, PhD Thesis, Harvard University, Cambridge, MA.

Flatters, C., Conway, R.G.: 1985, *Nature* **314**, 425.

Foley, A.R., Davis, R.J.: 1985, *Mon. Not. R. astr. Soc.* **216**, 679.

Fraix-Burnet, D., Nieto, J.-L.: 1988, *Astron. Astrophys.* **198**, 87.

Fraix-Burnet, D., Nieto, J.-L., Roques, S.: 1989, *Astron. Astrophys.* **217**, 387.

Gardner, F.F., Morris, D., Whiteoak, J.B.: 1969, *Aust.J.Phys.* **22**, 79.

Greenstein, J.L, Schmidt, M.: 1964, *Astrophys. J.* **140**, 1.

Guthrie, B.N.G., Napier W.McD.: 1975, *Mon. Not. R. astr. Soc.* **172**, 85.

Harris, D.E., Stern, C.P.: 1987, *Astrophys. J.* **313**, 136.

Hazard, C., Gulkis, S., Bray, A.D.: 1966, *Nature* **210**, 888.

Hazard, C., Gulkis, S., Bray, A.D.: 1966, *Nature* **212**, 461.

Hazard, C., Mackey, M.B., Shimmins, A.J.: 1963, *Nature* **197**, 1037.

Heavens, A.F., Meisenheimer, K.: 1987, *Mon. Not. R. astr. Soc.* **225**, 335.

Henry, J.P., Becklin, E.E., Telesco, C.M.: 1984, *Astrophys. J.* **280**, 98.

Hughes, M.P.: 1965, *Nature* **207**, 178.

Hutchings, J.B., Neff, S.G.: 1991, *Publ. Astron. Soc. Pacific* **103**, 26.

Keel, W.C.: 1988, *Astrophys. J.* **329**, 532.

Kössl, D.: 1988, PhD thesis, Technische Universität München.

Kron, G.E., Ables, H.D., Hewitt, A.V.: 1972, *Publ. Astron. Soc. Pacific* **84**, 303.

Lelièvre, G., Nieto, J.-L., Horville, D., Renard, L., Servan, B.: 1984, *Astron. Astrophys.* **138**, 49
Meisenheimer, K., Heavens, A.F.: 1986, *Nature* **323**, 419.
Morrison, P., Roberts, D., Sadun, A.: 1984, *Astrophys. J.* **280**, 483.
Morrison, P., Sadun, A.: 1992, *Mon. Not. R. astr. Soc.* **254**, 488.
Perley, R.A.: 1981, in *Proceedings of Second ESO/ESA Workshop on "Optical Jets in Galaxies"*, ESA SP-162, March 1981.
Perley, R.A.: 1984, in *"VLBI and Compact Radio Sources"*,, ed. R. Fanti *et al.*, IAU Symposium 110, p. 153.
Owen, F.N., Eilek, J.A., Keel, W.C.: 1990, *Astrophys. J.* **362**, 449.
Röser, H.-J., Meisenheimer, K.: 1986, *Astron. Astrophys.* **154**, 15.
Röser, H.-J., Meisenheimer, K.: 1991, *Astron. Astrophys.* **252**, 458.
Scarrott, S.M., Warren-Smith, R.F.: 1987, *Mon. Not. R. astr. Soc.* **228**, 35P.
Scarrott, S.M., Rolph, C.D.: 1989, *Mon. Not. R. astr. Soc.* **238**, 349.
Schmidt, G.D., Peterson, B.M., Beaver, E.A.: 1978, *Astrophys. J. (Lett.)* **220**, L31.
Schmidt, M.: 1963, *Nature* **197**, 1040.
Stockton, A.: 1980, in IAU-Symp. 92, "Objects of High Redshift", ed. G.O. Abell, P.J.E. Peebles, Dordrecht: Reidel, p. 89.
Tyson, J.A., Baum, W.A., Kreidl, T.: 1982, *Astrophys. J. (Lett.)* **257**, L1.
Unwin, S.C. *et al.*: 1985, *Astrophys. J.* **289**, 109.
Willingale, R.: 1981, *Mon. Not. R. astr. Soc.* **194**, 359.
Zensus, J.A., Pearson, T.J. (eds.): 1990, "Parsec-Scale Radio Jets", Cambridge University Press, Cambridge.
Zensus, J.A., Unwin, S.C., Cohen, M.H., Biretta, J.A.: 1990, *Astron. J.* **100**, 1777.

Spectral Ageing in Cygnus A

R.A. Perley [1], C.L. Carilli [1], and J.P. Leahy [2]

[1]National Radio Astronomy Observatory
Socorro, N.M. 87801, USA
[2]Nuffield Radio Astronomy Laboratories
Jodrell Bank
England

1 Observational Consequences of Spectral Ageing

The high-frequency downward curvature observed in the spectrum of many extragalactic radio sources has long been interpreted as a consequence of synchrotron ageing, whereby an isolated cloud of particles develops a bend in its spectrum at $\nu_B \sim B^{-3}t^{-2}$, where ν_B is the break frequency in GHz, B is the magnetic field in μGauss, and t is the age in Gyr. Below the break frequency, the spectrum remains power-law at the original index, while above the break frequency, the spectrum dramatically steepens. Over time, the break frequency decreases, allowing an estimate of the age of the source, presuming the magnetic field is known. This simple relationship has been used for many years to give estimates of the ages of synchrotron-emitting regions. However, interpretation of the results has been clouded by a number of complicating factors — our lack of knowledge of the magnetic fields, the poor resolution and sensitivity of our instruments, and the effects of complicating physics upon the spectral curvature. But, in fact, the effect of different physical processes is to subtly change the details of the spectrum around and above the break frequency. Accurate measures of the spectrum in these regions will give us unique information concerning the detailed physics of these regions.

This report summarizes work recently published by Carilli *et al.* (1991), reporting the results of a full VLA observational program on the prominent radio galaxy Cygnus A. This source, which has long been known to contain strong spectral curvature, is ideal for a detailed study of its spectrum, as it is both of large angular extent and high surface brightness.

1.1 Physical Processes and Spectral Curvature

The inclusion of more detailed physics into the basic model modifies the results only slightly – the interpretation of the break frequency is now a function of the processes and, most interestingly, the behaviour of the spectrum *above* the break frequency depends on the processes involved. We illustrate this point with some simple examples:

Consider first an isolated cloud into which a continuous injection of particles has been occuring for time t. The only energy loss mechanism considered is radiation, and assume that the source is unresolved. Under this scenario, the original power law spectrum of index α steepens to second power law spectrum of index $\alpha - 1/2$ above ν_B. This characteristic spectrum is abbreviated *CI*, for Continuous Injection.

A closely related model, resulting in an identical spectrum, is that of a hotspot with surrounding outflow, which is observed with a resolution which includes both the hot spot and the surrounding gas. In this case, a steady-state condition is set up, with particle input at the hot spot offset by particle outflow at the edge of the beam. The 'age' estimated by spectral curvature now refers to the particles at the 'edge' of the beam, and this can be used to derive the outflow velocity. (No account of adiabatic losses is made in this model, which assumes that particles lose energy only by radiation.)

The CI model is applicable only to unresolved sources, while in fact modern interferometers are constructed to resolve them. Consider an isolated cloud of gas with an initially isotropic distribution of pitch angles, into which there has been no particle input for time t, no reacceleration, and no pitch-angle scattering. In this scenario, the injection spectral index of value α becomes $4/3\alpha - 1$ above the spectral break. The break decreases with time, as before, following the same relationship. The form of spectrum is labelled *KP*, for Kardeshev-Pacholczyk.

Finally, consider the above situation, but allow continuous pitch-angle scattering of the radiating particles, resulting in an effective high-energy cutoff of the energy spectrum. The scattering mechanism can be collisions from photons (inverse Compton scattering) or from magnetic irregularities. The result is that the spectrum above the break frequency becomes exponential, with the same decline of break frequency with time.

Other processes are probably important in the lobes of radio sources. Continuing acceleration will retard the decline of the break frequency, causing the source radiation age to be too small, while adiabatic losses will lower the break frequency (and simultaneously decrease the intensity), giving a radiation age greater than the source age. These processes are probably important in the vicinities of the "hot spots", but are much less likely to occur in the extended lobes. Since we are here concerned with the lobes, we are not concerned about leaving out these processes from the model.

1.3 VLA Observations of Cygnus A

The spectral evolution of Cygnus A has been the object of numerous studies, notably that of Winter *et al.* (1980), and Alexander *et al.* (1984). Both of these studies were limited in their spectral coverage and spatial resolution. The unique feature of our study is the greatly increased sensitivity, resolution, and spectral coverage, allowing spectral fitting sensitive to the shape of the spectrum *above* the spectral break. To accomplish this, we observed Cygnus A in all six VLA observing bands with all VLA configurations. These data were used to construct intensity images, and theoretical spectra, appropriate for the cases discussed above, were fitted. The results of the spectral fitting gave the injection index (spectral index below the break), the resulting index (above the break), the break frequency, and the intensity scaling.

2 The Results from Cygnus A

2.1 The Hot Spots

Theoretical spectra were fitted to the observed flux densities of the A and D "hot spots" as seen with a 4.″5 beam. Nine frequencies (including non-VLA data from high-frequency

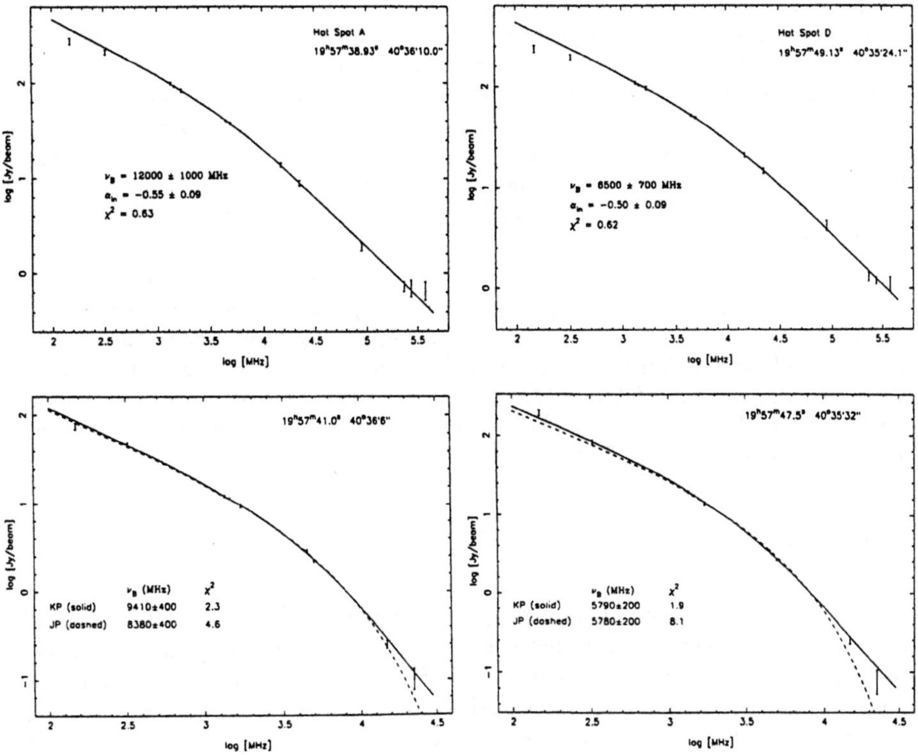

Fig. 1. The results of spectral fitting to several positions in the hot spots and lobes of Cygnus A. The upper pair of spectra show the fit of the CI form to the two main hot spots. The lower pair shows the JP and KP forms for two points in the lobes.

instruments) were used. The results show an excellent fit to the CI form, with break frequencies near 10 GHz, with an injection index for each hotspot of −0.55, in excellent agreement with the theory of non-relativistic shock acceleration. A low-frequency roll-off is also observed, which we interpret as a low-frequency cutoff in the electron energy spectrum, corresponding to a Lorentz factor of $\gamma \sim 450$. Using the 'equipartition' magnetic field estimate of $\sim 300\,\mu$G, and $s = 2.5$ kpc (corresponding to the beamsize at the redshift of Cyg A), a flow velocity of 0.08c is deduced which, when coupled with the presumed strong shock at the hot spot, indicates a jet velocity of 0.3c.

2.2 The Lobes

We fitted the simple JP and KP spectra described above, corresponding to the presence, or absence, or pitch-angle scattering, to the observed spectrum as seen with a $4\farcs5$ beam. Current opinion favors the presence of pitch-angle scattering, so that an exponentially decaying spectrum was expected above the break frequency.

The results of spectral fitting to the $4\farcs5$ images shows a strong and significant preference for the KP model, indicating a *lack* of pitch-angle scattering. Several examples are shown in Fig. 1. To confirm that this result is not due to blending of different JP spectra across the beam, we fitted spectra with higher resolution data where sufficient signal was present. The results were again the same, increasing our confidence that the spectra above the break are not exponentially falling. Furthermore, there is no evidence for a trend to JP spectral forms near the edge of the source, where blending due to different spectra along the line of sight is less. A second result from the spectral fitting is that the injection index is -0.7 throughout the lobes, in contrast to the value of -0.55 found in the hot spots.

The break frequency, ν_B, declines uniformly across the source, from greater than 50 GHz near the hot spots to 750 MHz in the small 'plume' rising above and below the central regions. The filaments show somewhat higher break frequencies, although this result's significance is hard to estimate. Using a representative value of the magnetic field of $50\,\mu$G, the age of Cygnus A is found to be 6 Myr, in good agreement with previous work. The 'age map' of the source is shown in Fig. 2. The average expansion velocity of the lobes is then found to be 0.06c.

3 Discussion

The expansion velocity found through spectral analysis is much higher than velocities determined from rough pressure balance arguments: the dynamic pressure balance at the lobes gives an expansion velocity $v_{exp} \sim (P/\rho)^{1/2}$ of ~ 0.005c, while even the high pressures inferred at the hot spots themselves require an expansion of only ~ 0.02c. (In these estimates, the pressures are derived from minimum energy arguments, while the mass density comes from analysis of X-ray emission). The simplest explanation for the discrepancy is to deduce the material velocity to be mainly backflow. However, backflow beyond what is necessary to fill the lobes must end up somewhere, and there is neither extensive plumes nor the bright central emission which would be expected to be the signature of backflow extending to the central regions of the source.

A more attractive option is to assume that the derived separation flow speed is too high, so that the source is older than indicated. This could conceivably be due to continuing acceleration throughout the lobes, or it could be a result of wrongly estimating

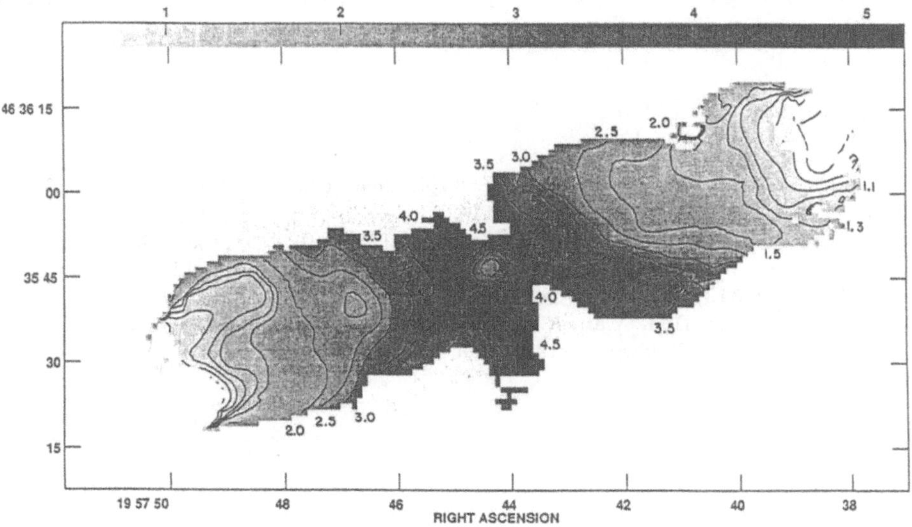

Fig. 2. The apparent age of the emitting regions of Cygnus A, using the derived break frequencies with equipartition magnetic fields. Note the curvature in the lines of equal age, indicating apparently older gas near the boundaries of the lobes

the value of the magnetic field. This latter explanation is especially attractive, as lowering the field will simultaneously lower the separation velocity (which is proportional to $B^{3/2}$), and raise the lobe advance velocity (which is proportional to $B^{-3/4}$). The rough age estimates from lobe expansion and flow velocity can then be made equal by lowering the field below equipartition by a factor of three. This factor is the same as that by Feigelson *et al.* (1987) for Virgo A, based on analysis of X-ray emission. Thus, the indication is that the expansion velocity of Cygnus A is close to 0.01c, with an age of 30 million years.

Two other results deserve mention. First, the break frequency is found to be higher on the edges of Cygnus A, indicating either that the material there is older than that near the middle, or that the magnetic fields are higher in these regions. The former effect could be due to the detailed flow back from the hot spots, which may primarily flow back along the axis, then towards the edges, while the latter explanation requires an enhancement of about 60% above the mean magnetic field, perhaps due to an edge interaction effect. Second, the derived injection index in the lobes of −0.7 is in disagreement with that observed in the hot spots. There is no easy explanation for this. It is possible that for this source, the effect of adiabatic expansion in the flow immediately behind the hot spots is just right to translate the broad curve of the spectrum of the gas in these regions to frequencies in the 100 MHz to 1 GHz range. Lower frequency observations must then see the rollover to flatter spectra. However, no evidence for this is seen, and furthermore, the fact that the injection index for all radio sources in general is in the range of −0.8

to −0.7 indicates that some other, unknown physical mechanism, operating just behind the hot spot acceleration region, is responsible for modifying the energy spectrum to give the observed steeper spectra.

References

Alexander, P., Brown, M.T., and Scott, P.F.: 1984, *Mon. Not. R. astr. Soc.* **209**, 851.
Burbidge, G.R.: 1956, *Astrophys. J.* **124**, 416.
Carilli, C.L., Perley, R.A., Dreher, J.W., and Leahy, J.P.: 1991, *Astrophys. J.* **383**, 554.
Feigelson, E.D., Wood, P.A.D., Schreier, E.J., Harris, D.E., and Reid, M.J.: 1987, *Astrophys. J.* **312**, 101.
Winter, A.J., Wilson, D.M., Warner, P.J., Waldram, E.M., Routledge, D., Nicol, A., Boysen, R.C., Bly, W., and Baldwin, J.: 1980, *Mon. Not. R. astr. Soc.* **192**, 932.

HST Observations of Jets and Radio Lobes

Philippe Crane

European Southern Observatory
Karl-Schwarzschildstrasse 2
D-8046 Garching, Germany

Abstract: The Faint Object Camera on the Hubble Space Telescope has observed the radio jets in 3C 66B, PKS 0521−36 and in M 87. The radio galaxy NGC 6251 which has a large radio jet has also been observed. In addition, the south radio lobe of 3C 33 has been observed with the Planetary Camera. Preliminary results from these observations are reported.

1 Introduction

As part of the Guaranteed Time Observing program of the Faint Object Team, several extragalactic jets and radio lobes have been and will be observed. The results from the first of these observations are presented here, and they clearly demonstrate the advantages of the high resolution provided by HST. The data show the enormous advantages of putting optical telescopes in space, since these images give a glimpse of what should have been and what will be achieved once the spherical aberration of HST is corrected.

The following sections present some images obtained, and refer to more detailed papers for further information. Table 1 gives a summary of the data obtained until early November 1991 and references to published works are given. The following paragraphs will concentrate on data and results which have not previously been reported. In particular, results for the nucleus of NGC 6251 and for the south radio lobe of 3C 33 are presented.

The results from the observations of M 87, PKS 0521−36, and 3C 66B have been published, and will not be reviewed in detail in this paper. The results for M 87 will be published by Boksenberg *et al.* (1992), and for PKS 0521−36 have been published by Macchetto *et al.* (1991a).

Table 1. Summary of Obervations

Object	Camera	Filter	Exposure	Date
NGC6251	FOC f/96	F342W	1200	17/5/91
		F502M	600	9/5/91
3C33	WFPC PC	F606W	2x900	11/6/91
PKS0521-36	FOC f/96	F430W	2x1500	25/8/90
		F320W	2x1500	25/8/90
M87Nuc	FOC f/96	F120M	2x1200	18/4/91
		F140W	4x1200	
		F220W	2x1200	
		F501N	4x1200	
	FOC f/96Z	F372M	1x1200	
M87-A	FOC f/96	F140W	2x1200	18/4/91
		F220W	2x1200	
		F430W	2x1200	
3C66B	FOC f/96	F320W	2x1500	23/9/90

2 Jet in 3C 66B

The nuclear jet in 3C 66B is shown in Fig. 1 and the details of this observation have been discussed by Macchetto *et al.* (1991b). The HST image clearly shows stranded structure in the outer parts of the jet. This structure corresponds quite closely to the structure seen in the radio maps. This morphological similarity is also seen even better in the M 87 data discussed by Boksenberg *et al.* (1992). The similarity between jets and radio lobes is discussed below and in more detail by Crane and Stiavelli (1992).

3 Nucleus of NGC 6251

The radio elliptical galaxy NGC 6251 was imaged with the FOC through a V-like filter, F502M, and through a U-like filter, F342W. NGC 6251 has a very strong one-sided radio jet (Perley, Bridle and Willis, 1984) that shows structure on scales up to several hundred kpc, and may be able to shed some light on the connection between emission in jets and in lobes. Figure 2 is a reproduction of the F342W image with the gray scale stretched to show a small extension from the nucleus indicated by the arrow in Fig. 2. Although it is tempting to imagine that this extension coincides with the nucleus, the position angle of this small extension is almost 90° from the position angle of the radio jet. This extension does not show up in the F502M image. More observations are obviously needed to understand the origin of this feature.

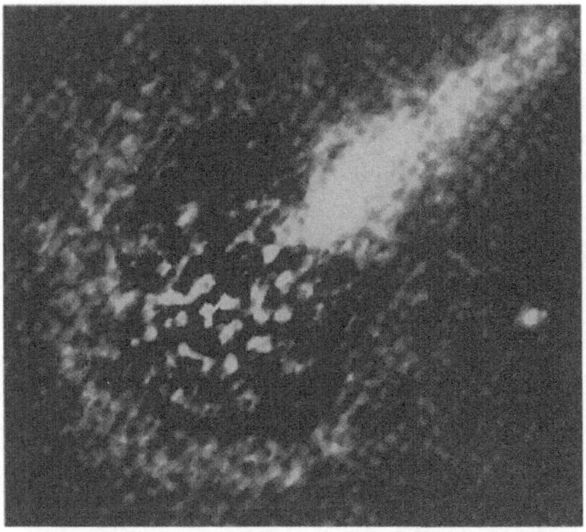

Fig. 1. FOC image of 3C 66B obtained with the f/96 camera and a F320W filter. The galaxy has been subtracted to reveal the details of the jet structure.

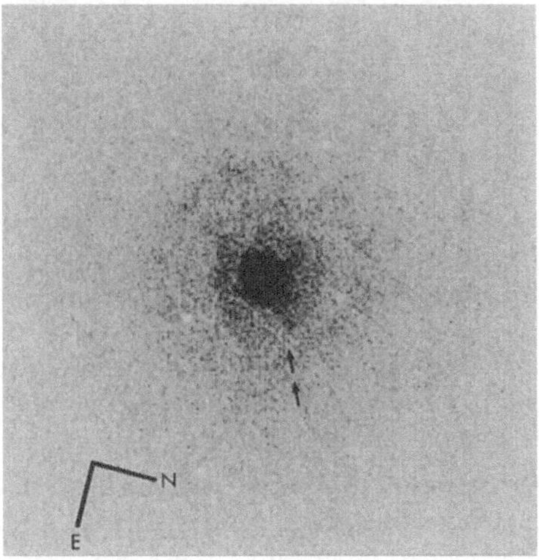

Fig. 2. FOC image of NGC 6251 obtained with the f/96 camera and a F342W filter. Note the slight extension to the lower right of the image. Careful comparison with point source images of other objects indicates that this is not due to an artifact of the HST point spread function.

4 3C 33 South Radio Lobe

The south radio lobe of 3C 33 is one of the best studied radio lobes with optical emission (Crane, and Stockton, 1988; Meisenheimer and Röser, 1986). Two 15 minute PC images were obtained of the radio hot spot region, and the sum with cosmic rays removed is shown in Fig. 3 (see Crane and Stiavelli, 1992 for more details). The optical counter part to the radio hot pot is resolved and shows a remarkable similarity to the radio map which is overlayed.

It seems that this radio hot spot shows many similarities to jets. The optical morphology is similar to that in the radio as found in the HST observations of jets mentioned above. The spectral index behaves like that found in jets and is indicative of a synchrotron origin for the radiation. Also, the polarization seen in the radio and optical matches and adds further evidence to the synchrotron origin of the radiation. (Meisenheimer *et al.* 1989)

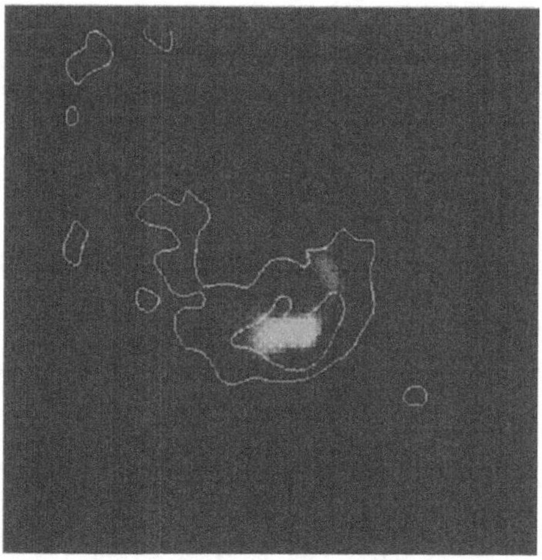

Fig. 3. Planetary Camera image of the radio hot spot in 3C 33 south. The image has been smoothed and deconvolved to a resolution to match the 2cm radio map which is overlayed.

The strong similarity between the observed characteristics of the radio lobe in 3C 33 south, and the jets in M 87 and others suggests that similar physical phenomena are responsible. However, the models proposed for these two cases rely on rather different scenarios. The charged particles responsible for the emission seen in jets is supposed to originate in the nucleus of the parent galaxy, whereas the charged particles responsible for the emission seen in the radio lobes are presumably accelerated *in situ*.

Acknowledgements

The data and analysis discussed here have been obtained in collaboration with the Faint Object Camera Investigative Definition Team, and the FOC Associates in Baltimore. It is a pleasure to acknowledge the contributions to this work of R. Albrecht, A. Boksenberg, C. Barbieri, J.M. Deharveng, M.J. Disney, P. Jakobsen, T.M. Kamperman, I.R. King, F.D. Macchetto, C.D. Mackay, F. Paresce, G. Weigelt, D. Baxter, P. Greenfield, R. Jedrzejewski, A. Nota, and W.B. Sparks.

References

Boksenberg, A., Macchetto, D.M., Albrecht, R., Barbieri, C., Blades, J.C., Crane, P., Deharveng, J.M., Disney, M.J., Jakobsen, P., Kamperman, T.M., King, I.R., Mackay, C.D., Paresce, F., Weigelt, G., Baxter, D., Greenfield, P., Jedrzejewski, R., Nota, A., Sparks, W.B.: 1992, *Astron. Astrophys.* **261**, 393

Crane, P., Stiavelli, M.: 1992, *Mon. Not. R. astr. Soc.* **257**, 17P.

Crane, P., Stockton, A.: 1988, in *Hot Spots in Extragalactic Radio Sources*, eds. K. Meisenheimer and H.-J. Röser, Heidelberg: Springer Verlag, p. 115.

Macchetto, F.D., Albrecht, R., Barbieri, C., Blades, J.C., Boksenberg, A., Crane, P., Deharveng, J.M., Disney, M.J., Jakobsen, P., Kamperman, T.M., King, I.R., Mackay, C.D., Paresce, F., Weigelt, G., Baxter, D., Greenfield, P., Jedrzejewski, R., Nota, A., Sparks, W.B.: 1991a, *Astrophys. J. (Lett.)* **369**, L55.

Macchetto, F.D., Albrecht, R., Barbieri, C., Blades, J.C., Boksenberg, A., Crane, P., Deharveng, J.M., Disney, M.J., Jakobsen, P., Kamperman, T.M., King, I.R., Mackay, C.D., Paresce, F., Weigelt, G., Baxter, D., Greenfield, P., Jedrzejewski, R., Nota, A., Sparks, W.B., Miley, G.: 1991a, *Astrophys. J. (Lett.)* **373**, L55.

Meisenheimer K., Röser, H.-J.: 1986, *Nature* **319**, 419.

Meisenheimer, K., Röser, H.-J., Schlötelburg, M.: 1989, in *Extranuclear Activity in Galaxies*, eds. E.J.A. Meurs, R.A.E. Fosbury, ESO Conference and Workshop Proceedings No. 32, p. 235.

Perley, R.A., Bridle, A.H., Willis, A.G.: 1984, *Astrophys. J. Suppl.* **54**, 291

3-D Hydrodynamical Simulations of Extragalactic Jets

Michael L. Norman and Dinshaw S. Balsara

Department of Astronomy and
National Center for Supercomputing Applications
University of Illinois at Urbana-Champaign
405. N. Mathews Street
Urbana IL 61801, USA

Abstract: We summarize the results of 3-D numerical hydrodynamical simulations studying the origin of morphological asymmetries in radio jets. Four cases are studied which relate to the interpretation of narrow-angle tailed radio sources, wide-angle tailed radio sources, and asymmetric radio hot spots in classical doubles. Our simulations are performed at high numerical resolution using the most accurate numerical algorithms available. They reveal a number of new and interesting features in 3-D jet hydrodynamics which lend further support to the fluid beam hypothesis.

1 Introduction

Many of the more interesting aspects of radio source morphology concern asymmetries of various sorts — wiggles, bends, one-sidedness, multiple hot spots, etc. —- which require 3-D simulations to study theoretically. Until the advent of large memory supercomputers, these topics have been practically out of reach to numerical modelers despite some valiant early attempts (Williams and Gull 1984,1985; Arnold and Arnett 1986). Two-dimensional simulations have shown the importance of numerical resolution in capturing essential fluid dynamical behaviour (Norman 1988; Kössl and Müller 1988) and dictate the use of numerical grids in 3-D of order one million zones or larger.

This paper summarizes the Ph. D. research of one of us (Balsara 1990) to carry out a series of 3-D numerical investigations into the physical mechanisms responsible for various radio source asymmetries, including jet bending, jet deflection and multiple radio hot spots in classical doubles. The guiding philosophy for this research was to seek general fluid dynamical explanations for morphological characteristics which are common within a class of radio sources. Although these simulations are purely hydrodynamical, and thus cannot be Stokes mapped for direct comparison with observations as in the work of Clarke (these proceedings), they were carried out with sufficient numerical resolution and accuracy so as to reveal a number of new and interesting phenomena that should also be present in 3-D MHD jets with passive fields.

These simulations were made possible by the installation of a CRAY-2 supercomputer with 128 million words of central memory at the NCSA, University of Illinois at Urbana-Champaign in early 1988. The simulations were carried out using the RIEMANN code, which implements the Piecewise Parabolic Method (PPM) of Colella and Woodward (1984) in its Lagrangian-remap formulation. Woodward and Colella (1984) have demonstrated that PPM is the most accurate algorithm available for ideal gas dynamical flows with strong shocks and contact discontinuities. This property as well as PPM's low numerical viscosity make it an ideal and efficient method to study supersonic jets. Details of RIEMANN and our code calibration suite can be found in Balsara (1990).

2 Narrow-Angle Tailed Sources

Narrow-angle tailed (NAT) radio sources are interpreted as double radio jets which are ram pressure-bent into their characteristic "C" shape by virtue of their parent active galaxy's peculiar motion through the intracluster medium. The radio galaxy NGC 1265 is a prototypical example of a NAT source, and has been sufficiently well mapped by O'Dea and Owen (1986, 1987a,b) using the VLA that several interesting morphological characteristics are seen for which we seek general dynamical explanations. These features include (1) the jet switches on at a distance of several kpc from the core; (2) the jets bend in discrete steps; (3) emission knots are commonly located at the bends; (4) the jets flare and disrupt after bending by 90 deg.

In this first investigation, we ignore the presence of an interstellar medium in the active galaxy as in the Jones and Owen (1979) model, and focus instead on pure ram pressure effects as in the Begelman, Rees and Blandford (1978, hereafter BRB) model. The calculation is performed on a rectangular domain of $183 \times 40 \times 165$ uniform zones. A symmetry plane is introduced at the midplane of the jet to conserve zones. A Mach 1.3 cross wind of unit density and pressure is injected along the Z axis. A Mach 3.8 (internal) jet is injected along the X axis with a density of one-fifth the ambient value. The jet radius is resolved by 8 zones. These parameters are close to those derived by O'Dea and Owen (1987a) as a best fit for NCG 1265.

Figure 1 shows a slice through the jet midplane at an advanced time and Fig. 2 shows a schematic summarizing of the salient features. The low density cocoon is blown downstream by the strong crosswind, leaving the upwind surface of the jet in direct contact with the ambient medium. We find that jet bending is mediated by a network of oblique internal shock waves and rarefaction waves. This network is established as the jet attempts to come into pressure balance with the cross wind, and is unavoidable so long as (1) the jet is supersonic; and (2) the jet encounters the cross wind suddenly–conditions generally met in radio jets. High pressure regions form downstream of the bending shocks, and are a likely mechanism for forming emission knots. We find that the jet radius of curvature R is closely approximated by the simple scaling relation $R/r_{\rm jet} =$

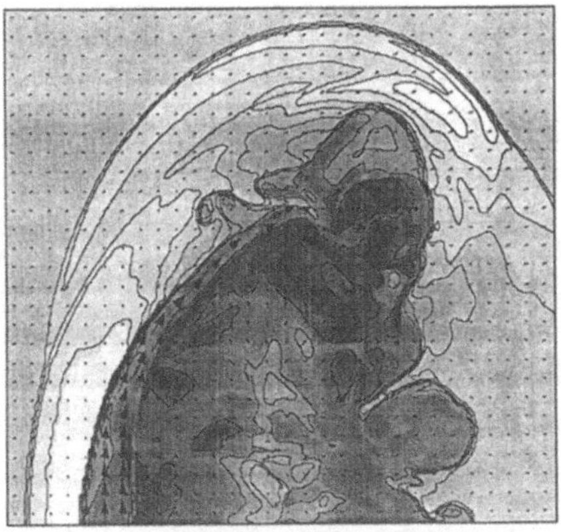

Fig. 1. Numerical simulation of a NAT radio source. A Mach 3.8, $\eta = .2$ jet is injected vertically into a Mach 1.2 left-to-right crosswind. Density (greyscale, contours) and velocity (arrows) distributions are displayed on the central slice plane.

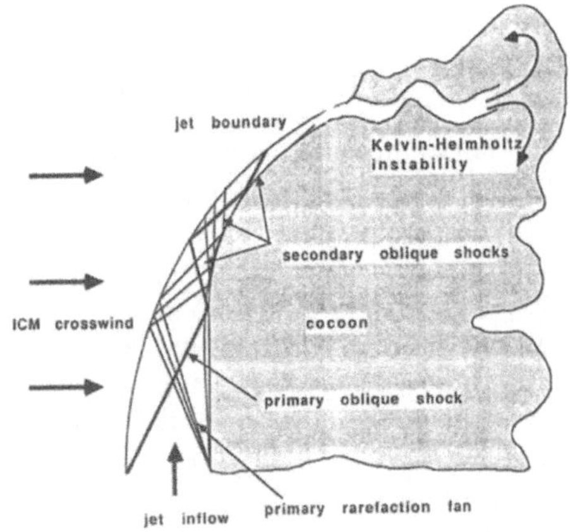

Fig. 2. Schematic diagram of the bending shock and rarefaction waves in the jet channel found in the numerical simulation.

$\rho_{jet} v_{jet}^2 / \rho_{amb} V_{gal}^2$, in stark contrast to the finding of De Young (1991). Furthermore, we find the jet profile is well described by the analytic solution of BRB.

We find that the jet is disrupted after it has bent by 90 deg by a combination of Rayleigh-Taylor (RT) and Kelvin-Helmholtz (KH) instabilities. The ram pressure of the cross wind flattens the jet into a ribbon with major axis transverse to the cross wind. The RT instability acts to filament the ribbon. The filaments are then susceptible to the fundamental kink mode of the KH instability (e.g., Norman and Hardee 1989).

3 Wide-Angle Tailed Sources

Burns (1986) has discussed the general morphological characteristics of tailed radio sources and proposed a tripartite classification scheme based on their physical size. The smallest ones have gradual bends and are associated with the more rapidly moving cluster galaxies. He concludes that ram pressure is the likely bending mechanism in this class which we designate as NATs. The ones of intermediate size are also of intermediate power between FRI and FRII sources. Their jets bend and flare suddenly but quite symmetrically with respect to the core in the vicinity of the galaxy's edge. The angle between the two jets is generally greater than 90 deg, hence the designation wide-angle tailed (WAT). The largest tailed radio sources are asymmetrical, and have sharp bends, sometimes as large as 90 deg, occurring far beyond the galaxy's edge. Because WATs are associated with cD galaxies which are nearly stationary with respect to the cluster center, Eilek et al. (1984) have concluded that it is unlikely that ram pressure bends the jets. The bending mechanism(s) for WATs has remained elusive, and motivates the simulations described next.

3.1 Jet Bending and Disruption at External Shocks

The O'Donohue, Owen and Eilek (1990) catalog of WAT sources of intermediate size display three striking changes in jet behaviour which are spatially coincident with the galaxy edge: (1) sudden flaring; (2) sudden bending; and (3) onset of jet disruption via what appears to be mass entrainment and fluid instability. Norman, Burns and Sulkanen (1988; NBS) showed that jet disruption and flaring could be induced by the jet's passage through an external shock in the ICM. Such a shock could be produced by a galactic wind contacting the ICM or the merger of cluster subclumps. Because their simulations were 2-D, jet bending could not be addressed. We studied the NBS model in 3-D and found that all three WAT characteristics could be reproduced quite naturally. Fig. 3 shows a typical case.

The calculation used a grid of $100 \times 100 \times 100$ uniform zones. Initially, an oblique shock inclined at an angle of 75 deg to the jet axis is established on the grid. The Mach number of this shock is 2.9. A helically perturbed jet is injected along the X axis with $\eta_{\text{jet}} = .02$ and $M_{\text{jet}} = 2$. NBS showed that a jet will flare and undergo a subsonic transition provided $M_{\text{jet}} < M_{\text{shock}}$. Our parameters satisfy this constraint, and the same results are found here in 3-D. The jet bends upon crossing the shock because all streamlines bend away from the shock normal. The helical perturbation, as well as small scale perturbations, grow rapidly on the post-shock jet as it is now subsonic. As in NBS, we find that the flaring angle is a sensitive function of M_{jet} and M_{shock} but is insensitive to η, whereas the growth of the helical perturbation is sensitive to η. Provided a suitable analytic model for these quantities can be found, we may be able to derive useful constraints on jet parameters from easily observable quantities.

3.2 Jet Deflection and Disruption by Cloud Collisions

Burns (1986) proposed that the largest scale WATs can be explained as the result of a jet-cloud collision in a clumpy ICM. This explanation is attractive on both statistical and physical grounds. Statistically, the longer the jet, the more chances it has to hit a clump if they are randomly distributed. Physically, cosmological N-body simulations are beginning to show that galaxy clusters form by gravitational clustering of sub-galactic clumps for a wide range of initial conditions (e.g., Bertschinger and Gelb 1991). Moreover, recent ROSAT observations of rich and poor clusters containing WATs reveal clumps (Burns, private communication).

An apparent stumbling block of this picture is how to explain the large (up to 90 deg) deflection angles observed in sources like 1919+479 (Burns et al 1986); it is well known that a single oblique shock cannot bend a supersonic flow by more than about 37 deg without rendering it subsonic. Yet many WAT jets are observed to propagate stably for many jet diameters *downstream* of the deflection point. This would seem to require that supersonic conditions are either maintained or reestablished somehow. We have investigated the physics of a jet-cloud collision in 3-D and have identified a fluid dynamical mechanism which naturally overcomes these problems for all deflection

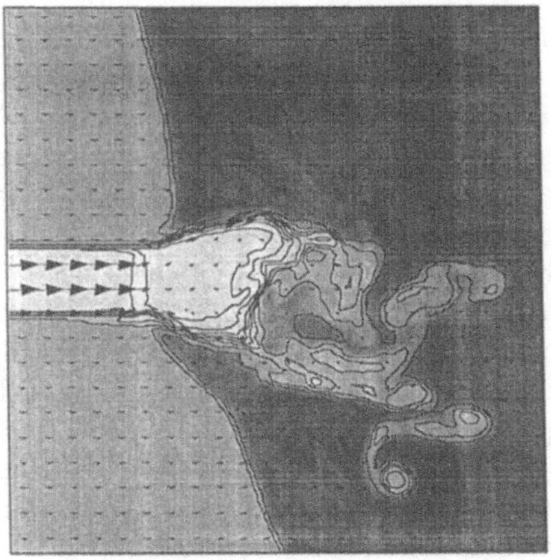

Fig. 3. Numerical simulation of a WAT radio source. A Mach 2, η = .02 jet is injected vertically into a Mach 2.8 shock inclined at 70 deg to the jet axis. Density (greyscale, contours) and velocity (arrows) distributions are displayed on the central slice plane. The jet bends, flares and disrupts suddenly as it crosses the shock wave.

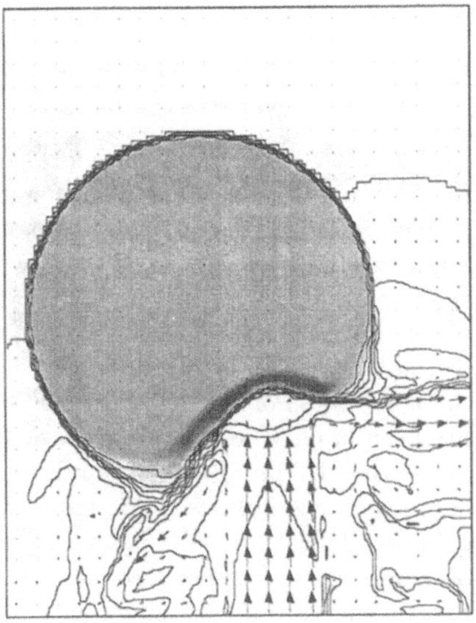

Fig. 4. Numerical simulation of a jet-cloud collision. A Mach 4, η = .2 jet is deflected through 90 deg by a η = 50 cloud. Density (greyscale, contours) and velocity (arrows) distributions are displayed on the central slice plane. The jet kinetic energy is thermalized at the point of impact and is subsequently reaccelerated to Mach 4.5 by a de Laval nozzle mechanism.

angles 0 deg $\leq \Theta \leq$ 90 deg; namely, our old friend the de Laval nozzle. Fig. 4 shows a particularly interesting case.

The calculation used a grid of $130 \times 70 \times 103$ zones to cover the upper half plane centered on the cloud (*i.e.*, reflection symmetry is assumed about Y=0). The jet and cloud parameters were taken to be $\eta_{\rm jet} = .2, M_{\rm jet} = 4, \eta_{\rm cloud} = 50, R_{\rm cloud} = 3.5 r_{\rm jet}$. The jet hits the cloud with an impact parameter of $1.5 r_{\rm jet}$ producing a nearly head-on collision. A high pressure cavity is created at the point of jet impact bounded on the one side by the shock driven into the cloud and on the other side by a working surface shock in the jet. The high pressure cavity is lens-shaped geometrically, and is filled with thermalized jet material. This high pressure gas is accelerated to supersonic speeds through one or two de Laval nozzles depending upon the angle of incidence; Fig. 4 shows an asymmetric two nozzle case. Quite generally, the nozzles are found to possess a flattened cross section, and thus the jets emerge as slab jets. For the cases we explored, we find the exit Mach number of the dominant channel exceeds the input Mach number by 5—10% and thus the reflected jet inherits the stability properties of the original jet.

4 Wall Jets as a Model for Multiple Hot Spots

At the last Ringberg conference, Mike Wilson (1988) presented an interesting paper on the topic of wall jets and their potential relevance to multiple hot spot formation. In particular he presented some laboratory results which show that a supersonic jet impinging obliquely on a rigid wall creates an exit jet which flows along the wall — hence the name wall jet — rather than rebounding from it as in a specular reflection. One could summarize this result by saying that the momentum component perpendicular to the wall is completely thermalized. As Wilson pointed out, this calls into question the correctness of the "ricochet hot spot" model of Williams and Gull (1985), which has been applied to the interpretation of the multiple hot spots in Cygnus A (Carilli, Perley and Dreher 1988). However, as further discussed by Wilson, radio jets do not impinge on rigid plates in the lobes of radio galaxies but rather on a deformable fluid interface. Thus the possibility of a piece of the jet splashing back is not ruled out. Wilson concluded by presenting a couple of suggestive cartoons on how wall jets may operate to form the multiple hot spots seen in many radio doubles (e.g., Laing 1988).

Motivated by these ideas, we have simulated the 3-D interaction of a supersonic jet with a contact discontinuity for a variety of incidence angles and density ratios to explore how radio jets interact with the leading edge of their radio lobes. So far, we have considered only planar discontinuities, and thus curvature effects as in Wilson's Fig. 3 are not addressed. Our simulations do directly address, however, Wilson's Fig. 4. The initial conditions for the simulations are an oblique contact discontinuity of density ratio $\eta_{\rm wall} = \rho_{\rm lobe}/\rho_{amb}$ inclined at an angle $\Theta_{\rm wall}$ to the jet axis. A jet is injected at t=0 with parameters $\rho_{\rm jet} = \rho_{\rm lobe}$ and internal Mach number $M_{\rm jet}$. Although the numbers vary, we employed a grid of roughly $100 \times 100 \times 100$ for the six cases considered.

We find that, depending upon the parameter regime, both kinds of interactions are possible. Fig. 5 shows a wall jet ($\eta_{\rm wall} = 0.03, \Theta_{\rm wall} = 50\,{\rm deg}, M_{\rm jet} = 5$), while Fig. 6 shows a splash-back jet ($\eta_{\rm wall} = 0.03, \Theta_{\rm wall} = 20\,{\rm deg}, M_{\rm jet} = 5$). In all cases studied, a wall jet is created at early times. However, the persistence of the wall jet depends critically on $\eta_{\rm wall}$ and $\Theta_{\rm wall}$. If $\eta_{\rm wall}$ is too large (>0.05) or $\Theta_{\rm wall}$ is too small ($< 45\,{\rm deg}$), the jet penetrates the denser medium as in Wilson's Fig. 3c. The splash-back jet is then formed as the usual sort of backflow encountered in axisymmetric jet simulations, except now it is highly asymmetric. Interestingly, we find that the exit Mach number of splash-back jets is approximately the incident Mach number; the exit Mach number of the splash-back jet in Fig. 6 is 5–10% larger than the incident Mach number. The physics for this reacceleration is simply the de Laval nozzle mechanism as discussed above in yet another context.

The astronomical implications of these findings are threefold: (1) the bright rims seen downstream of the primary hot spot may be wall jets created by the highly oblique interaction of a low density jet with its lobe. In such cases, one would not expect to see a very pronounced hot spot protrusion; (2) the radio filaments that appear to connect the primary and secondary hot spots as in Cygnus A west may indeed be splash-back

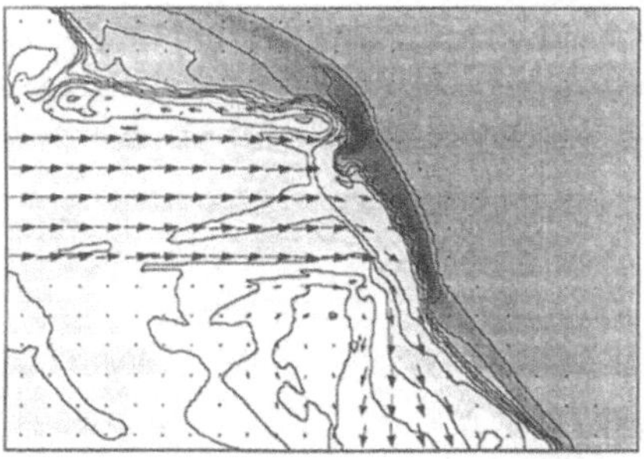

Fig. 5. Numerical simulation of a wall jet. A Mach 5, $\eta = .03$ jet impinges on a contact discontinuity inclined at 50 deg to the jet axis. Density (greyscale, contours) and velocity (arrows) distributions are displayed on the central slice plane. The jet kinetic energy is thermalized at the point of impact and is subsequently reaccelerated to Mach 5.3 by a de Laval nozzle mechanism.

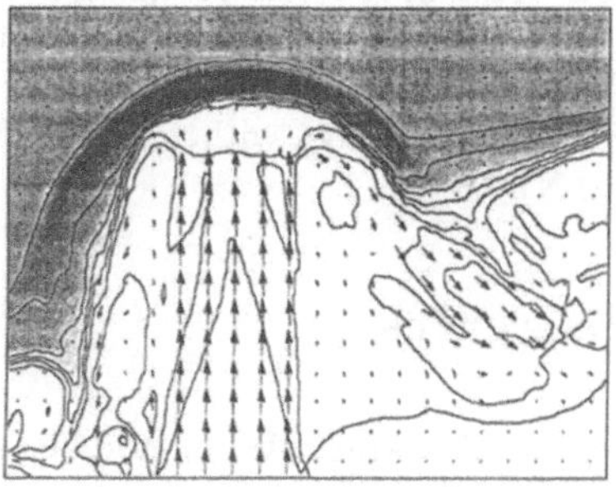

Fig. 6. Numerical simulation of a splash-back jet. A Mach 5, $\eta = .03$ jet impinges on a contact discontinuity inclined at 20 deg to the jet axis. Density (greyscale, contours) and velocity (arrows) distributions are displayed on the central slice plane. The jet kinetic energy is thermalized at the point of impact and is subsequently reaccelerated to Mach 5.5 by a de Laval nozzle mechanism.

jets that have been accelerated to supersonic and perhaps relativistic speeds by a de Laval nozzle located at the primary hot spot. In such cases, one would expect to see a protruding hot spot, as is indeed observed in Cyg A west; (3) the observed length scale of these features (bright rims, filaments) implies much lower jet densities than are assumed in the "standard model" of Norman (1990) — *i.e.*, $\eta = 0.1$. A detailed comparison of observations and simulations of the sort described here should provide some useful constraints on jet density from purely fluid dynamical considerations.

5 Conclusions

The simulations described here just scratch the surface of the new view of radio jet fluid dynamics that is beginning to unfold with the advent of high resolution 3-D numerical simulations. From the standpoint of either supporting or disproving the *fluid beam hypothesis* for radio galaxy structure and evolution as first set forth by Scheuer (1974) and Blandford and Rees (1974), there were two *a priori* possibilities: either the 3-D simulations would look *more* like the real world, or they would look *less* like the real world as compared to the 2-D models. As interesting and suggestive as the 2-D studies have been (e.g., Norman 1990; Burns, Norman and Clarke 1991), one was always comparing apples and oranges. As we hope we have shown here, the 3-D simulations are indeed beginning to look more like the real world. All of the fluid dynamical processes found in the 2-D studies carry over to 3-D: cocoons, internal shock waves, vortices, interfacial instabilities, etc. However, it appears that asymmetric flow structures and shock patterns are the rule rather than the exception. The one new piece of physics to come out of our work is the discovery that the de Laval nozzle mechanism may operate in the extremities of radio galaxies rather than in their cores.

The calculations reported here were carried out using the CRAY-2/4128 system at the National Center for Supercomputing Applications at the University of Illinois, Urbana-Champaign with partial support from NSF grant AST-9012353 to one of us (MLN).

References

Arnold, C.N., Arnett, W.D.: 1986, *Astrophys. J. (Lett.)* **305**, L57.
Balsara, D.S.: 1990, Ph.D. Dissertation, University of Illinois at Urbana-Champaign.
Begelman, M.C., Rees, M.J., Blandford, R.D.: 1984, *Rev. Mod. Phys.* **56**, 255.
Bertschinger, E., Gelb, J.M.: 1991, *Computers in Physics* **3/4**, 164.
Blandford, R.D., Rees, M.J.: 1974, *Mon. Not. R. astr. Soc.* **169**, 395.
Burns, J.O.: 1986, *Can.Jour.Phys.* **63**, 373.
Burns, J.O., Norman, M.L., Clarke, D.A.: 1991, *Science* **253**, 485.
Carilli, C.L., Perley, R.A., Dreher, J.H.: 1988, *Astrophys. J. (Lett.)* **334**, L73.
Colella, P., Woodward, P.R.: 1984, *J.Comp.Phys.* **54**, 174.
De Young, D.S.: 1991, *Astrophys. J.* **371**, 69.
Eilek, J.A., Burns, J.O., O'Dea, C.P., Owen, F.N.: 1984, *Astrophys. J.* **278**, 37.
Kössl, D., Müller, E.: 1988, *Astron. Astrophys.* **206**, 204.
Jones, T.W., Owen, F.N.: 1979, *Astrophys. J.* **234**, 818.
Laing, R.A.: 1988, in *Hot Spots in Extragalactic Radio Sources*, eds. K. Meisenheimer , H.-J. Röser, Lecture Notes in Physics 327, Heidelberg: Springer Verlag, p. 27.
Norman, M.L.: 1988, *ibidem* , p.193.
Norman, M.L., Burns, J.O., Sulkanen, M.E.: 1988, *Nature* **335**, 146.
O'Dea, C.P., Owen, F.N.: 1986, *Astrophys. J.* **301**, 841.
——: 1987a, *Astrophys. J.* **316**, 95.
——: 1987b, *Astrophys. J.* **316**, 113.
O'Donohue, A.A., Owen, F.N., Eilek, J.A.: 1990, *Astrophys. J.* **72**, 75.
Scheuer, P.A.G.: 1974, *Mon. Not. R. astr. Soc.* **166**, 513.
Williams, A.G., Gull, S.F.: 1984, *Nature* **310**, 33.
——: 1985, *Nature* **331**, 34.
Wilson, M.J.: 1988, in *Hot Spots in Extragalactic Radio Sources*, eds. K. Meisenheimer , H.-J. Röser, Lecture Notes in Physics 327, Heidelberg: Springer Verlag, p. 215.
Woodward, P.R., Colella, P.: 1984, *J.Comp.Phys.* **54**, 115.

Numerical Simulation of a Hypersonic Jet[21]

Chris Loken [1], Jack O. Burns [1],

David A. Clarke [2], Michael L. Norman [2]

[1]Department of Astronomy
New Mexico State University
Las Cruces, NM 88003, USA

[2]National Center for Supercomputing Applications
University of Illinois
Urbana-Champaign, IL 61801, USA

Abstract: Luminous extragalactic radio jets associated with classical double radio sources often appear to be overpressured with respect to the ambient medium. This overpressuring can be easily explained if the jet is hypersonic. Here we present the results of a hydrodynamical simulation of a jet with an initial Mach number of 150. We find that the hypersonic jet is highly dynamic in nature with varying head velocity and shape. The jet remains well-collimated while the region behind the bow shock is overpressured by a factor of $\sim 1,600$ with respect to the quiescent ambient medium, thus establishing the validity of the ram-pressure confinement scheme.

1 Introduction

An important, yet unresolved, problem regarding extragalactic radio sources concerns the mechanism for confining luminous jets. Potash and Wardle (1980) were the first to recognize that radio jets associated with quasars are unlikely to be transversely confined by thermal gas surrounding the QSOs. For many classical double radio sources, it has now been demonstrated that the minimum internal pressure of the luminous jet exceeds the external pressure (calculated assuming thermal bremsstrahlung emission and using X-ray luminosities) (*e.g.* Burns *et al.* 1984; Miller *et al.* 1985). As an example of an overpressured jet, we present a new VLA map of the source 3C 208 in Fig. 1. From the

[21]Paper presented by J.O. Burns

radio and X-ray data, we estimate that the jet is at least 40 times more highly pressured than the ambient medium.

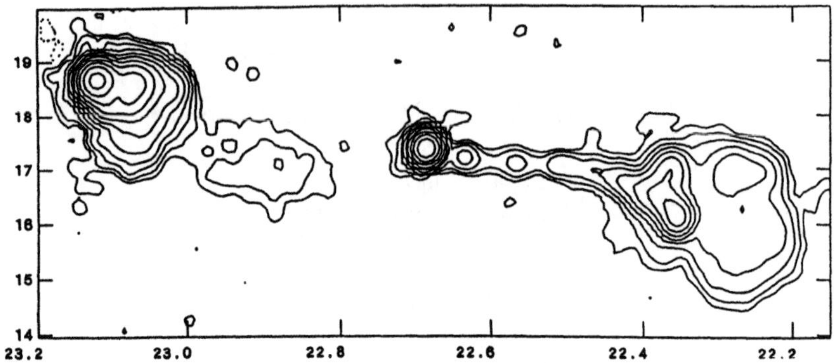

Fig. 1. VLA map of the radio source 3C 208 at 6 cm from Bridle *et al.* (1992). The high dynamic range (4800:1) reveals a continuous jet from the core to the western lobe.

In this paper, we investigate a strictly hydrodynamical explanation for overpressured jets. A hypersonic jet will produce an extensive but narrow bow shock that may stretch back to the nuclear region of the galaxy or quasar. The jet will be in pressure balance with the cocoon material that lies behind the bow shock and thus can be at a pressure considerably higher than that of the ambient medium (cocoon pressure is proportional to ${M_\mathrm{h}}^2$). Such hypersonic jets may be said to be ram pressure confined. This hydrodynamical solution to the problem of collimating a strongly overpressured jet was first proposed by Begelman and Cioffi (1986) and Norman (1987) and further refined by Burns *et al.* (1987) and Begelman and Cioffi (1989). A more detailed account of the current research appears in Loken *et al.* (1992).

2 Theory

The basic idea behind ram-pressure confinement can be simply illustrated by treating the bow shock surrounding a supersonic jet as a steady, oblique shock. The pressure jump across such a shock is given by (*e.g.* Landau and Lifshitz, 1959):

$$\frac{P_\mathrm{sa}}{P_\mathrm{ua}} = \frac{5{M_\mathrm{h}}^2 \sin^2\phi - 1}{4} \tag{1}$$

where P_ua and P_sa are, respectively, the pressures on the upstream (unshocked ambient) and downstream (shocked ambient) sides of the shock. M_h is the Mach number of the upstream material (measured in the rest frame of the shock), or equivalently, the Mach number of the head of the jet (*i.e.*$M_\mathrm{h} = v_\mathrm{h}/c_\mathrm{ua}$ where v_h is the velocity of the head of the jet and c_ua is the sound speed in the undisturbed ambient medium). ϕ is the angle between the shock and the velocity vector of the upstream material and we have assumed an ideal gas with $\gamma = 5/3$. It is clear from this expression that very large pressure jumps can arise for hypersonic jets.

The sideways expansion of the jet can be determined (following Begelman and Cioffi, 1989) by balancing the pressure of the cocoon (P_c) with the ram pressure of the ambient medium ($\rho_{ua} v_c{}^2 \sim \rho_{ua} r_c{}^2/t^2$). Expressing the pressure in terms of the jet kinetic luminosity (assumed constant), we find the cocoon radius as a function of time to be given by the expression:

$$r_c = \left(\frac{\epsilon_L}{2\epsilon_V}\right)^{1/4} (r_h + r_j\sqrt{\eta})^{1/4} r_j{}^{1/4} \left(\frac{K}{\sqrt{\eta}}\right)^{1/4} M_j{}^{1/2} c_{ua}{}^{1/2} t^{1/2}. \tag{2}$$

Here ϵ_L is an efficiency factor for the conversion of jet kinetic luminosity into thermal energy of the cocoon while ϵ_V is a volume factor ($\epsilon_V = 1$ if the cocoon is cylindrical, and $\epsilon_V \sim 1/3$ for the more likely case of a bi-conical geometry). M_j is the jet Mach number (with respect to the speed of sound in the jet), K is the ratio between the jet and undisturbed ambient pressures ($K = P_j/P_{ua}$), η is the ratio between the jet and undisturbed ambient densities ($\eta = \rho_j/\rho_{ua}$) and c_{ua} represents the sound speed in the undisturbed ambient medium. The jet radius is r_j while r_h denotes the effective head radius (the effective working surface of the jet is actually larger then r_j since shocked jet material expands transversely to the jet axis at the head of the jet).

Although the above analysis suggests that hypersonic, overpressured jets could explain several important observed properties of jets in classical doubles, the inherently non-linear character of the problem necessitates the use of numerical simulation techniques to supplement our simple analysis.

3 Numerical Method

The three-dimensional MHD code ZEUS-3D (Clarke, 1990) was used in its two-dimensional mode to model an axisymmetric jet in cylindrical polar coordinates (Z,R,ϕ). ZEUS-3D is an Eulerian, finite-difference MHD code under development at the National Center for Supercomputing Applications (NCSA). Detailed discussion of the numerical method used for the hydrodynamics may be found in Stone and Norman (1992).

Flow-in boundary conditions inject a cylindrically symmetric jet (radius r_j) with constant density (ρ_j), pressure (P_j) and velocity (v_j) along the Z-axis. The computational grid was chosen to have $Z_{max} = 160\, r_j$, $R_{max} = 60 r_j$, and was divided into 1200 zones axially and 350 zones radially. The jet was resolved by 15 zones in the radial direction.

Only three dimensionless parameters are needed to characterize our jet; the simulation described here corresponds to a light ($\eta = .01$), initially overpressured ($K = 10$), hypersonic ($M_j = 147$) jet.

4 Results

The hypersonic jet discussed above was evolved until it reached the end of the computational grid ($\sim 160 r_j$), corresponding to a scaled-time of $t = .725$ (*i.e.*material moving at the sound speed of the undisturbed ambient medium would have travelled nearly 3/4 of a jet radius). The simulation required a total of $\sim$ 47 CPU hours on the Cray Y-MP at NCSA.

A visualization of the density distribution at the final epoch is shown in Fig. 2. At this point the bow shock radius at $Z = 0$ is about 51 jet radii yielding a final length to width ratio of 3:2 for the entire object. A great deal of complicated structure is apparent in this image. A well-collimated jet is evident along the central axis. A bow shock separates the undisturbed ambient medium from the shocked ambient medium while a Mach disk separates the jet from the shocked cocoon gas at the working surface. Both the cocoon and the shocked ambient have been shock-heated to high pressures and are separated by a contact discontinuity. Notice that the cocoon is extensive and occupies a large fraction of the region behind the bow shock. This is in contrast with lower Mach number jets but in keeping with the correlation between cocoon size and jet Mach number (Norman *et al.*, 1983).

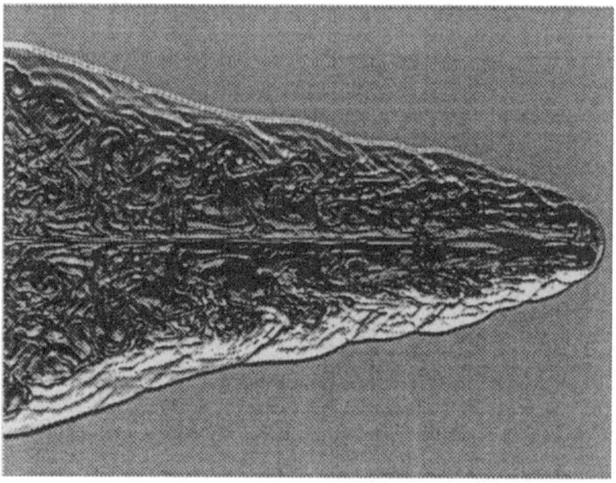

Fig. 2. "Gradient-shade" visualization of the density for the final state of the hypersonic jet.

Most of the time-dependent structure occurs in the cocoon where we see numerous turbulent eddies of extremely low density material. These eddies drive oblique shock waves into the jet and have almost managed to pinch off the jet (there is a region of strong backflow in the jet at $Z \sim 110 - 114 r_j$) giving rise to a strong rarefaction zone further downstream (the jet density drops by almost 2 orders of magnitude in the region $\sim 116 \leq r_j \leq 140$). This situation resembles a "restarting jet"(Clarke and Burns, 1991) during the phase just after the beam is turned off and before it is turned back on. The head of the jet also changes its appearance during the evolution. The working surface (just behind the leading bow shock) resembles the flat end of a cylinder at some epochs (as in Fig 2), whereas it is much more conical at other times. This continual change in shape is probably responsible for the observation that the jet advance speed is not constant with time.

Our (2) predicts the cocoon radius will grow as $t^{1/2}$. A first-order polynomial fit to the cocoon position *vs.* time data yields a slope of $0.512 \pm .003$ which is in good agreement with the predicted value. The fitted value for the intercept, along with (2) enables us to solve for the effective radius of the head, resulting in $r_h/r_j = (7.43 \pm .10)\epsilon_V/\epsilon_L - 0.1$. The measured average advance speed of the head (using the jet length at the end of the simulation) is $\bar{v}_h = 218 c_{ua}$. This implies (by balancing the ram pressures of the jet and ambient medium at the working surface) that $r_h \sim 2.0\, r_j$ which may be combined with the above value for r_h to yield $\epsilon_V/\epsilon_L = .29$. Then, for example, $\epsilon_L = 0.9$ implies $\epsilon_V = .26$. The simulation shows the cocoon to be roughly conical in shape but because of the blunt head, undulating outer boundary and the portion occupied by the jet itself, we certainly expect $\epsilon_V < 1/3$. Nevertheless, ϵ_V could not be much smaller than 1/3 and therefore we conclude that, on average, the conversion of jet kinetic energy to thermal energy is quite efficient.

The fact that the jet and the cocoon are truely overpressured with respect to the ambient medium can be demonstrated by examining the pressure along various cuts through the jet. The pressure profile along the axis of the jet remains fairly constant (initially $P_j = 10 P_{ua}$) until a strong shock is encountered after $\sim$ 10 jet radii. Several more shocks occur along the jet and the average pressure remains more than 1000 times greater than the ambient pressure after this point. Profiles of the pressure along several lines through the computational grid perpendicular to the jet symmetry axis all show a sudden drop by a factor of about 1,600 between the cocoon pressure and the ambient pressure at the bow shock.

5 Summary and Conclusions

We have presented a purely hydrodynamical solution to the problem of confining over-pressured jets which invokes a hypersonic jet. In order to verify the validity of this mechanism and study multi-dimensional effects as well as their time evolution, we performed a numerical simulation of a hypersonic jet with a Mach number of $M_j = 147$. This Mach number is much greater than that of any previously published calculation and the jet was evolved further than in previous simulations. The overall structure is stable and the jet remains well collimated over the course of the lengthy time evolution. The magnitude of the jet overpressuring, the cocoon size and even the cocoon shape were shown to agree reasonably well with our simple, analytic predictions. The effective radius of the head was determined to be $\sim$ 2.1 jet radii, indicating that the forward momentum of the jet is effectively spread out over an area larger than just a single jet radius. The radial extent of the cocoon was found to grow at a rate $\propto t^{1/2}$, as predicted analytically.

This work was supported by a grant from the U.S. National Science Foundation (AST-9012353) to J.O.B. and M.L.N.

References

Begelman, M.C., Cioffi, D.F.: 1986, *Bull.AAS* **18**, 909.
Begelman, M.C., Cioffi, D.F.: 1989, *Astrophys. J. (Lett.)* **345**, L21.
Bridle , A.H., Hough, D.H., Burns, J.O., Laing, R.A., Lonsdale, C.J.: 1992, in preparation.
Burns, J.O., Basart, J.P., De Young, D.S., Ghiglia, D.C.: 1984, *Astrophys. J.* **283**, 515.
Burns, J.O., Clarke, D.A., Norman, M.L.: 1987, in *13th Texas Symposium on Relativistic Astrophysics*, ed. M.P. Ulmer (Singapore: World Scientific), p. 341.
Clarke, D.A.: 1990, *Bull.AAS* **22**, 1302.
Clarke, D.A., Burns, J.O.: 1991, *Astrophys. J.* **369**, 308.
Landau, L.D., Lifshitz, E.M.: 1959, *Fluid Mechanics*, Pergamon Press, p. 336.
Loken, C., Burns, J.O., Clarke, D.A., Norman, M.L.: 1992, *Astrophys. J.* **392**, 54.
Miller, L., Longair, M.S., Fabbiano, G., Trinchieri, G., Elvis, M.: 1985, *Mon. Not. R. astr. Soc.* **215**, 799.
Norman, M.L.: 1987, *Bull.AAS* **19**, 651.
Norman, M.L., Winkler, K.-H.,A., Smarr, L.: 1983, in *Astrophysical Jets*, eds. A. Ferrari and A.G. Pacholczyk, Dordrecht: Reidel, p.227.
Potash, R.I., Wardle, J.F.C.: 1980, *Astrophys. J.* **239**, 42.
Stone, J.M., Norman, M.L.: 1992, *Astrophys. J. Suppl.* **80**, 753.

3-D MHD Simulations of Extragalactic Jets

David A. Clarke

National Center for Supercomputing Applications
The Beckman Institute
The University of Illinois
Urbana, IL 61801, USA

Abstract: With the advent of fast, large memory supercomputers such as the Cray 2, it is now possible to investigate the dynamics of extragalactic radio sources in three dimensions. In this contribution, results from two simulations are presented which use the equations of ideal MHD to model extragalactic radio sources. Among other things, these calculations reveal a simple hydrodynamical origin for the formation of total intensity filaments in the extended regions of well-resolved extragalactic radio sources, and point to important observational differences between extragalactic radio sources with active and passive magnetic fields.

1 Introduction

Blandford and Rees (1974) describe what has become the standard model for extragalactic radio sources. This model posits that jets are supersonic collimated plasma beams emanating from a central engine which may consist of an accretion disc surrounding a $\sim 10^9$ $M_\odot$ black hole situated at the centre of the parent galaxy (*e.g.* Lovelace *et al.* 1991). The jet material is hot and light relative to the ambient intergalactic medium (IGM). At the "working surface", the jet material is shocked and heated to great pressures, inflating an extended lobe or cocoon surrounding the jet. Note that an extended lobe would not result if the jet material were heavy relative to the IGM (*e.g.* Norman *et al.* 1982). This model and its variants grossly describe a large number of observed Fanaroff and Riley (1974) class II (hereafter FR II) extragalactic radio sources and even some Fanaroff and Riley class I (hereafter FR I) extragalactic radio sources, but there remain many unresolved questions. For one, observational features such as knots along the jet, total intensity filaments in the extended structure, and the polarisation properties are not addressed by the Blanford and Rees model. Second, many fundamental physical attributes of extragalactic radio sources such as the origin of the inferred magnetic field energy density are unexplained.

At the last Ringberg workshop, Norman (1989) presented 2-D axisymmetric numerical simulations of extragalactic radio sources from which emissivity images were derived.

These were compared with radio images of selected extragalactic radio sources. It is the purpose of this investigation to "upgrade" those results with fully three dimensional MHD simulations and to cast light on some of the remaining questions regarding the dynamics, structures, and classification of extragalactic radio sources.

2 Numerical Procedures

The computer simulations are performed using ZEUS-3D, a computer program designed to solve the equations of ideal magnetohydrodynamics (MHD) on a three dimensional discretised grid. The numerical techniques for solving these equations are beyond the scope of these proceedings and the interested reader is referred to Stone and Norman (1992a, 1992b).

Boundary conditions are maintained which are consistent with the propagation of a jet into the desired ambient medium. When the simulation is complete, the flow variables (density ρ, velocity $\mathbf{v}$, internal energy e, and magnetic induction $\mathbf{B}$) are known as a function of three spatial coordinates and time. These are used to compute the synchrotron emission properties of the simulation (*e.g.* Longair 1981). Since the relativistic electrons are not computed independently, assumptions have to be made about their distributions. The simplest prescription is to set:

$$\int N(E)\,\mathrm{d}E \propto \rho; \qquad \int EN(E)\,\mathrm{d}E \propto e \tag{1}$$

where $N(E)\,\mathrm{d}E$ is the number of electrons with energy between E and $E+\mathrm{d}E$. For $N(E) \sim E^{-2}$, the Stokes I, Q, and U parameters can be written as (e.g. Clarke 1988):

$$I \propto \int e\,(B\sin\psi)^{3/2}\,\mathrm{d}l \tag{2}$$

$$Q \propto \int e\,(B\sin\psi)^{3/2}\cos 2\chi\,\mathrm{d}l \tag{3}$$

$$U \propto \int e\,(B\sin\psi)^{3/2}\sin 2\chi\,\mathrm{d}l \tag{4}$$

where:

ψ = angle between local $\mathbf{B}$ and line of sight
χ = position angle of local $\mathbf{B}$ projected in the plane of the sky
$\mathrm{d}l$ = increment along line of sight

Note that Stokes I represents the total intensity, while the linear polarisation distribution is given by:

$$F = \frac{\sqrt{Q^2+U^2}}{I}; \qquad \chi_0 = \frac{1}{2}\tan^{-1}\frac{U}{Q} \tag{5}$$

where:

F = fraction of emission that is (linearly) polarised
χ_0 = apparent position angle of polarisation pseudo-vector

The variables I, F, and χ_0 constitute the "numerical observations" of the computation and can be directly compared with radio observations of extragalactic radio sources. As shall be seen, similar structures appear in the numerical and radio observations. Thus, the computation can be probed for the physical agents responsible for the feature, thereby gaining insight into the physics of the radio source itself.

3 Numerical Experiments

3.1 Jet with a Passive Toroidal Magnetic Field

The original intent of this calculation was to investigate the possibility that a jet-cloud collision could be responsible for WAT sources such as 1919+479 (Burns *et al.* 1986). To this end, a light ($\rho_{\rm jet}/\rho_{\rm amb} \equiv \eta = 0.1$), supersonic ($M = 4$) jet propagated into an ambient medium which was uniform except for a massive cloud along the jet path 30 jet radii ($r_{\rm jet}$) from the jet orifice. A helical perturbation was applied to the jet velocity at the jet orifice to break the azimuthal symmetry. For the purpose of computing the synchrotron properties of the simulation, a dynamically negligible (i.e. passive) toroidal magnetic field was transported by the jet. This magnetic field was 100% *ordered*: no random component was specified at the jet orifice.

The calculation was done on the Cray-2 at the NCSA, and consists of some three million zones fixed on a Cartesian grid. The grid dimensions in units of $r_{\rm jet}$ are $40\times40\times24$ and there are 16 zones across the jet diameter at the orifice (i.e. 200 zones per jet cross-section).

The jet-cloud collision itself is incidental to the present discussion, and will not be considered further. Instead, the observational characteristics of a 3-D jet transporting a passive magnetic field will be the focus of this subsection. Figure 1 shows various displays of the final epoch of the calculation. Figure 1a is a cartoon highlighting the salient features. Figure 1b is a superposition of the polarisation **B**-vectors [length proportional to F, orientation given by χ_0 (equation 5)] and a grey-scale representing the total intensity (equation 2). The grey-scale is logarithmic with a dynamic range (value represented by white divided by value represented by the darkest grey) of 5,000. The remaining panels are grey-scale representations of magnetic pressure (Fig. 1c), and velocity shear (Fig. 1d) integrated along the line of sight. The latter was evaluated by summing the squares of all the off-diagonal (shear) components in the tensor $\nabla\mathbf{v}$.

The striking aspect of Fig. 1b is the filamentary nature of the total intensity and the alignment of the polarisation **B**-vectors with the long axis of the filaments. By viewing the emission pattern at various angles, the total intensity filaments are seen to be one-dimensional features (as opposed to sheets being viewed end-on) with elliptical cross sections. Filamentation (with aligned apparent magnetic field) is commonly observed in well resolved, high dynamic range images of extragalactic radio sources, irrespective of FR class (*e.g.* Hercules A, Dreher and Feigelson 1984; Cygnus A, Perley *et al.* 1984; Centaurus A, Clarke 1988; Clarke, Burns, and Norman 1992; M 87, Hines *et al.* 1989; 3C 219, Clarke *et al.* 1992). However, until now, the origin of total intensity filaments was not known (*e.g.* Hines *et al.* 1989).

While there is no evidence of filamentation in the density and pressure distributions (not shown), the magnetic field is clearly arranged in "bundles" (Fig. 1c) which correspond in a one to one fashion with the total intensity filaments (Fig. 1b). "Intermittency"

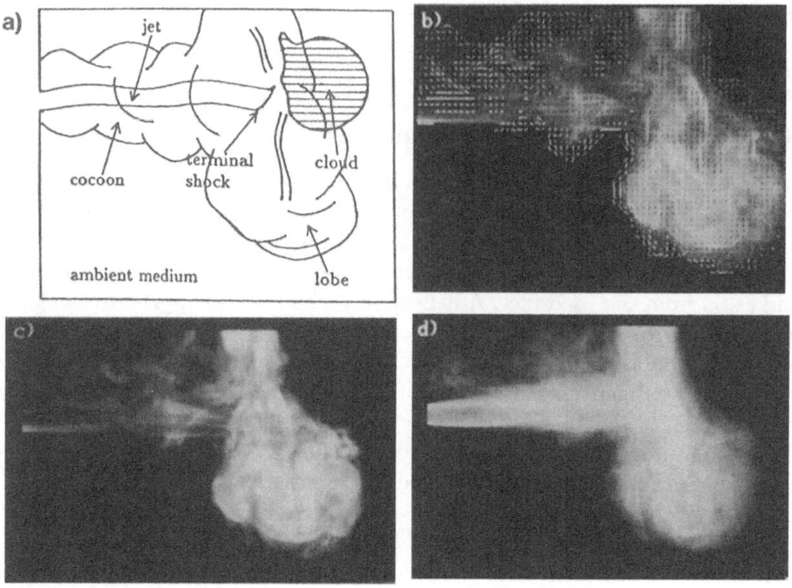

Fig. 1. Final epoch of the jet-cloud simulation. a) Cartoon depicting salient features, b) Grey-scale of computed total intensity with polarisation **B**-vectors superposed, c) Grey-scale of magnetic pressure ($\propto B^2$) integrated along the line of sight, d) Grey-scale of velocity shear modulus integrated along the line of sight.

in the magnetic field distribution was previously observed by Meneguzzi *et al.* (1981). Since the magnetic field is passive, there must be a dynamical variable which causes it to form bundles. The *only* dynamical variable found to have any evidence of filamentation is the velocity shear (Fig. 1d). While the velocity shear filaments do not correspond in a one to one fashion with the magnetic field bundles, there is a strong correlation. Thus, it is concluded that velocity shear is ultimately responsible for the formation of total intensity filaments in extragalactic radio sources.

The nature of the mechanism is illustrated in Figs. 2 and 3. Eddies are generated in the lobe by the flapping (Kelvin-Helmholtz unstable) jet. Now, the surface of each eddy is a shear layer. Figure 2 illustrates how a seed magnetic field threading a shear layer is amplified via a "kinematic dynamo". Formally, as applied to the conditions in Fig. 2, the induction equation $[\partial \mathbf{B}/\partial t = \nabla \times (\mathbf{v} \times \mathbf{B})]$ reduces to:

$$\frac{\partial B_x}{\partial t} = B_y \frac{\partial v_x}{\partial y} \quad \Rightarrow \quad \frac{B_x}{B_y} \sim \frac{\delta t \, 2 v_x}{w} \tag{6}$$

where B_x/B_y represents the amplification of the seed magnetic field and δt is a measure of the lifetime of the eddy. Taking $\delta t \sim \lambda/c_s$ (the circumference of the eddy divided by the sound speed) and $v_x \sim c_s$ (transonic flow in the lobe), then (10) becomes

$$\frac{B_x}{B_y} \sim \frac{2\lambda}{w} \tag{7}$$

where λ is now interpreted as the length of the filament. Thus, the amplification of the seed magnetic field is roughly the aspect ratio of the filament, which can be of order ten's for the longest features. This rough calculation is confirmed by the simulation, in which the magnetic field strength within the bundles is typically several and rarely 100 times greater than in the inter-bundle region. [Field amplification in the lobes of extragalactic radio sources has been discussed semi-analytically by De Young (1980)]. Note that demonstrating the ubiquity of shear amplification of magnetic fields helps to explain the relatively high energy density of magnetic fields inferred in the lobes of extragalactic radio sources.

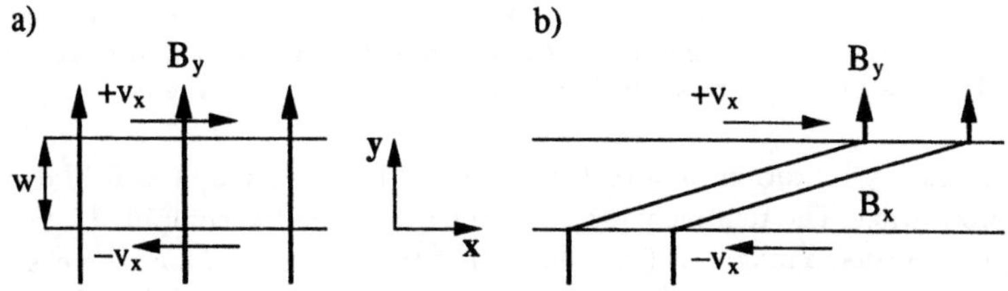

Fig. 2. The "Kinematic Dynamo". a) A seed magnetic field (along $\hat{\mathbf{y}}$) initially penetrates a shear layer (along $\hat{\mathbf{x}}$). b) Shearing B_y in the x-direction gives rise to B_x which can be significantly greater than the original B_y.

The degeneration of shear layers into shear filaments (vortex tubes) is part of the process by which organised flow cascades into turbulence (*e.g.* Stone and Norman 1992c). In short, shear layers are Kelvin-Helmholtz unstable to the formation of vortex tubes (Fig. 3). Figure 3a shows a perturbed shear layer broken up into two components: the original shear layer plus vortex tubes (columns of rotating gas) running along its length. As illustrated in Fig. 3b, vortex tubes possess at least one property of electrical currents. By Bernoulli's principle, parallel vortex tubes attract while antiparallel vortex tubes repel. For sufficiently low viscosity (i.e. sufficiently high Reynolds numbers), parallel vortex tubes will migrate and merge within a dynamical timescale (comparable to δt above), driving the instability. This is the nature of the Kelvin-Helmholtz instability and the fundamental process which drives the total intensity filamentation in the simulation.

Finally, a word about the distribution of fractional polarisation (F). In many FR I and FR II sources (*e.g.* 3C 219, Clarke *et al.* 1992), the distribution of F may be best described as "mottled", with little resemblance to the total intensity distribution. Typically, $F \sim 0.6$ at spots around the circumference of the source, and 0.2 to 0.3 in most of the interior. Qualitatively, the computed distribution of F in this simulation is *identical*.

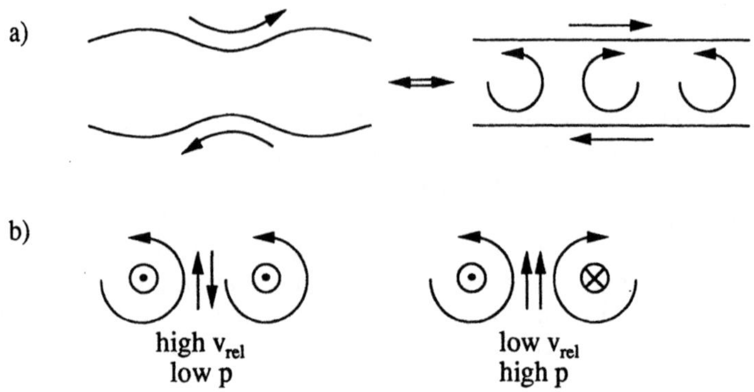

Fig. 3. a) A sinusoidal perturbation of a shear layer can be decomposed into two components: the original unperturbed shear layer plus a series of vortex tubes. b) Between parallel vortex tubes, the relative velocity is high yielding (by Bernoulli's principle) a low pressure region. Thus parallel vortex tubes attract. Similarly, antiparallel vortex tubes repel.

This is a necessary condition for a realistic and well-resolved magnetic field distribution in the computation. The relatively low values of F pervade the computed image despite the completely ordered magnetic field specified at the jet orifice. Thus, there is a net loss in the apparent ordering of the global magnetic field. In a separate ZEUS-3D simulation (still in progress), a jet is launched with a passive, totally *random* magnetic field and yet yields a similar distribution of F. In this case, there is a net gain in the apparent ordering of the global magnetic field. The total intensity distribution of this simulation is also filamentary. It would seem, therefore, that memory of the structure of a *passive* magnetic field at the jet orifice (or the AGN for real sources) is completely lost in the extended emission of the lobe and perhaps in the jet emission as well. Note that this contradicts some of the conclusions of previous 2-D studies (Norman 1989; Clarke *et al.* 1989).

3.2 Jet with an Active Toroidal Magnetic Field

Clarke *et al.* (1986) present the first axisymmetric simulation of a magnetically confined jet. In this calculation, an $M = 6$, $\eta = 0.1$ jet was launched with an active toroidal magnetic field ($\beta \equiv 2\mu_0 p/B^2 = 0.4$). In Fig. 4, this jet (panel a) is compared with an $M = 6$, $\eta = 0.1$ hydrodynamical ($\beta = \infty$) jet (panel b) after comparable evolution times. In both cases, a Mach disc is formed at roughly the same distance from the jet orifice (labelled). For the hydrodynamical jet, material shocked at the Mach disc is heated and then expands laterally to inflate an extended cocoon surrounding the jet (the Blandford and Rees model). Conversely, the shocked jet material in the magnetically confined jet is launched ahead of the Mach disc by the $\mathbf{J} \times \mathbf{B}$ forces forming a "nose-cone" instead of a cocoon.

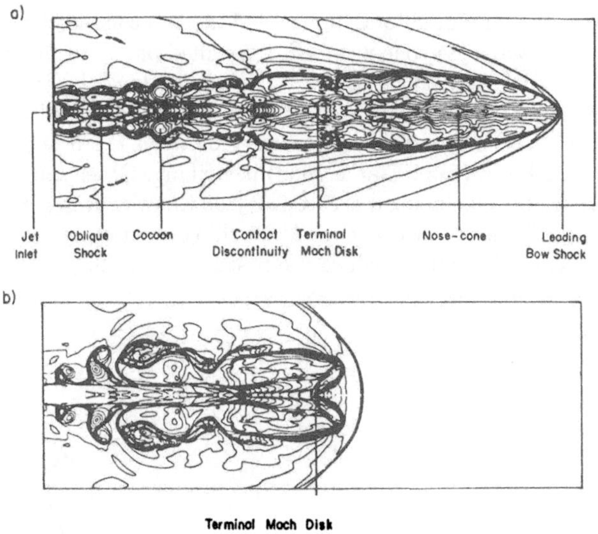

Fig. 4. a) An axisymmetric magnetically confined jet. b) An axisymmetric hydrodynamical jet with the same Mach number and density ratio. Both jets were evolved to comparable times (from Clarke *et al.* 1986).

Observationally, there are no convincing examples of extended nose-cones known to the author. This suggests two possibilities. Either magnetic confinement of jets does not occur in nature, or the imposed axisymmetry has lead to unrealistic results. It was therefore decided to repeat the calculation of Clarke *et al.* (1986) in three dimensions and to re-examine the observations in light of these new results.

The computation consists of some three million zones fixed on a Cartesian grid. The grid dimensions in units of $r_{\rm jet}$ are $50 \times 30 \times 30$ with the long direction along the jet axis. There are 20 zones across the jet diameter at the orifice (i.e. 314 zones per jet cross-section).

In 3-D, the nose-cone is not a stable structure. A magnetically confined column of plasma is subject to pinch and kink instabilities (*e.g.* Jackson 1975). While the former can be observed in the axisymmetric calculation, the latter are manifestly three-dimensional. As the amplitude of the kink instability grows along the length of the nose-cone, the axis of the nose-cone becomes misaligned with the axis of the jet, and the structure gets sloughed aside. The sloughing events prevent the nose-cone from growing into the large structures seen in 2-D, and allow the material in the nose-cone to be ingested by the narrow cocoon surrounding the jet. The toroidal geometry of the magnetic field is preserved all along the length of the jet, and even along the growing nose-cone. However, when the nose-cone is sloughed, the toroidal loops of magnetic field are twisted into the poloidal plane, giving rise to a predominantly axial magnetic field in the cocoon. Note, however, that the magnetic field energy density in the cocoon is still comparable to the internal and kinetic energy densities of the thermal plasma. Thus, the magnetic field effectively resists being tangled and remains ordered. That is to say, the random component of an active magnetic field is small.

Figure 5 shows the computed total intensity of the 3-D confined jet at its final epoch. At this time, the second nose-cone was being sloughed off. Note the knotty jet with a small lobe confined near the tip of the jet. The knots along the jet are generated by the pinching mode instability while the small lobe may be identified with the nose-cone being sloughed off. The morphological similarity between Fig. 5 and the radio quasar 1007+417 shown in Fig. 6 (from Clarke 1988) is striking. Other sources which may be candidates for magnetic confinement are 3C 273 (see Conway, Röser, these proceedings) in which the "blister" may be evidence for a sloughing event some time in the past, and 0800+608 (Jackson *et al.* 1990), in which the first nose-cone may still be forming. Note that both of these sources are also quasars.

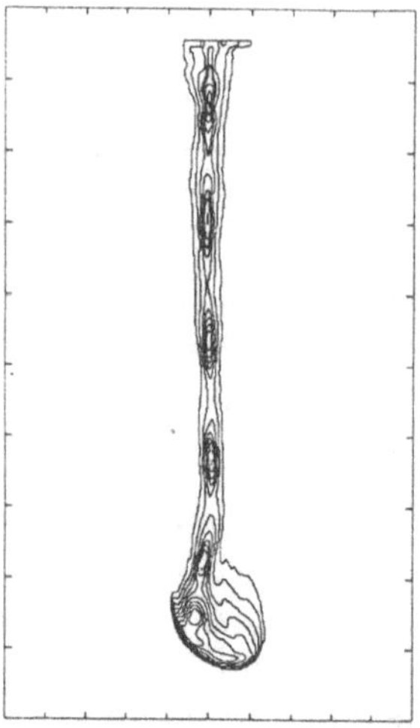

Fig. 5. Computed total intensity image of the 3-D magnetically confined jet at the final epoch.

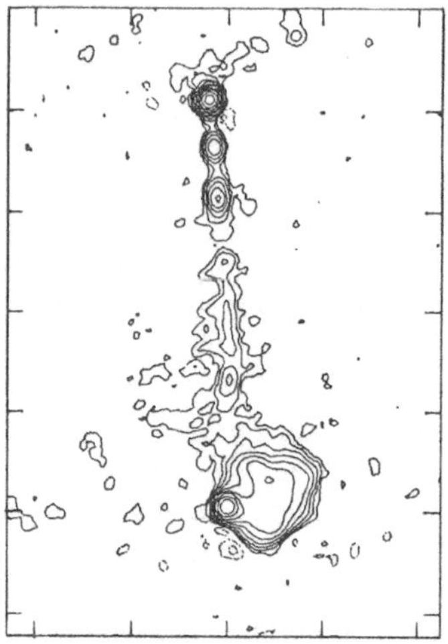

Fig. 6. VLA total intensity image at 6 cm of the quasar 1007+417 (from Clarke 1988).

What of the polarisation properties? First, because an active magnetic field resists twisting, there are large regions of relatively high F on the computed polarisation image (not shown). Second, the polarisation **B**-vectors will always be transverse on the axis of a jet with a toroidal magnetic field, even when the Stokes Q and U parameters are poorly resolved. However, if the magnetic field were helical with a sufficiently low pitch angle, the axial component of the magnetic field could yield axial polarisation **B**-vectors along the jet. Such a calculation is being performed at the time of this writing.

4 Summary

Table 1 summarises the important observational characteristics of the two simulations presented herein. The majority of FR I and FR II sources seem to be consistent with the numerical observations of an extragalactic radio source with a passive magnetic field. Specifically, the extended, filamentary lobe emission surrounding relatively weak radio jets typical of many FR II sources (*e.g.* Cygnus A) is *inconsistent* with an active magnetic field. However, a minority of FR II sources (*e.g.* 3C 273, 0800+608, 1007+417) do exhibit properties of an object with a magnetically confined jet.

Table 1. Observational Characteristics of Passive and Active Magnetic Fields

Passive Magnetic Field	Active Magnetic Field
Extended lobe	Confined Lobe
Filamentary	Few Filaments
Low FPOL	High FPOL
Confused Jet Emission	Knotty, Naked Jet

The addition of magnetic fields and the extension to three dimensions has brought numerical modelling of extragalactic radio sources to a golden age. At last, numerical models are beginning to resemble nature in a meaningful way. At the very least, this should give us confidence that ideal MHD is the basis of a usable physical model for the dynamics of extragalactic radio sources. More optimistically, we can begin to investigate the origin of observable features such as total intensity filaments and even reproduce the polarisation properties of these sources. Indirectly, physical properties such as the relative importance of the magnetic field are being revealed.

References

Blandford, R.D., Rees, M.J.: 1974, *Mon. Not. R. astr. Soc.* **169**, 395.
Burns, J.O., O'Dea, C.P., Gregory, S.A., Balonek, T.J.: 1986, *Astrophys. J.* **307**, 73.
Clarke, D.A.: 1988, Ph.D. thesis, University of New Mexico.
Clarke, D.A., Bridle, A.H., Burns, J.O., Perley, R.A., Norman, M.L.: 1992, *Astrophys. J.* **385.**,
Clarke, D.A., Burns, J.O., Norman, M.L.: 1992, *Astrophys. J.* **395**, 444.
Clarke, D.A., Norman, M.L., Burns, J.O.: 1986, *Astrophys. J. (Lett.)* **311**, L63.
——: 1989, *Astrophys. J.* **342**, 700.
De Young, D.S.: 1980, *Astrophys. J.* **241**, 81.
Dreher, J.W., Feigelson, E.D.: 1984, *Nature* **308**, 43.
Fanaroff, B.L., Riley, J.M.: 1974, *Mon. Not. R. astr. Soc.* **167**, 31P.
Hines, D.C., Owen, F.N., Eilek, J.A.: 1989, *Astrophys. J.* **347**, 713.
Jackson, J.D.: 1975, *Classical Electrodynamics*, New York: Wiley.
Jackson, N., Browne, I.W.A., Shone, D.L., Lind, K.R.: 1990, *Mon. Not. R. astr. Soc.* **244**, 750.

Longair, M.S.: 1981, *High Energy Astrophysics*, Cambridge: Cambridge University Press.
Lovelace, R.V.E., Berk, H.J., Contopoulos, J.: 1991, *Astrophys. J.* **379**, 696.
Meneguzzi, M., Frisch, U., Pouquet, A.: 1981, *Phys.Rev.Letters* **47**, 1060
Norman, M.L.: 1989, in *Hot Spots in Extragalactic Radio Sources*, eds. K. Meisenheimer and H.-J. Röser, Heidelberg: Springer Verlag, p 193.
Norman, M.L., Smarr, L., Winkler, K.-H.A., Smith, M.D.: 1982, *Astron. Astrophys.* **113**, 285.
Perley, R.A., Dreher, J.W., Cowan, J.J.: 1984, *Astrophys. J. (Lett.)* **285**, L35.
Stone, J.M., Norman, M.L.: 1992a, *Astrophys. J. (Lett.)* **390**, L78.
——: 1992b, *Astrophys. J. Suppl.* **80**, 753.
——: 1992c, *Astrophys. J. Suppl.* **80**, 791.

Radio Galaxies and Their Environment

Wil van Breugel

Institute of Geophysics and Planetary Physics
L-413, Lawrence Livermore National Laboratory
P.O. Box 808
Livermore, CA 94551-9900, USA

"What you see is what you get"

Wild Bill Hickock

1 Introduction

The relationships between radio galaxies and their environment are varied, complex, and evolve with cosmic epoch. Basic questions are what role the environment plays in triggering and fuelling (radio) galaxy activity, what the effects of this activity are on its environment, and how radio galaxies and environment evolve. Clearly, this could be the topic of a workshop all in itself and the scope of this review will necessarily be limited. A review of the connections between environment and galaxy activity in general has been given by Heckman (1990).

First, I will briefly summarize the relationships between parent galaxy and cluster environments, and radio galaxies. A more detailed discussion of various aspects of this will be given elsewhere by F. Owen, J.O. Burns and R. Perley. I will then discuss the current status of investigations of extended emission-line regions in radio galaxies, again referring elsewhere in this volume for more detailed discussions of some particular aspects (kinematics and ionization mechanisms by K. Meisenheimer; polarization and spectral index lobe asymmetries by G. Pooley). I will conclude with a brief discussion of the current status of observations of high redshift radio galaxies.

2 Parent Galaxies, Neighbourhoods and Cluster Environments

Optical observations have shown that a large fraction (30% to 50%) of nearby ($z < 0.5$), *powerful* ($P_{1400,\mathrm{rest}} > 10^{33}$ ergs Hz^{-1} s^{-1})[1] FR-II type radio galaxies exhibit peculiar optical features (tails, fans; Heckman *et al.* 1986; Hutchings 1987) and unusually blue, extra-nuclear colors (Smith and Heckman 1989). While their $r^{-1/4}$ surface brightness profiles still classify them as ellipticals, the morphological distortions strongly suggest

[1] $H_0 = 50$, $q_0 = 0.0$

that they are interacting or merging with nearby gas rich companions (see Toomre and Toomre 1972). Support for this also comes from stellar dynamics studies which show that powerful radio galaxies which are morphologically peculiar exhibit more stellar rotation than other radio galaxies and normal ellipticals, presumably because of recent merging with a disk galaxy (Smith *et al.* 1990).

On the other hand, low power ($P_{1400,\mathrm{rest}} < 10^{33}\,\mathrm{ergs\,Hz^{-1}\,s^{-1}}$), FR I type radio galaxies appear associated with relatively normal elliptical galaxies, albeit generally of the D or cD type (*e.g.* Owen and Laing 1989). In addition to this morphological difference between the parent galaxies of low and high power radio sources, investigations reported at this workshop by F. Owen further suggest that the radio FR I / FR II type break may be a function of optical luminosity as well.

Several studies have been made of the galaxian/cluster environments of radio galaxies (*e.g.* Heckman, Carthy and Bothun 1985; Yates, Miller and Peacock 1989; Smith and Heckman 1990; Hill and Lilly 1991). These studies have shown that, at low redshifts, powerful FR II type radiogalaxies inhabit relatively sparsely populated regions with average galaxy densities a factor 2–3 lower than those of low power FR I type radio galaxies. At larger redshifts ($z \gtrsim 0.5$), the cluster environments of FR II type radio galaxies appear to be richer, while no change in environment has been found for FR I types (Hill and Lilly 1991). This suggests that FR II activity may somehow be related to the evolution of its cluster environment (intergalactic medium, galaxy richness, or both). Incidentally, the cluster environments of powerful radio galaxies and quasars are very similar (Smith and Heckman 1990; Hill and Lilly 1991), suggesting that these types of objects are also closely related.

While much about the detailed interpretation of these results remains unclear, the difference in parent galaxy types, environments, and stellar kinematics suggests that the two classes of radio sources do not simply evolve into one another, at least not on typical FR II timescales which may be at most a few times 10^7 years (Leahy 1991). The evidence strongly suggests that the interaction and merging of elliptical galaxies with gas rich companions plays an important role in triggering or fuelling powerful, FR II type radio source activity. The resulting decrease in galaxy density surrounding the radio galaxies would be consistent with the observed decrease in cluster richness with redshift. The importance of galaxy interactions in connection with galaxy activity and central star formation is being recognized in a variety of galaxy types, including star burst galaxies, Seyferts and quasars, as well as radio galaxies (see Heckman 1990). The processes of galaxy activity and star formation may occur simultaneously, or may even be causaly connected, when dense gas rapidly accumulates in the central regions of the interacting galaxies (Norman and Scoville 1988; Hernquist 1989; Shlosman, Begelman and Frank 1990).

The lower radio luminosities of the less disturbed parent galaxies of FR I type radio sources suggest a different origin and evolution of their activity. In these cases the "gas-supply" to their galaxy nuclei may be due to infall from their halos or clusters (cooling flows), and is presumably more sedate, continuous and of a lower level than is provided my the rapid merging processes thought to occur in FR IIs.

3 Interaction of Radio Galaxies with Their Environment

3.1 Extended Emission-Line Regions

Extended emission-line regions (EELRs) in radio and other active galaxies are being

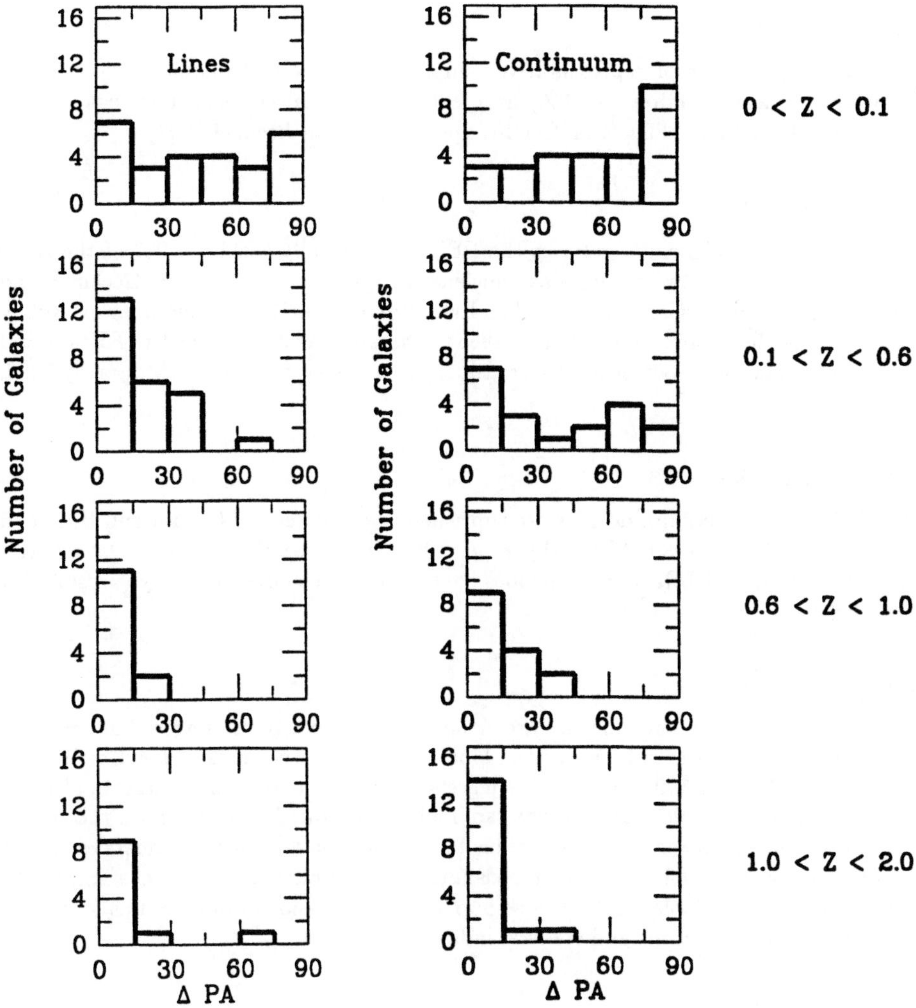

Fig. 1. These histograms show the inceasingly better alignment of EELR (left panels) and optical continuum emission (right panels) as function of redshift/radio power in the 3CR sample studied by McCarthy (1989).

studied by many groups. Important new developments since the last Ringberg workshop in 1988 are the realization that emission-line "cones" may be ubiquitous in active galaxies (Seyferts and radio galaxies), and the extremely good correlation between EELR and radio lobe asymmetries. I briefly summarize the current status and refer for a more detailed discussion of many relevant aspects to my previous review (van Breugel 1988), and to the proceedings of the ESO workshop on "Extranuclear Activity in Galaxies" (1989).

3.1.1 Occurrence

Narrow-band surveys of representative samples of nearby and distant radio galaxies show that a large fraction ($\sim 70\%$) have extended optical emission line regions (*e.g.* Baum and Heckman 1989*a*, *b*; McCarthy, Spinrad and van Breugel 1992).

3.1.2 Alignments

In relatively powerful radio galaxies the EELRs are usually within $\sim 20°$ of the radio axes, in most cases between the galaxy nucleus and radio hotspots (Baum, Heckman and van Breugel 1990; McCarthy, Spinrad and van Breugel 1992) but sometimes extending beyond radio hot spots (van Breugel 1988). The alignment between the EELRs and radio sources becomes better for high redshift / more powerful objects (McCarthy 1989; Fig. 1).

3.1.3 Luminosities

The EELR and radio luminosities are suprisingly well correlated (Baum and Heckman 1989*a*, *b*; McCarthy 1989; McCarthy *et al.* 1991) This, together with 3.1.2., strongly suggests that the EELRs and radio lobes have a common source of energy (presumably the nucleus).

3.1.4 Asymmetries

EELR and radio lobe asymmetries are strongly correlated: the strongest line emission is always found on the side where the lobe is closest to the nucleus (McCarthy, van Breugel and Kapahi 1991; Fig. 2). It suggests that the gaseous environments of radio galaxies are asymmetric, and that the radio lobes are slowed down most on the side of the densest gas. As a corollary, this means that radio lobe distance asymmetries cannot be used to constrain lobe advance speeds and radio source ages. I am referring to the contribution by G. Pooley at this workshop for the discussion of observed asymmetries in radio lobe polarizations and spectral indices.

3.1.5 Kinematics

The kinematics of EELRs in nearby radio galaxies appears to be dominated by gravitational forces, including rotation of gaseous disks acquired through merging with gas-rich companions or accretion from halos or cooling flows (*e.g.* Baum *et al.* 1991; Tadhunter 1991). However, radio jets/lobes can significantly affect the kinematics of emission-line gas at locations where they interact with ambient gas. Recent examples are Cygnus A and 3C 265, with EELR velocities of as much as 1800 km s^{-1} (Tadhunter 1991). Other

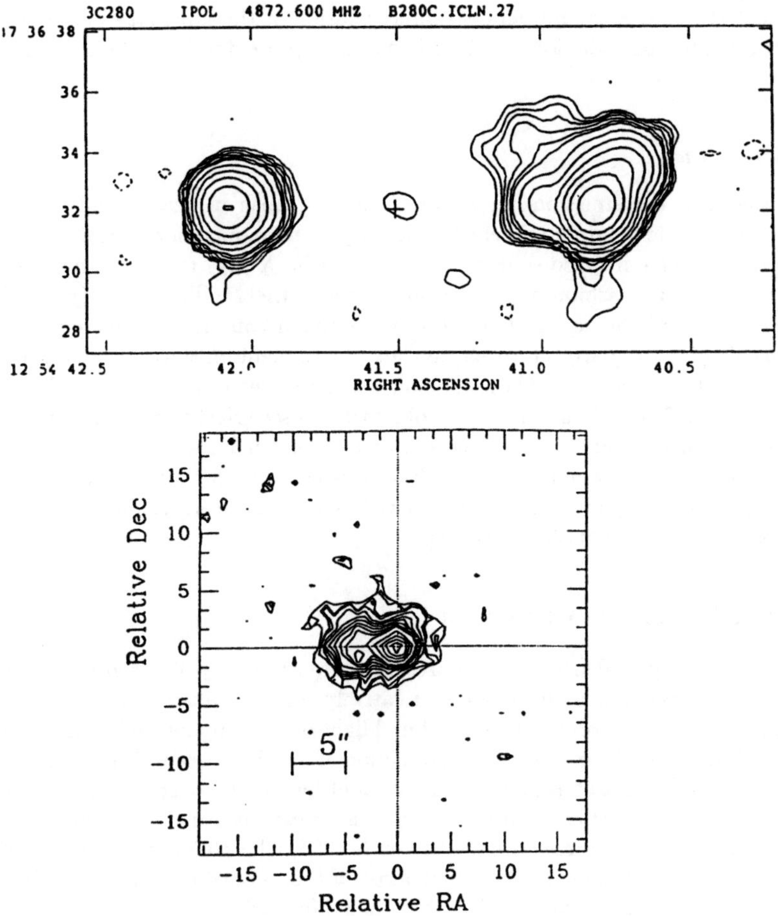

Fig. 2. An example of the correlation between EELR and radio lobe asymmetries (McCarthy, van Breugel and Kapahi 1991): shown are a 6 cm, 1″ resolution VLA map of 3C 280, $z = 0.99$ (top), and a continuum subtracted narrowband [OII]λ3727 image (bottom). Note that EELR is on the side with the lobe nearest to the nucleus (lobe distance ratio 1/1.27).

examples are 3C 277.3, 4C 29.30 and 3C 321, and it is worth repeating (see van Breugel 1988) that the velocities of emission-line gas seem to be nearly always blue-shifted on the side of the brightest jets. This might mean that these jets are Doppler boosted and pointed towards us. The question remains, however: where are the redshifted EELRs near radio knots of unseen (counter-) jets, which in such models should also exist and be pointed away from us? It is clearly of interest to continue studies of the kinematics of EELRs associated with radio jets. The emission-line kinematics near hotspots may also be interpreted in terms of gas entrainment by lobes propagating through dense gas (*e.g.*

van Breugel *et al.* 1985*a* [3C 277.3]; Wilson and Ulvestad 1987 [NGC 1068]; Pedlar *et al.* 1989 [Seyfert's]). A nice and detailed interpretation of the kinematics of emission-line gas entrained by the radio lobes of 3C 368 has been presented by K. Meisenheimer (this workshop).

3.1.6 Ionization

Quite generally, the emission-line ratios of EELRs, both in nearby and distant radio galaxies, can be adequately modelled with simple photoionization models in which the radio galaxy nuclei are also sources of ionizing (UV, X-ray) radiation directed or collimated in cone-like fashion along the same axes as their bulk outflow (jets, winds). I consider it beyond the scope of this review to give a complete discussion of the pros and cons of this "standard interpretation" except to mention that 1) the low ionization EELRs and their associated (?) low power radio sources in cooling flow clusters may well require a different dominant source of ionization see (Heckman *et al.* 1986), 2) close scrutinity of emission line ratios reveals some inconsistencies which require either more sophisticated AGN photoionization models, or leave room for alternative explanations (i.e. shock ionization) in specific cases. Two such cases are discussed in detail by K. Meisenheimer (3C 368) at this workshop.

3.2 Extended Optical Continuum

There is increasing evidence that the presence of very blue optical continuum emission associated with radio galaxies may be a relatively common phenomenon. The origin of this continuum may be young stars, scattered light or synchrotron emission, all of which are related to different aspects of the interaction of a radio source with its environment.

This is perhaps most dramatically illustrated by the discoveries that distant radio galaxies exhibit blue optical continuum which is nearly always aligned with the axes of their associated radio sources (McCarthy *etal.* 1987; Chambers *et al.* 1987; McCarthy *et al.* 1990, 1991). While it is likely that most of this optical emission is probably due to young stars (see for example Chambers and McCarthy 1990), optical polarization observations have shown that at least some of the UV excess may be due to scattered light from the nucleus (Scarrott *et al.* 1990; Januzzi and Elston 1991). Furthermore, as emphasized by Meisenheimer at this workshop, objects with very bright EELRs such as 3C 368 may significantly "contaminate" broadband images if strong emission-lines such as [OII]λ3727 fall in the filter band pass. This is really only an issue for objects in the redshift range 0.6 – 1.1 (for example with the Lick/"Spinrad" R_s filter covering 6000 – 7700 Å), and even in this redshift range emission-line free "intermediate broadband" images have confirmed the basic alignment effect (Rigler *et al.* 1992).

Also, in several *nearby* radio galaxies blue continuum emission has been found. There are the wellknown cases of optical synchrotron knots associated with radio jets (M 87, 3C 66B, 3C 277.3, PKS 0521−36 etc.) and lobes (3C 33, Pictor A and others; see elsewhere in these proceedings and those of the first Ringberg workshop). Young stars and scattered continuum are not as commonly found, although some spectacular examples are known (Minkowski's Object, van Breugel *et al.* 1985*b*; PKS 2152−69, di Serego Alighieri *et al.* 1988).

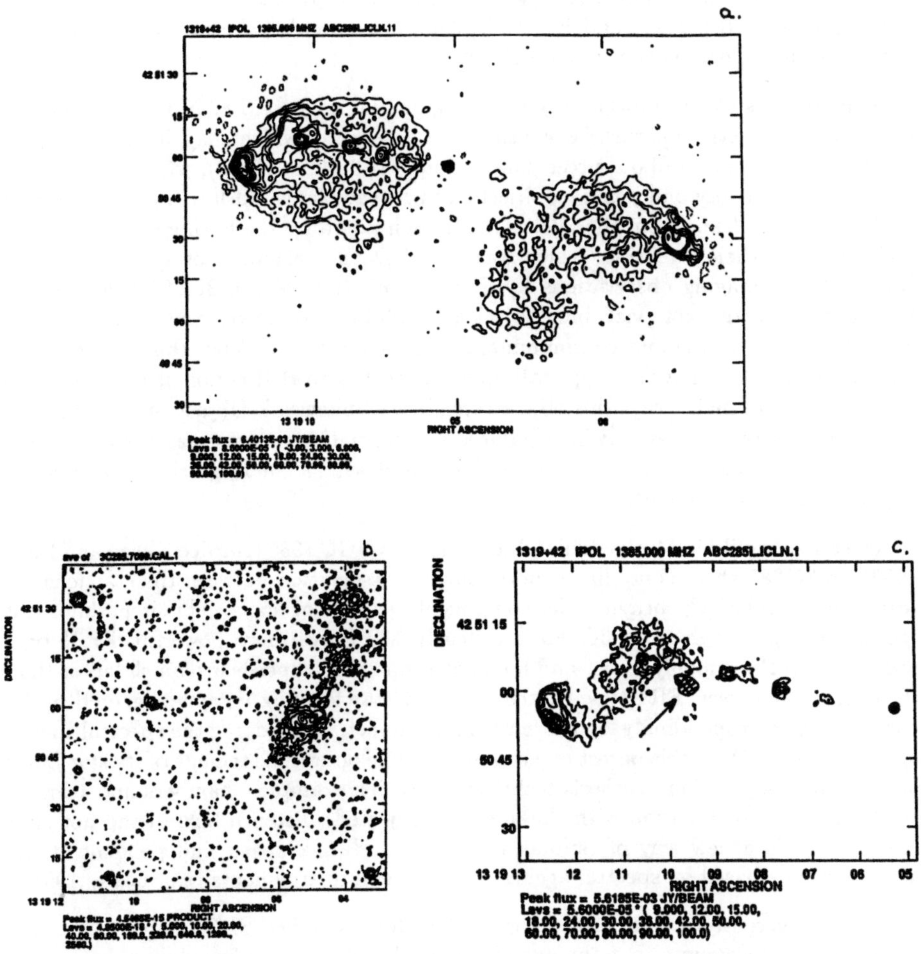

Fig. 3. Radio and optical images of 3C 285 showing the location of a star forming region behind the eastern hot spot. Fig. 3a shows a smoothed 2.″1 resolution 21 cm image of 3C 285. Fig. 3b shows a narrow-band image centered at the Hα emission-line (Baum *et al.* 1990; continuum not subtracted). Note the small, slightly extended EELR at $\sim 45''$ east of the parent galaxy. Lick spectra have shown that this EELR is an HII region at the redshift of 3C 285. Its location relative to the radio hot spot is shown in Fig. 3c (see arrow).

Motivated by the rather general presence of UV continuum associated with distant radio galaxies, and by evidence that similar effects — on a more modest scale — also seem to occur in nearby, less powerful radio galaxies, A. Dey (UC Berkeley) and I are conducting a systematic optical U- and R-band survey of a well defined sample of relatively nearby radio galaxies selected from the 3CR and PKS catalogs. All objects are constrained to have redshifts between 0.085 – 0.200 so that no bright emission-lines fall in the U- ([OII]λ3727) and R- ([OIII]λ5007) passbands (when convolved with the detector/telescope/atmosphere throughput).

Our first results have been very encouraging and suggest that in as many as 30% of the objects extra-nuclear UV continuum knots exist which may in some way be associated with their radio sources. One example is 3C 285 (Fig. 3), which was already known to have a very blue compact object along its radio axis (Saslaw, Tyson and Crane 1978), at some 60 kpc from the nucleus. Although noted by the authors it was not the main focus of their attention, which was aimed at the eastern "hot spot" of 3C 285. Narrow-band imaging observations by Baum *et al.* 1990 of the 3C 285 field showed, by chance, that this compact, blue object also exhibits extended optical line emission, with the brightest line and continuum emission facing the nucleus. Our spectroscopic observations at Lick have subsequently shown that it is an H-II region at the redshift of 3C 285. Thus, in many ways this object resembles Minkowski's Object, suggesting that the expanding radio lobes may have induced star formation in dense gas along the radio axis (possibly a cloud of cold gas or gas associated with a small, gasrich galaxy which happened to be in the path).

Of course, unlike in Minkowski's Object and also NGC 7385 (Hardee, Owen and Eilek 1980), in 3C 285 there is no direct morphological association between the star forming region and a relatively bright radio component (jet or hot spot). This is perhaps not too surprising though since 3C 285 is a nearly 100 times more powerful, FR II type radio source. Presumably its jet and hot spot propagation speeds are much larger than those of lower power, FR I type sources and any star forming regions triggered by its passage will correspondingly lag further behind. In fact, by studying the spectral energy distribution (SED) of this object one can in principle determine the age of its star burst using stellar population synthesis techniques (see for example Chambers and Charlot 1990). This in combination with the known (projected) distance behind the hot spot (43 kpc) yields a new way of estimating the lobe advance speed, independent of the usual methods based on spectral ageing or ram pressure arguments (*e.g.* Leahy 1991).

We are currently still lacking sufficient data for a detailed SED analysis, but it is probably safe to assume that the age of the starburst is somewhere between 10^7 and 10^8 years. Assuming, for example, an age of 5×10^7 years, we derive a hot spot advance speed of 900 km/s. Allowing for projection effects, which may led to underestimate this velocity by a factor of 2 to 3, and an uncertainty of a factor of 3 or so in the starburst age, which may go either way, I would conclude at this point that the hot spot speed is probably more than 1000 km/s, but probably less than 10,000 km/s. The upper estimate is consistent with the expansion speed derived for 3C 285 from spectral ageing arguments (8400 km/s; Alexander and Leahy 1987). Studies of H-II regions behind radio hot spots in nearby FR II sources may thus provide unique new constraints for radio galaxy models.

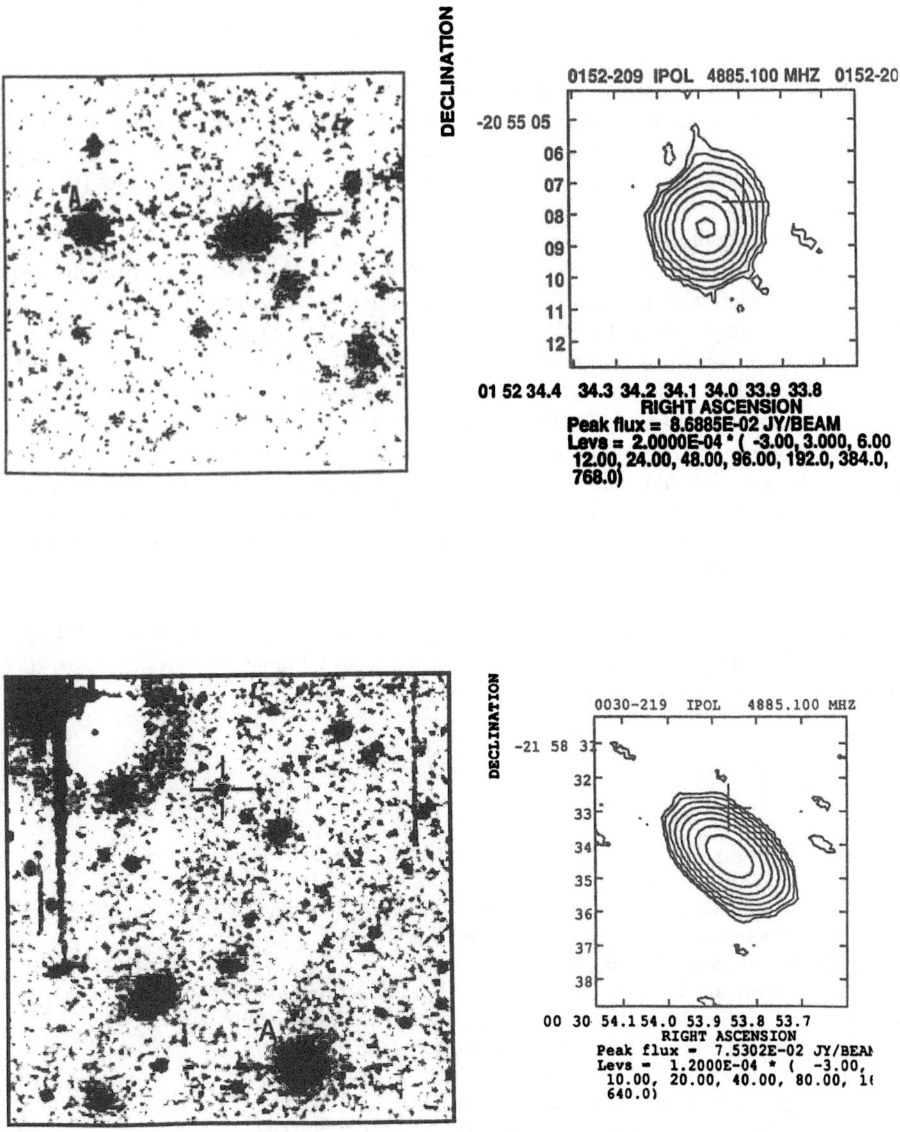

Fig. 4. Two examples of high redshift radio galaxies with very compact parent galaxies (left hand panels; MRC 0030−219 and 0152−209 with $z = 2.168$ and $z = 1.890$ respectively) and equally compact steep spectrum radio sources (right hand panels; from McCarthy *et al.* 1990, 1991).

4 Quasi-Stellar Distant Radio Galaxies

While high redshift ($z > 1$) radio galaxies generally have multi-modal, elongated morphologies more or less directed along their radio axes, it has also become clear from ungoing searches for distant radio galaxies that some of them can be rather compact or even stellar in good seeing images. Two such examples are the Molonglo sources MRC 0152-209 (McCarthy *et al.* 1991) and MRC 0030-219 (McCarthy *et al.* 1990; Fig. 4). Interestingly, in both cases the associated radio sources are also very compact ($0\farcs6$ and $< 0\farcs3$ respectively). Their optical spectra show relatively narrow Ly-α, indicating that these objects are galaxies, not quasars.

Another example is MRC 0349−211 ($z = 2.310$; McCarthy *et al.* 1991). In this case, the galaxy is relatively compact but the radio source is fairly large ($7\farcs4$; 94 kpc), and has a relatively strong core. There is also evidence that the CIV 1549 line is fairly broad. Together these observations suggest that MRC 0349−211 might be an example of a high redshift "broad-line radio galaxy" (this class includes nearby sources such as 3C 390.3 and 3C 445, and their parent "N-type" galaxies usually have strong nuclear continua). I would conclude from these ongoing investigations that there is increasing evidence that distant radio galaxies may exhibit a wide variety of properties and it will be interesting to compare these with those of quasars at similar redshifts to investigate a possible connection, evolutionary or otherwise (*e.g.* Heckman *et al.* 1991; Neff and Hutchings 1990; Barthel 1989).

Acknowledgements: This work was performed under the auspices of the U.S. Department of Energy at the Institute of Geophysics and Planetary Physics, Lawrence Livermore National Laboratory, under contract W-7405-Eng-48.

References

Alexander, P., Leahy, J.P.: 1987, *Mon. Not. R. astr. Soc.* **225**, 1.
Barthel, P.D.: 1989, *Astrophys. J.* **336**, 606.
Baum, S.A., Heckman, T.M., van Breugel, W.J.M.: 1990, *Astrophys. J. Suppl.* **74**, 389.
Baum, S.A., Heckman, T.M.: 1989a, *Astrophys. J.* **336**, 681.
——: 1989b, *Astrophys. J.* **336**, 702.
Baum, S.A., Heckman, T.M., van Breugel, W.J.M.: 1992, *Astrophys. J.* **389**, 208.
Chambers, K.C., Miley, G.K., van Breugel, W.J.M.: 1987, *Nature* **329**, 604.
Chambers, K.C., McCarthy, P.J.: 1990, *Astrophys. J. (Lett.)* **354**, L9.
Chambers, K.C., Charlot, S.: 1991, *Astrophys. J. (Lett.)* **348**, L1.
di Serego Alighieri, S. *et al.*: 1988, *Nature* **334**, 591.
Hardee, P.E., Owen, F.N., Eilek, J.A.: 1980, *Astrophys. J.* **242**, 502.
Heckman, T.M., Carty, T., Bothun, G.D.: 1985 *Astrophys. J.* **288**, 122.
Heckman, T.M.: 1986, *Astrophys. J.* **311**, 526.
——: 1990, in IAU Colloq. 124 *Paired and Interacting Galaxies*, NASA Conf. Publ. 3098, eds. J.W. Sulentic, W.C. Keel and C.M. Telesco.
Heckman, T.M., Lehnert, M.D., van Breugel, W.J.M., Miley, G.K.: 1991, *Astrophys. J.* **370**, 78.
Hernquist, L.: 1989, *Nature* **340**, 687.

McCarthy, P.J.: 1989, Ph.D. Thesis, UC Berkeley.
McCarthy, P.J., van Breugel, W.J.M., Spinrad, H. Djorgovski, S.: 1987, *Astrophys. J. (Lett.)* **321**, L29.
McCarthy, P.J., Kapahi, V.K., van Breugel, W.J.M., Subrahmanya, C.R.: 1990, *Astron. J.* **100**, 1014.
McCarthy, P.J., van Breugel, W.J.M., Kapahi, V.K., Subrahmanya, C.R.: 1991, *Astron. J.* **102**, 522.
McCarthy, P.J., van Breugel, W.J.M. Kapahi, V.K.: 1991, *Astrophys. J.* **371**, 478.
McCarthy, P.J., Spinrad, H., van Breugel, W.J.M.: 1993, *Astrophys. J. Suppl.*, in press.
Neff, S.G., Hutchings, J.B.: 1990, *Astron. J.* **100**, 1441.
Norman, C.A., Scoville, N.: 1988, *Astrophys. J.* **332**, 124.
Owen, F.N., Laing, R.: 1989, *Mon. Not. R. astr. Soc.* **238**, 357.
Pedlar, A.: 1989 *Mon. Not. R. astr. Soc.* **238**, 863.
Rigler, M.A.: 1992, *Astrophys. J.* **385**, 61.
Saslaw, W.C., Tyson, J.A., Crane, P.: 1978, *Astrophys. J.* **222**,, 435.
Scarott, S.M., Rolph, C.D., Tadhunter, C.N.: 1990, *Mon. Not. R. astr. Soc.* **243**, 5P.
Shlosman, I., Begelman, M.C., Frank, J.: 1990, *Nature* **345**, 679.
Smith, E.P., Heckman, T.M.: 1989, *Astrophys. J.* **341**, 658.
———: 1990, *Astrophys. J.* **348**, 38.
Smith, E.P., Heckman, T.M., Illingworth, G.D.: 1990, *Astrophys. J.* **356**, 399.
Tadhunter, C.N.: 1991, *Mon. Not. R. astr. Soc.* **251**, 46P.
van Breugel, W.J.M.: 1985, *Astrophys. J.* **290**, 496.
———: 1985, *Astrophys. J.* **293**, 83.
———: 1988, in *Hot Spots in Extragalactic Radio Sources*, Lecture Notes in Physics 327, eds. K. Meisenheimer and H.-J. Röser, Heidelberg: Springer Verlag, p. 121.
Toomre, A., Toomre, J.: 1972, *Astrophys. J.* **178**, 623.
Wilson, A.S., Ulvestad, J.S.: 1987, *Astrophys. J.* **319**, 105.
Yates, M.G., Miller, L., Peacock, J.A.: 1989, *Mon. Not. R. astr. Soc.* **240**, 129.

The Emission Line Lobes of 3C 368 ([23])

Klaus Meisenheimer and Hans Hippelein

Max-Planck-Institut für Astronomie
Königstuhl 17
D-6900 Heidelberg 1
Germany

Abstract: We present a detailed study of the morphology and kinematics of the emission-line lobes around the high-redshift radio galaxy 3C 368 ($z = 1.13$). The observations have been obtained with a new imaging Fabry-Pérot Spectrometer at the Calar Alto 3.5 m telescope. We have achieved a spatial resolution of $1\farcs33$ (FWHM) and a velocity resolution of 440 kms^{-1}. Our results confirm the close connection between the radio and emission-line lobes reported previously. However, our line-free continuum images taken between 700 and 900 nm show that any correlation between the continuum light distribution and the radio source is loose – if present at all.

The spatial and kinematic structure of the southern emission-line lobe is well described by an expanding conical shell of gas around the radio emitting plasma. It seems to be located in between the outer bow shock and the hot cocoon of the radio jet. A quantitative model in which the emission-lines originate from cooling material having passed the bow shock accounts well for the observed line intensity, the apparent lag behind of the emission-line region with respect to the radio source, and the velocities found in the gas. For the southern lobe we infer an external density $n_{ex} = 0.13 \times 10^{-6}\,\mathrm{m}^{-3}$ (in front of the bow shock), an advance speed $V_{bow} = 1170$ kms^{-1}, and a kinematic age of the radio source of 4.5×10^7 years. The pronounced north-south asymmetry of both the radio and emission-line lobes is easily explained by a moderate density asymmetry $n_{ex}^N : n_{ex}^S = 1 : 1.3$. The emission-line gas seems to be excited by shocks which are generated by the jet and its hot cocoon. No indications for the presence of a nuclear source of photo-ionization could be found.

[23]This paper appeared in *Astron. Astrophys.* **264**, 455–471 (1992).

Jet Disruptions at the Cores of Rich Galaxy Clusters

Jack O. Burns [1], Chris Loken [1], Kurt Roettiger [1],

Michael L. Norman [2], David A. Clarke [2]

[1]Department of Astronomy
New Mexico State University
Las Cruces, NM 88003, USA
[2]National Center for Supercomputing Applications,
University of Illinois
Urbana-Champaign, Urbana, IL 61801, USA

Abstract: The abrupt flaring of jets in wide-angle tailed radio sources and the newly recognized class of amorphous radio sources associated with dominant cluster galaxies may be understood in terms of various hydrodynamic jet instabilities. We present the results of numerical hydrodynamic simulations which duplicate many of the characteristics of these sources. In particular, the first simulation of a two-dimensional, perturbed jet in a cooling flow may have application to the amorphous sources.

1 Introduction

The cores of rich clusters of galaxies are interesting and dynamic places. Radio observations indicate that galaxies near the cores of rich clusters, especially the dominant galaxies, are very often AGNs (Brown & Burns 1992; Burns 1990). In such a dynamic, high gas pressure environment, the evolution of the radio plasma appears to be dramatically affected by its surroundings. Radio jets suddenly, yet symmetrically, disrupt and form diffuse tails of emission within the halos of some central dominant galaxies. The jets and tails bend under the (yet poorly understood) influence of pressure gradients in the intracluster medium (ICM).

Over the past few years, we have begun to explore the physical mechanisms which shape the radio morphology at the cores of clusters by means of a series of numerical hydrodynamic simulations which focus on jet disruption. We discuss two of these simulations in this paper. The first involves the jet/tail transition in wide-angle tailed (WAT) radio sources. The second involves jet instabilities in cooling inflows.

2 Jet Disruption in Wide-Angle Tailed Sources

WATs are generally the most luminous radio sources found in nearby rich clusters. As shown in Fig. 1, the large-scale morphology of WATs consists of two edge-darkened tails which can be up to a Mpc long. Generally, two straight and narrow jets emerge from the galaxy core, extend out to distances of 10-50 kpc (for $H_0 = 75$ km s^{-1} Mpc^{-1}) from the nucleus, and then symmetrically transform into diffuse tails. Often, the jets bend near the disruption point. WATs are associated with dominant galaxies (usually D or cD) at the optical and X-ray centroids of non-cooling flow clusters.

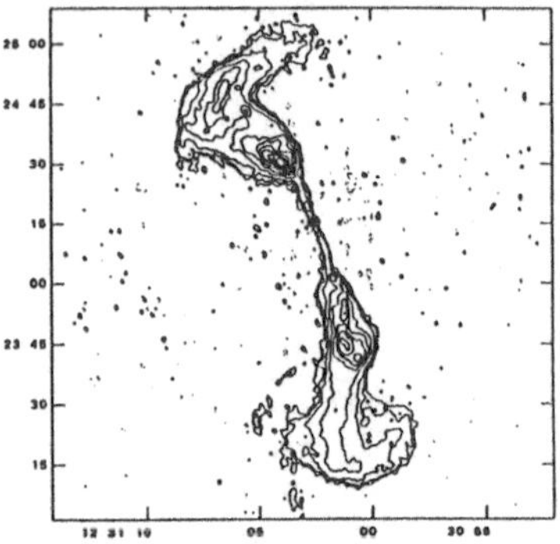

Fig. 1. VLA map of the wide-angle tailed radio source 1231+674 in Abell cluster 1559 from O'Donoghue *et al.* (1990). Note the sudden, and symmetric, transition from jet to tail.

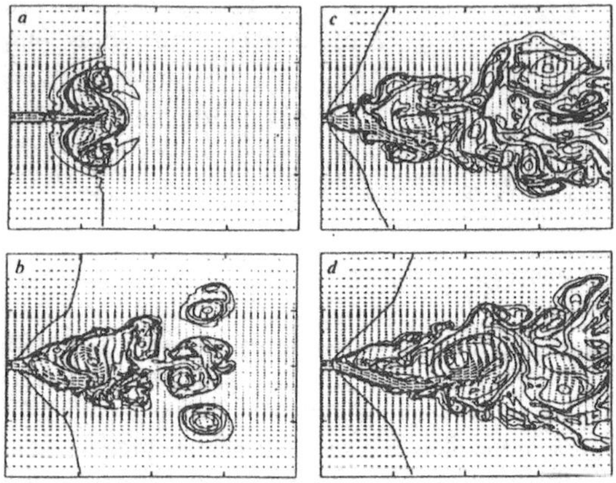

Fig. 2. Evolution of a 2-D, shock disrupted jet. Density contours are superposed on velocity vectors. Initial Mach numbers (with respect to jet sound speed) of the external wind and the jet are 2.9 and 2.5, respectively. Jet to ambient density ratio is 0.01.

We believe that the symmetrical disruption of jets in the halos of WAT galaxies is caused by the sudden change in the characteristics of the ambient medium, which occurs approximately at the expected interface between the interstellar medium (ISM) and the ICM. This jump in ambient gas conditions may be a contact discontinuity or a shock. Numerical simulations, shown in Fig. 2, demonstrate that a shock in the external gas will produce a corresponding Mach disk within the jet (Norman *et al.* 1988). According to Bernouilli's equation, the post-shock jet flow must decelerate and become subsonic. Such a fluid flow is well known to be highly turbulent and will entrain ambient gas as in Fig. 2. In order to reach pressure equilibrium with the surrounding medium, the shocked jet material must expand. The net result is a sudden jet/tail transition which looks strikingly similar to WATs like that in Fig. 1.

The origin of the discontinuity at the ISM/ICM interface is still unclear. It may arise from a cosmic ray driven supersonic wind which loosely couples with the ISM (and thus does not drive out all the galaxy gas and produce a hole in the X-ray emission) and causes a shock at the ISM boundary. Alternatively, a partial galaxy wind, predicted from time-dependent calculations of galaxy gas evolution (*e.g.*, Lowenstein & Mathews 1987), may produce the same effect. Another possibility is that a triaxial galaxy potential (consistent with the ellipticity common in cDs) could develop a contact discontinuity at the boundary of the ISM and the spherical distribution of cluster gas. We are actively pursuing these alternatives and will test them against current X-ray and radio data.

3 Unstable Jets in Cooling Inflows

Recent surveys of Abell clusters suggest that clusters with dominant galaxies and X-ray cooling inflows are radio-loud far more often than clusters without cooling cores (*e.g.*, Zhao *et al.* 1989). Burns (1990) found that $\approx 70\%$ of cDs with cooling inflows are radio-loud whereas $< 25\%$ of cDs in non-cooling flow rich clusters are radio-loud. This striking correlation suggests an intimate link between the central cluster environment and the generation of radio emission.

Many of the radio sources associated with dominant galaxies in cooling inflows are weak and compact. However, there is a sizable subset that are relatively luminous and

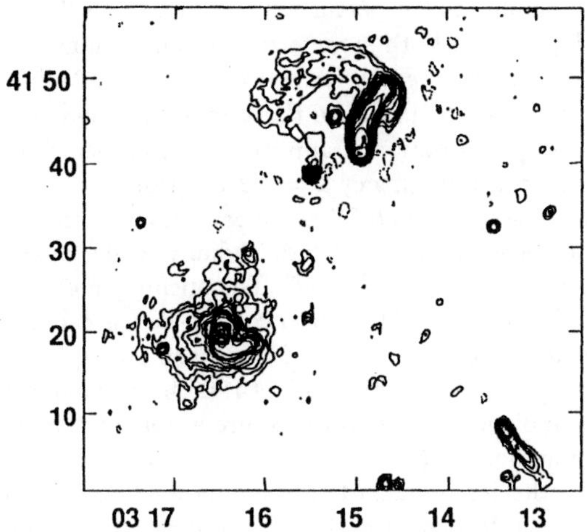

Fig. 3. VLA image of the Perseus cluster at 90 cm. 3C 84 is located in the west-central part of the map and has an amorphous, core-halo morphology. The well-known head-tail source, NGC 1265, is in north-central portion of the map and another head-tail, IC 310, is to the SW.

amorphous. A good example of this second class is 3C 84 in the Perseus cluster shown in Fig. 3. These amorphous sources are < 100 kpc in diameter, have little (< 1 kpc long) or no jets, have steep radio spectra, and have high rotation measures ($10^3 - 10^4$ rad m^{-2}). Burns *et al.* (1992) suggest that such sources seem to be trapped by the high pressure cooling inflows, where diffusion of electrons and magnetic fields may be important. The high rotation measures also indicate that a strong magnetic field is a crucial part of the physics of these clusters.

A common characteristic of radio sources in cooling flows is the small or absent radio jet. It appears that the cooling inflows produce an environment which causes the jets to become unstable and decollimate very near the nucleus (< 1 kpc). (It is worth noting that the sonic point in a transonic cooling flow occurs at about this same distance (Fabian *et al.* 1984) with an accompanying jump in temperature and pressure.) The amorphous radio emission may then result from diffusion of the relativistic particles into the dense ICM or via a slow bulk flow of the poorly collimated radio plasma (Zhao 1990). We have begun to investigate the disruption of radio jets in cooling inflows via 2-D numerical experiments using ZEUS-3D (Clarke 1990), an Eulerian finite-difference MHD code under development at the National Center for Supercomputing Applications which is directly descended from ZEUS-2D (Stone and Norman 1992).

The linear theory of instabilities in slab jets has been developed and compared with numerical simulations by Hardee and Norman (1988) and Norman and Hardee (1988). Propagating slab jets were considered in Hardee and Norman (1990), while slab jets in atmospheric gradients were modelled by Hardee *et al.* (1991). Axisymmetric (pinch) instabilities of jets in cooling flows have been discussed by Sumi and Smarr (1984) and simulated by Sumi and Norman (1987), while limited 3D models of jets in cooling flows were published by Soker and Sarazin (1988). Here we present new results of sophisticated 2D simulations of perturbed jets in cooling flows in a geometry which allows for the development of both symmetric and antisymmetric instabilities. We use spherical coordinates and assume axisymmetry about the rotation axis. A steady-state cooling flow solution is used to initialize the atmosphere and then a jet is injected at $\theta = 90^0$. A small (2%) perturbation is applied to the jet velocity in the perpendicular (θ) direction at the inlet at a frequency $\omega = 0.21\omega^*$, where ω^* is the resonant frequency (Hardee and Norman 1988) calculated for the conditions at the inner boundary.

The computational grid spans 90 jet radii radially (400 zones) and covers 50° in the θ direction (320 zones), centered on $\theta = 90°$. The opening angle of the jet is 2° and it is well-resolved axially by 40 zones. The initial cooling-flow is calculated as in Sulkanen *et al.*(1989) for parameters applicable to A2634 and has a temperature of $T_s = 2.1 \times 10^6$ at the sonic point which is located about 19 jet radii from the inner boundary. The inlet values of the jet-to-ambient density and pressure are $\eta = 0.3$ and $K = 1$, respectively, while the jet Mach number is $M_j = 6$.

At first (Fig. 4a), the jet remains fairly symmetric. At the time of Fig. 4b, a pronounced asymmetry has developed as the working surface is moving at a substantial angle with respect to the initial direction of propagation. The jet head has moved about 42 jet radii by this time. The working surface subsequently flaps about and the protrusion in Fig. 4b is swept backwards by the ram pressure of the cooling flow (Fig. 4c). The jet stagnates after the epoch shown in Fig. 4b and has hardly moved any further by the end of the simulation (Fig. 4d). In terms of the e-folding length 1^*, the jet in Fig. 4c has moved about $3.4l^*$ whereas linear analysis in the case of a uniform ambient medium predicts that jet disruption will occur at about 6-7 e-folding lengths (Hardee and Norman 1988). The positive temperature gradient in the cooling flow therefore appears to decrease jet stability in agreement with the results of Hardee *et al.* (1991).

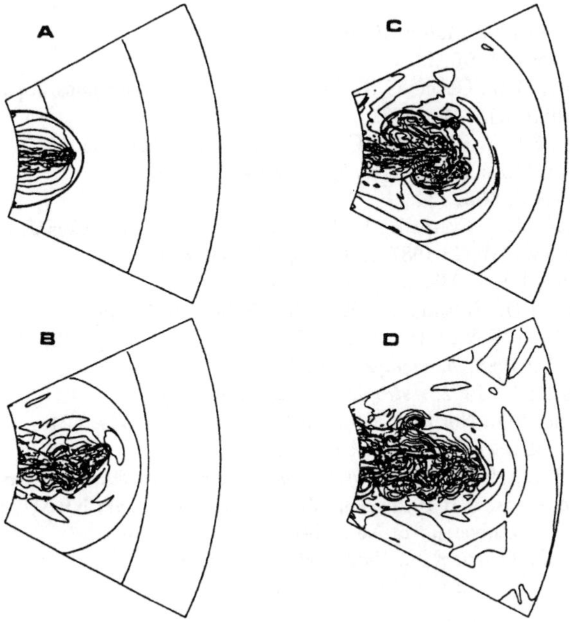

Fig. 4. 2-D numerical simulation of a perturbed jet in a cooling flow at four epochs. Equally spaced contours in log of density are shown. Initial Mach number of the jet is 6 (with respect to internal sound speed). Jet to ambient density ratio is 0.3.

4 Conclusions

Numerical simulations have now reached the point where we can begin to understand complex, distorted radio structure in terms of hydrodynamic instabilities. We have shown that a shock at the ISM/ICM interface can account for the symmetrical disruption of WATs. We also suggest that the amorphous radio sources may result from jet disruption in a cooling flow and have presented the first numerical simulation of such a scenario.

Acknowledgements: This work was supported in part by a grant from the U.S. National Science Foundation (AST-9012353) to J.O.B. and M.L.N.

References

Brown, D., Burns, J.O.: 1992, *Astron. J.*, in press.
Burns, J.O.: 1990, *Astron. J.* **99**, 14.
Burns, J.O., Sulkanen, M.E., Gisler, G.R., Perley, R.A.: 1992, *Astrophys. J. (Lett.)*, in press.
Clarke, D.A.: 1990, *Bull.AAS* **22**, 1302.
Fabian, A.C., Nulsen, P.E.J., Canizares, C.R.: 1984, *Nature* **311**, 733.
Hardee, P.E., Norman, M.L.: 1988, *Astrophys. J.* **334**, 70.
——: 1990, *Astrophys. J.* **365**, 134.
Hardee, P.E., Norman, M.L., Koupelis, T., Clarke, D.A.: 1991, *Astrophys. J.* **373**, 8.
Lowenstein, M., Mathews, W.G.: 1987, *Astrophys. J.* **319**, 614.
Norman, M.L., Hardee, P.E.: 1988, *Astrophys. J.* **334**, 80.
Norman, M.L., Burns, J.O., Sulkanen, M.E.: 1988, *Nature* **335**, 146.
O'Donoghue, A.A., Owen, F.N., Eilek, J.A.: 1990, *Astrophys. J. Suppl.* **72**, 75.
Soker, N., Sarazin, C.L.: 1988, *Astrophys. J.* **327**, 66.
Stone, J.M., Norman, M.L.: 1992, *Astrophys. J. Suppl.* **80**, 753.
Sulkanen, M.E., Burns, J.O., Norman, M.L.: 1989, *Astrophys. J.* **344**, 604.
Sumi, D.M., Norman, M.L.: 1987, *Bull.AAS* **19**, 1049.
Sumi, D.M., Smarr, L.L.: 1984, *Physics of Energy Transport in Extragalactic Radio Sources*, NRAO Workshop 9, eds. A.H Bridle, and J.A. Eilek, Greenbank: NRAO, p. 168.
Zhao, J.-H.: 1990, thesis, University of New Mexico.
Zhao, J.-H., Burns, J.O., Owen, F.N.: 1989, *Astron. J.* **98**, 64.

Steps Toward a Radio H-R Diagram

Frazer N. Owen

National Radio Astronomy Observatory
Socorro, New Mexico 87801, USA

Abstract: We discuss the possibility of establishing a set of observational parameters for radio galaxies which could be used to study radio source evolution in the same way that the Hertzsprung-Russell diagram is used to understand stellar evolution. As a first step towards a radio H-R diagram, we show that the FR I/II break is a function of both radio and optical luminosity. This result combined with the radio linear size/luminosity diagram suggests that the FR I/II break is very sharp and the two morphological classes probably have an evolutionary connection.

1. Introduction

For some time the jet or beam model has been accepted as the basic physical picture which describes powerful extragalactic radio sources. For almost a decade this model has been studied numerically with increasing physical and mathematical sophistication. In virtually all versions of the model, both analytic and numerical, the model describes a rapidly evolving structure. Furthermore, arguments based on the reasonable parameters for the model — on particle lifetimes, on estimated velocities for the advance of the hot spots or on their rapid cosmological evolution — suggest that the sources have lifetimes short with respect to the evolution of stars and/or galaxies.

Interpretation of radio source images are usually made by assuming that the source being studied is in some particular stage of its evolution. However, the standard picture would suggest that the radio sky is made up of sources in all stages of evolution. If this is true then we should be able to create some analogy to the stellar H-R diagram which we can use to understand radio source evolution and to test our existing theoretical ideas. Of course, the measured parameters may not be quite the same. Four come to mind that might be useful in relating to the standard model and that are practical to obtain: absolute radio luminosity, radio size, integrated radio spectral index, and optical luminosity of the parent galaxy. The radio luminosity and size are most directly predicted by jet models, although practically they must be measured at a particular frequency. The spectral index can also be predicted, although particle losses and acceleration need to be included in the model. The luminosity of the parent galaxy seems likely to be important since it gives an idea of the mass of the parent system and possibly the environment into which the radio jet is injected.

Ideally, these four quantities or any other set one cares to study can be formed into an n-dimensional cube. For a given optical parent galaxy magnitude, the density of points in each n-dimensional plane should be proportional to the percentage of the time each galaxy stays in that state.

2. The VLA Abell Cluster Survey

My coworkers and I are beginning to study the distribution of these four parameters for a sample of radio sources in Abell clusters of galaxies. We have completed all clusters in Abell's northern survey with measured redshifts less than 0.09. We have surveyed most of the 353 clusters down to a limiting flux density at 20cm of 10mJy using the VLA C array. For the rest we have used published Westerbork observations of the clusters with similar limiting flux densities. A second sample of about 2000 Abell clusters has been surveyed down to 200mJy with the 300 foot telescope at 20cm. In both cases we have included sources out to 0.5 Abell radii for clusters with redshifts greater than 0.047. At lower redshifts, we are complete out to 20 arcmin from the cluster at the 10mJy level and to 100mJy out to 0.5 Abell radii. All of the sources in the 200mJy sample have been mapped with the VLA C and/or B arrays.

In order to isolate the true cluster sources and remove the large number of background sources from the sample we have identified all the sources on the Palomar Sky Survey E prints using the NRAO MANN measuring machine in Socorro. We have kept only sources with absolute magnitudes R brighter than $M = -20.5$, assuming $H_0 = 75$ km s^{-1} Mpc^{-1} if they were cluster members. The resulting sample has 370 objects with $z \leq 0.09$ and 110 more objects in the 200mJy sample.

From this work we have statistically complete samples with which one can study radio source evolution. A graduate student (Michael Ledlow) and I are in the process of measuring radio sizes by adding VLA A and B configuration data at 20cm and optical R-band magnitudes for a subset of about three hundred galaxies from these two samples. Radio spectra will be added at a later stage. In the next few years we should be able to produce our H-R three or four dimensional cube and test the predictions of various forms of the models.

3. The Current Situation

However, right now we can begin to look at projections of the cube for various samples and begin to see some trends. One such projection is the radio luminosity function for E and S0 galaxies. This function has been studied by many authors. Auriemma *et al.* (1977) produced the first bivariate luminosity function by studying the probability distribution of radio luminosity as a function of optical absolute magnitude. Recently, Sadler *et al.* (1989) and de Ruiter *et al.* (1990) have studied both the univariate and bivariate functions. While the latter two studies are at very different frequencies, 5000 MHz and 408 MHz respectively, and disagree about the low luminosity part of the function, both conclude that there is a break in the slope of the univariate function near $\log P_{tot}$(1400 MHz) = 10^{25} W Hz^{-1}. This break is interestingly close to the change in morphological type between FR I and FR II structures (Fanaroff and Riley, 1974). However, the change in morphological type takes place over at least an order of magnitude in radio luminosity and is not quite at the break frequency in the univariate luminosity function.

Another projection of our hypothetical cube is the linear size versus radio luminosity plane. Baldwin (1980) discussed this diagram for FR II sources and concluded that some modification of the standard model was necessary to understand radio source evolution. He suggested that taking into account the radial decrease in density as the jet propagated outward produced a more plausable set of evolutionary tracks than a jet propagating into a constant density. Wiita and coworkers have explored a similar idea theoretically (Wiita *et al.* 1990 and references therein). They include propagation of the jet through an interstellar/intergalactic medium boundary. In this way they can explain a number of the features of the radio luminosity function and the change of the radio source population with epoch.

De Ruiter *et al.* (1990) give an up-to-date discussion of this projection for FR I and FR II sources using 3CR and B2 radio galaxies. In Fig. 9 of their paper, one can see that the linear sizes of their sample of B2 and 3CR radio galaxies increases from 10^{22} to 10^{25} W Hz^{-1} and then levels off or increases more slowly. This plot shows that the lower luminosity FR I radio galaxies are generally smaller than the FR II sources and that the size-luminosity trend is continuous. This plot makes it hard to imagine that FR II radio galaxies simply evolve into FR Is as they get older and bigger. However, the continuity suggests some connection between the two classes.

4. The Bivariate FR I/II Correlation

A further clue to the evolution is found in the results of groups which have studied the parent galaxy luminosity as a function of FR type (Lilly and Prestage, 1987; Owen and Laing, 1989; Owen and White, 1991). Others have looked at FR type versus local galaxy density and evolution of this relation with redshift (Prestage and Peacock, 1988, Yates *et al.* 1989, Hill and Lilly 1991). These studies have concluded that locally, (*i.e.* $z < 0.2$), FR I sources are associated with more luminous galaxies and denser environments than FR II sources. At higher redshifts ($z \sim 0.5$), FR IIs are also found in brighter galaxies in rich environments. Thus the environments in which FR IIs are found appear to have evolved since $z \sim 0.5$.

Another projection of our proposed cube is the absolute optical/radio luminosity plane. Owen and Laing (1989) and Owen and White (1991) have discussed this diagram. The latter paper reports that the FR Is and IIs seem to be separated by a diagonal line in this diagram . Below we show an updated version of this figure. The 1s and 2s represent FR I and II morphology respectively.

The figure has been created by combining the optical photometry and radio morphological types from Lilly and Prestage (1987), Smith and Heckman (1989), Baum *et al.* (1988), Owen and Laing (1989), Hill and Lilly (1991), Owen and White (1991), and Owen *et al.* (1992). A few morphogical types have estimated from maps provided by van Breugel (private communication). Objects with uncertain morphological classifications

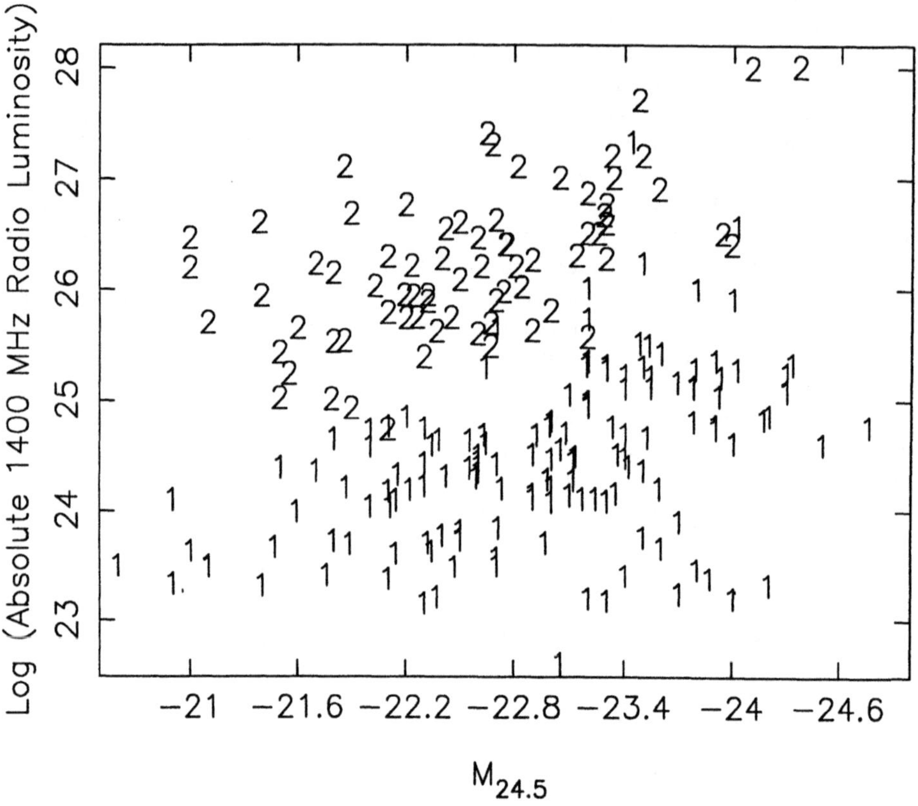

Fig. 1. The log of the 1400 MHz absolute radio luminosity is plotted against the isophotal R magnitude of the parent galaxy, $M_{24.5}$. Units for the radio luminosity are W Hz^{-1}. 1s and 2s refer to FR I and FR II morphological types, respectively.

or uncertain redshifts have not been included in the figure. In cases where the measurements were not on the R-band $M_{24.5}$ scale defined by Owen and Laing, the results have been transformed to $M_{24.5}$ using mean colors for galaxies and/or the relation between the Gunn-Oke absolute magnitudes and $M_{24.5}$ displayed in Fig. 3 of Owen and White.

Cumulative errors in the transformation range from 0.1 to 0.5 magnitudes and thus direct measurements should be made to improve the reliability of the diagram. However, since the figure is a log – log plot, even with these large errors the trend reported by Owen and White is clearly visible and stands out more clearly with the much larger number of data points. This plot is not based on a complete sample so the density of points does not reflect the probability distribution of any complete parent sample. The next step will be to study more complete samples. However, the current version should show us what parts of the diagram is occupied by which classes of sources. Given the errors and the selection, this diagram should be looked on as a preliminary result and a sign of the potential of such studies. However, we can already gain some insight from the current version.

As one can see, there appears to be a linear boundary between FR I and FR II sources which depends on radio and optical luminosity. This two dimensional diagram shows that the width of the FR I/II break is due to projecting the sharp two dimensional transition between the two classes into one dimension, leaving only radio luminosity as the variable. It seems likely that the bivariate luminosity function has a break frequency that depends on optical luminosity and drops off rapidly at the FR I/II break. Initial studies of the VLA cluster sample seem consistent with this hypothesis. However, this speculation needs to be tested by more quantitative work.

The FR I/II dividing line goes roughly as $L_{radio} \propto L^2_{optical}$. The relative sharpness of the dividing line suggests that FR I and FR II sources have some relation, perhaps an evolutionary relation. Given the size luminosity/diagram, perhaps one should conclude that FR I sources evolve into FR II sources rather than the other way around. Alternatively, FR IIs may just be the signature of a higher power state in the jet than FR Is.

In order to construct the current diagram and draw our conclusions, we have mixed sources inside and outside of rich clusters. One might expect to see the effect of environment. However, as shown in Owen and Laing and Owen and White, no evidence of a difference inside and outside clusters can be found. De Ruiter *et al.* also do not find clear differences inside and outside cataloged clusters. Furthermore, analysis of the VLA cluster survey agrees with an earlier result of Fanti (1984) which shows that the radio luminosity function inside the cores of Abell clusters is the same as one finds for radio galaxies not selected to be in rich clusters. However, the lack of any effect of galaxy density may mean that environment has no effect on radio properties or that we simply do not understand the environments of radio galaxies.

Since the points from Hill and Lilly are included in Fig. 1, there is little evidence that a change in this relation has occurred out to $z = 0.5$, although it is hard to rule out the possibility absolutely. The data seem consistent with a dependence of morphology on radio and optical luminosity rather than on epoch. If more optically more luminous galaxies occur more often in rich environments, then the apparent change in environment at redshifts near 0.5 may also be related to this correlation. However, further work with good optical magnitudes and complete samples is necessary to clarify this point and to lead us closer to radio H-R diagram.

References

Auriemma, C., Perola, G.C., Ekers, R.D., Fanti, R., Lari, C., Jaffe, W.J., Ulrich, M.H.: 1977, *Astron. Astrophys.* **57**, 41.

Baldwin, J.E.: 1980, in *Extragalactic Radio Sources, IAU Symp. 97*, Dordrecht: Reidel, p21.

Baum, S.A., Heckman, T., Bridle, A., van Breugel, W., Miley, G.: 1988, *Astrophys. J. Suppl.* **68**, 714.

de Ruiter, H.R., Parma, P., Fanti, C., Fanti, R.: 1990, *Astron. Astrophys.* **227**, 351.

Lilly, S.J., Prestage, R.M.: 1987, *Mon. Not. R. astr. Soc.* **225**, 531.

Hill, G.J., Lilly, S,J.: 1991, *Astrophys. J.* **367**, 1.

Fanaroff, B.L., Riley, J.M.: 1974, *Mon. Not. R. astr. Soc.* **167**, 31.

Fanti, R. (1984) in *Clusters and Groups of Galaxies*, Dordrecht: Reidel, p185.
Owen, F.N., Laing, R.A.: 1989, *Mon. Not. R. astr. Soc.* **238**, 357.
Owen, F.N., White, R.A.: 1991, *Mon. Not. R. astr. Soc.* **249**, 164.
Owen, F.N., White, R.A., Burns, J.O.: 1992, *Astrophys. J. Suppl.* **80**, 501.
Prestage, R.M., Peacock, J.A.: 1988, *Mon. Not. R. astr. Soc.* **230**, 131.
Sadler, E.M., Jenkins, C.R., Kotanyi, C.G.: 1989, *Mon. Not. R. astr. Soc.* **240**, 591.
Smith, E.P., Heckman, T.M.: 1989, *Astrophys. J. Suppl.* **69**, 494.
Wiita, P.J., Rosen, A., Norman, M.L.: 1990, *Astrophys. J.* **350**, 545.
Yates, M.G., Miller, L., Peacock, J.A.: 1989, *Mon. Not. R. astr. Soc.* **240**, 129.

Where Have All the Old Radio Sources Gone?

Jean A. Eilek

New Mexico Institute of Mining and Technology
Socorro, NM 87801, USA

Abstract: The standard picture of radio sources, in which a steady jet supplies the extended lobes with energy and mass, predicts distributions of source properties which contradict observations. The most dramatic contradiction is the non-detection of the large numbers of steep spectrum sources predicted by the theory. It seems to be time for a new approach to the basic source model.

1. The Problem

Most work on radio galaxies shares a common picture: that a steady or quasi-steady jet transports energy and mass to the large-scale radio lobes. The interaction of this outflowing plasma with the external medium results in the radio source we see. This picture makes strong predictions for the dynamical and energetic evolution of the sources. In the context of this model, we can easily predict the evolution with time of the linear size (D) of a source. With a judicious choice of microphysics, we can also predict the evolution of radio power at a given frequency, p_ν, and of the spectral index, $\alpha = -d\ln p_\nu / d\ln \nu$. The most important parameter in such a model is the total jet power, P_{jet}; the predictions turn out to be less sensitive to other factors such as the ambient density and pressure.

The dynamic and energetic lifetimes of the sources are probably short compared to a Hubble time. If radio sources are being born at a steady rate, an observed sample will contain sources of all ages, young through to old. The observed distribution of source properties — $N(p_\nu, D, \alpha)$ — will be determined by the evolution of individual sources (in p_ν, D and α), as well as by the intrinsic spread of basic parameters (such as P_{jet}).

Comparing $N(p_\nu, D, \alpha)$ to real data tests the basic model (*e.g.* Baldwin 1982). Type I and type II sources should be treated separately in this exercise. In addition to current models which suggest that the underlying physics of the sources are different, we know that these two source types differ in their (p_ν, D, α) distributions.

One basic statistic is the luminosity function, $N(p_\nu)$. When derived for type I and II sources separately (Eilek, 1990), $N_{II}(p_\nu)$ peaks at a high p_ν value, while $N_I(p_\nu)$ continues to low powers as $\propto p_\nu^{-s}$, where $s \sim 0.2$.

Another statistic is the distribution of sources in the (D, p_ν) plane. This distribution is given in de Ruiter *et al.* (1990); the general trend is for higher-power sources to be larger. A rough fit to the data in that paper would be, $D \propto p_\nu^{0.5}$ for B2 sources (mostly type I), and $D \propto p_\nu^{0.2}$ for 3CR sources (mostly type II); but the data show a great deal of scatter around this trend.

A third statistic is the distribution of sources in the spectral index. Available spectral information is easy to describe: both the B2 and 3CR data sets have mean spectral indices $\langle\alpha\rangle \sim 0.75 - 0.80$, with a dispersion $\sigma_\alpha \sim 0.3$ (de Ruiter *et al.* 1990).

2. Models: How Might Sources Evolve?

The dynamical evolution of the sources is predicted in a straightforward way by applying momentum and energy conservation to the interaction of the jet with the ambient medium. These predictions depend only on the macrophysics of the source, and should therefore be robust.

Following Scheuer (1974), momentum balance at the end of the jet or tail can be used to find the advance speed of the radio plasma into the ambient gas. From this the linear size of the source can be predicted (see also Gopal-Krishna and Wiita 1987). I modelled type II sources as supersonic beams moving into a medium with background density dropping going away from the source. Type I sources were modelled as transonic flows moving into a constant density medium. The volume of type II sources is probably determined by a self-similar overpressure expansion (Eilek and Shore 1989); for type I sources, the data suggest that the aspect ratio is independent of source size, predicting $V(t) \propto D^3(t)$.

The mean magnetic field within the source is then found from the total magnetic energy divided by the source volume. If magnetic energy is supplied by the jet at a rate $\propto P_{jet}$ (which can include *in situ* growth of field energy from dynamos or shocks), the mean magnetic field will *decay* with time; the jet cannot supply energy fast enough to overcome the faster-than-linear volume growth.

The next step is the prediction of $p_\nu(t)$. In order to do this, we need to know the electron spectrum. The synchrotron power at frequency ν can be related to the magnetic field and electron spectrum by

$$p_\nu(t) = \frac{C_o}{\nu} N(e_{obs}, t) e_{obs}(t)^3 B(t)^2 \tag{1}$$

where C_o contains fundamental constants and $e_{obs}(\nu, t) \propto (v/B(t))^{1/2}$ is the electron energy which dominates the radiation at frequency ν.

To understand $N(e_{obs}, t)$, we need the microphysics. Since we do not have a trustworthy microphysical model for real radio sources, I chose simple models to investigate the effects of $N(e,t)$ in (1). The simplest model is that of steady injection, with no *in situ* reacceleration. (For instance, this might take place at the hot spot of a type II source). Injection of a power law plus synchrotron aging results in $N(e,t) \propto te^{-s}$ for $e < e_c$, and $e^{-(s+1)/(1-m)}$ for $e > e_c$, if $m = -d\ln B/d\ln t$ (Eilek and Shore 1989). The cutoff frequency decays as $e_c(t) \propto t^{(2m-1)}$. The time when $e_c(t_{sy}) = e_{obs}(t_{sy})$ is a useful definition of the synchrotron lifetime: $t_{sy} \sim 3 \times 10^7 \nu_{GHz}^{1/2} B_{10\mu G}^{3/2}$ yrs.

A variant of the simple ageing model is one in which the jet is turned off, say at $t \sim t_{sy}$. After turnoff, the lobe will expand sightly into pressure balance with the external medium. Past this time, the source volume and magnetic field will stay constant, and the electrons will age with $e_c(t) \propto t^{-1}$, in the usual manner.

There is, of course, a good chance that the electrons are reaccelerated in the lobes, for instance by turbulence or by local shocks. The resultant $N(e,t)$ depends on the details of the acceleration process, so that no simple and generally applicable model is available. Based on current models of reacceleration, one might pick a "best case" model: namely, that $N(e,t) \propto te^{-s}$ for $e < e_c$, with e_c a constant, and $N(e,t)$ falling exponentially above e_c (*e.g.*, Borovsky and Eilek 1986). That is, in good conditions, reacceleration might maintain a power law spectrum with a constant high-energy cutoff.

All of these models predict that $p_\nu(t)$ rises for $t \lesssim t_{sy}$ (young sources), and falls for $t \gtrsim t_{sy}$ (old sources); $\alpha(t)$ stays constant for young sources and steepens for old sources. The details of the $p_\nu(t)$ and $\alpha(t)$ behavior depend on the details of the electron spectrum, and on the source type; but these general trends are shared by all models I have considered.

3. Statistics: Do these Models Work?

The short answer is, not very well. I have investigated what these models predict for $N(p_\nu, D, \alpha)$. To do this, I assumed that the sources are born at a constant rate, with a spread $F(P_{jet})$ of intrinsic jet power. I followed the evolution of a sample of sources in linear size and in radio flux density at two frequencies; the latter gives the evolution of the two-frequency spectral index.

I first considered the luminosity function, $N(p_\nu)$. If individual sources are created at a rate $S(p_\nu, t)$, they will populate $N(p_\nu)$ as they evolve:

$$\frac{\partial N}{\partial t} + \frac{\partial}{\partial p_\nu}\left(N\frac{dp_\nu}{dt}\right) = S(p_\nu, t) \tag{2}$$

(*e.g.*, Cavaliere *et al.*, 1971). Thus, single-source evolution could be responsible for the shape of $N(p_\nu)$. However, this is not the answer for the sources we see. The slow decay of $p_\nu(t)$ which the models predict for older sources is not compatible with the broad spread of p_ν values observed, even for type II sources. The observed $N(p_\nu)$ must be due to the distribuiton of intrinsic jet power, $F(P_{jet})$, rather than to single-source evolution. $N(p_\nu)$ for type II sources can be reproduced with a Gaussian distribution for $F_{II}(P_{jet})$. This is reminiscent of the optical luminosity function of the parent galaxies (large ellipticals; *e.g.*, Binggelli *et al.* 1988). However, the extended, low-power tail of $N_I(p_\nu)$ requires a power-law $F(P_{jet})$. Since type I sources also have large elliptical galaxies as their parent galaxies, the relation of intrinsic jet power to galaxy size must be more complex for these objects.

I next used these $F(P_{jet})$ distributions to investigate the expected behavior of sources in the (D, p_ν) plane. Individual sources will follow a double-valued track in this plane: they continue to grow in size, while their radio power first brightens, then decays. Thus, a single P_{jet} value will produce a curved track in this plane. However, as Fig. 1 shows, the distribution of sources in this plane is dominated by the $F(P_{jet})$ distrubution *of the oldest sources* in the sample. Thus, the $F(P_{jet})$ distribution is more important than single-source evolution in this statistic as well.

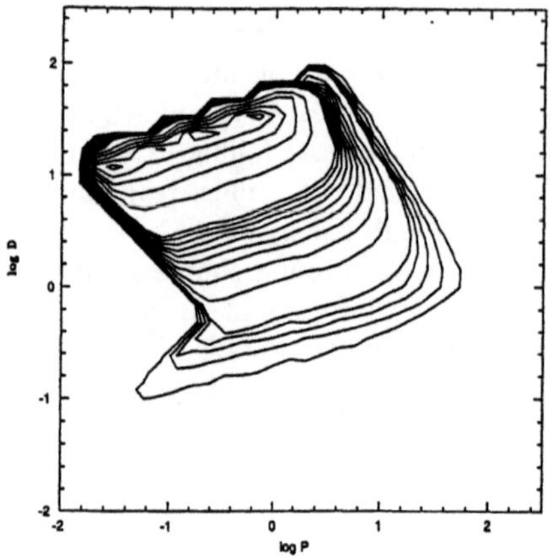

Fig. 1. The distribution of sources in the (D, p_ν) plane, assuming type II sources with a Gaussian $F(P_{jet})$ distribution, at an age $100t_{sy}$. The contours, in sources per log-log bin, are: .1 to .5 by .1; 1 to 5 by .5; 5 to 30 by 5.

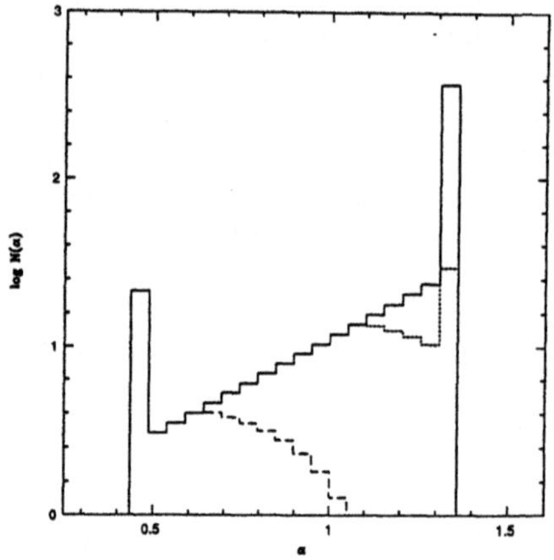

Fig. 2. The spectral index distribution of type II sources, at times 3, 10 and $30t_{sy}$. The injected spectrum was taken to be $N(e) \propto e^{-2}$, to set the initial $\alpha = 0.5$. Note the *log* of the number of sources per bin has been plotted.

The third area I investigated is the distribution of sources in spectral index. This is the one which shows the most dramatic conflict between the standard model and the data.

A typical prediction for the spectral index statistics in the standard, injection-only model is shown in Fig. 2. It is clear that this model is nonsense. The fraction of sources which are steep spectrum grows with time; by a time $\sim 30t_{sy}$ less than one percent of the sample are flat-spectrum.

Facing this problem, one is tempted to ask what happens if the jet turns off. I investigated jets which turn off at $\sim t_{sy}$. I found that this makes very little difference to this problem. The relativistic electrons age at the usual synchrotron loss rate, but the mean magnetic field stays constant; this leads to only a slow decay in $p_\nu(t)$. Thus, the predictions of this model are not much different from those of the injection-only model: we find far too few steep-spectrum sources for this model to be right.

To summarize, these simulations find that the $N(p_\nu)$ and $N(D, p_\nu)$ statistics can be explained by the standard model if the $F(P_{jet})$ distributions are right (although the model does not address why $F_I(P_{jet})$ and $F_{II}(P_{jet})$ should be different). However, these simple models predict far too many steep-spectrum sources compared to the numbers actually found in the sky.

4. Resolution: What Might We Do?

This work has re-emphasized an old problem: the theory predicts far too many steep-spectrum sources. Can we find any hints for the resolution of this problem?

One obvious extension of the picture was mentioned above: *in situ* electron reacceleration might solve the problem, *if* it keeps the high-energy cutoff of the electron spectrum from decaying. I investigated the "best case" acceleration model numerically, in which the cutoff of the electron spectrum is maintained at a constant energy. In this case, the photon spectrum will "age" only due to the decay of the magnetic field, and the steep-spectrum problem will be minimized. The simulations do bear this out. Most of a sample can be maintained at flat spectra for times up to $100t_{sy}$, for instance, if e_c is chosen a factor of ten above the initial e_{obs}.

Another likely extension of the standard picture could involve fast dissipation of magnetic energy. This would be a variant of the "dying source" problem; the simulations done so far show that the timescale for field or current dissipation must be much shorter than the mean synchrotron lifetime. Thus, overpressure expansion of the entire source, or diffusion of the relativistic particles, are too slow to deal with the problem. It might be that the currents and fields are localized in regions small compared to the source size – such as thin current sheets. The dissipation time for these features will scale as the size of the feature; overpressure expansion, shock dissipation or local reconnection might then provide a rapid decay of the radio power.

The two ideas above may well turn out to provide the answer, when they are developed in sufficient detail to be testable. However, it may be that a quite different model will turn out to be profitable.

One tempting possibility is the idea that the sources flare: that their radio power does not increase slowly, as in these steady jet-fed models, but rather that the energy from the jet is stored, and released suddenly (as in a stellar flare, perhaps). If the brightening is strong enough, our statistics might reflect only the sources in their "flare" stage.

In addition, one might question whether type I and II sources remain separate throught their lives. Recent observational work (described in Owen, these proceedings) suggests that the connection between type I and II sources is more complex than had been thought. In particular, the old idea that one type evolves into the other may yet turn out to be right. If this is the case, we need to figure out the physical mechanisms by which this evolution takes place.

References

Baldwin, J.E.: (1982), in *Extragalactic Radio Sources, IAU Symp. 97*, Dordrecht: Reidel, p.21.
Binggelli, B., Sandage, A. and Tammann G.A.: (1988), *Ann. Rev. Astron. Astrophys.* **26**, 509.
Borovsky, J.O. and Eilek, J.A.: 1986, *Astrophys. J.* **308**, 929.
Cavaliere, A., Morrison, P. and Wood, K.: 1971, *Astrophys. J.* **170**, 223.
de Ruiter, H.R., Parma, P., Fanti, C., Fanti, R.: 1990, *Astron. Astrophys.* **227**, 351.
Eilek, J.A.: (1990), *Bull.AAS* **22**, 801.

Eilek, J.A. and Shore, S.N.: 1989, *Astrophys. J.* **342**, 187.
Gopal-Krishna and Wiita, P.J.: (1987), *Mon. Not. R. astr. Soc.* **226**, 531.
Scheuer, P.A.G.: (1974), *Mon. Not. R. astr. Soc.* **166**, 513.

Asymmetries in Powerful Radio Galaxies

Guy Pooley

Mullard Radio Astronomy Observatory
Cavendish Laboratory
Madingley Road
Cambridge CB3 0HE, U.K.

Summary: Several recent studies have shown that most powerful radio galaxies which show extended emission-line regions have correlated optical and radio asymmetries. McCarthy, van Breugel & Kapahi (1991) showed that, in cases where the emission-line region is extended, it is brighter on the side with the shorter radio arm-length in the great majority of cases. Liu & Pooley (1991a) also confirmed that in 4 cases there is strong spatial association of the emission-line material with depolarization of the radio emission.

Liu & Pooley (1991b) demonstated a clear relation between spectral index and depolarization in powerful 3C sources: the steeper radio spectrum is associated with the stronger depolarization.

These correlations appear to be related to unsymmetrical distributions of gas around the host galaxies; this is in contrast to the depolarization asymmetry for sources with one-sided radio jets, which is most easily explained in terms of the orientation of the sources to the line of sight (Garrington & Conway 1991).

References

Garrington, S.T., Conway, R.G.: 1991, *Mon. Not. R. astr. Soc.* **250**, 198.
Liu, R., Pooley, G.: 1991a, *Mon. Not. R. astr. Soc.* **253**, 669.
——: 1991b, *Mon. Not. R. astr. Soc.* **249**, 343.
McCarthy, P.J., van Breugel, W., Kapahi, V.K.: 1991, *Astrophys. J.* **371**, 478.

Ultrahigh Energy Cosmic Rays from Fanaroff Riley Class II Radio Galaxies[24]

Jörg Rachen and Peter L. Biermann

Max-Planck-Institut für Radioastronomie
Auf dem Hügel 69
D5300 Bonn-1, Germany

Abstract: The hot spots of Fanaroff Riley class II radio galaxies are argued to be the sources of the ultrahigh energy component in cosmic rays. We present calculations of cosmic ray transport in an evolving universe, taking the losses against the microwave background properly into account. As input we use the models for the cosmological radio source evolution derived by radio astronomers (mainly Peacock 1985). The model we adopt for the acceleration in the radio hot spots has been introduced by Biermann and Strittmatter (1987), and Meisenheimer *et al.* (1989) and is based on first order Fermi theory of particle acceleration at shocks (see *e.g.* Drury 1983). As an unknown the actual proportion of energy density in protons enters, which together with structural uncertainties in the hot spots should introduce no more than one order of magnitude in uncertainty: we easily reproduce the observed spectra of high energy cosmic rays. It follows that scattering of charged energetic particles in intergalactic space must be sufficiently small in order to obtain contributions from sources as far away as the nearest radio galaxies.

1 Radiogalaxy Hot Spots

1.1 Acceleration of Particles

The spectra of the radio hot spots of those radio galaxies that have a characteristic double structure (see *e.g.* Miley 1980) often show a spectrum that extends into the optical range and then usually cuts off rather abruptly. Similar properties are found for some red quasars and jets. Biermann and Strittmatter (1987) proposed that such spectra could be understood as arising from electrons accelerated in shocks, that are mediated by energetic protons. The role of the protons is here crucial: they provide the messenger for the upstream gas to react to the shock downstream, thereby excite turbulence and so provide the wavefield for the electrons to scatter on both sides of the shock; this repeated scattering is essential for electrons to gain energy on their repeated passages through the shock. A critical assumption is here that this wavefield is of Kolmogorov

[24]Paper presented by P. Biermann.

character; however, from plasma simulations, from the interstellar medium, as well as from the solar wind it is now known that Kolmogorov turbulence appears to be one standard mode of turbulent motion in a plasma outside the dissipation scales. A first order Fermi theory approach was also used by Meisenheimer *et al.* (1989) to actually fit observed hot spot spectra. The fit is quite good and suggests that the model has merit. For many of the hot spots they discuss, the mean free path of scattering can be estimated and so it can actually be checked whether a Kolmogorov spectrum is consistent with the data: it is. From the argument of Biermann and Strittmatter (1987) it then follows that protons exist in the hot spots that reach rather high energies, in magnetic field environments of 10^{-4} Gauss energies up to 10^{12} GeV. It can be ascertained that the space available and the competing losses from proton-proton collisions do not normally win over the synchrotron and photon-interaction losses which limit the proton energy that can be reached. The spectra of these energetic protons and heavier nuclei can be estimated from the electron population: these spectra are very close to E^{-2}, as shown by the $\nu^{-0.5}$ synchrotron spectra observed. Thus, these arguments provide clear evidence that radio hot spots harbour a population of very high energy protons, and, by analogy, heavier nuclei. Such radio hot spots are situated far outside their central galaxy in most cases, and may well be considered to be embedded in intergalactic space. Thus, it it an interesting question to ask to what degree such hot spots might contribute to the observed high energy cosmic rays, since their expected particle energy range is very close to that observed.

1.2 Radio Source Evolution

Radio surveys provide information on the cosmological evolution of powerful radio sources. Peacock (1985) has provided a number of models that describe the evolution of the radio source population to explain the radio data. We will use these radio source evolution models as input to discuss the sources of cosmic rays as a function of cosmological epoch.

2 The Evolution of the Injected Spectrum of Energetic Particles

2.1 The Injection

When observing radio sources, the energy density of the various relativistic components can be estimated from the requirement that the observed radio emission be reproduced in the observed space. The luminosity of synchrotron emission is clearly proportional to some power of the magnetic field, depending on the spectrum of the emission. The required particle energy density in relativistic electrons then is inversely related to the assumed magnetic field strength. The energy density of the magnetic field is obviously proportional to the square of the field strength. The assumption that the sum is near the minimum in real environments then leads to a unique relation between magnetic field, particle energy density and total energy density. The relative proportion of proton — and heavier nuclei — and electron populations of relativistic particles k_p remains an unknown here; in the interstellar medium this ratio is 100 for particle energies above

1 GeV. In radio lobes and hot spots this number could well be anywhere between unity and more than 1000. This enters below into the fudge factor, the factor given in the figures.

2.2 The Evolution

Interaction of the high energy particles with the microwave background is a strong limiting factor in determining for how long the injected particles can travel in intergalactic space (Greisen 1966, Stecker 1968). The dominant loss processes in this interaction are first of all pion production, due to the extremely strong increase of the proton-photon interaction cross section at pion energies (the Delta resonance), then the corresponding electron-positron pair production (Blumenthal 1970), and last, the adiabatic losses due to the expansion of the universe. Obviously, we do include the cosmological evolution of the microwave background itself in energy density and temperature. We treat here the proton-photon interaction as a continuous loss process, following Berezinsky and Grigor'eva (1988). They considered this approximation and demonstrated that it does

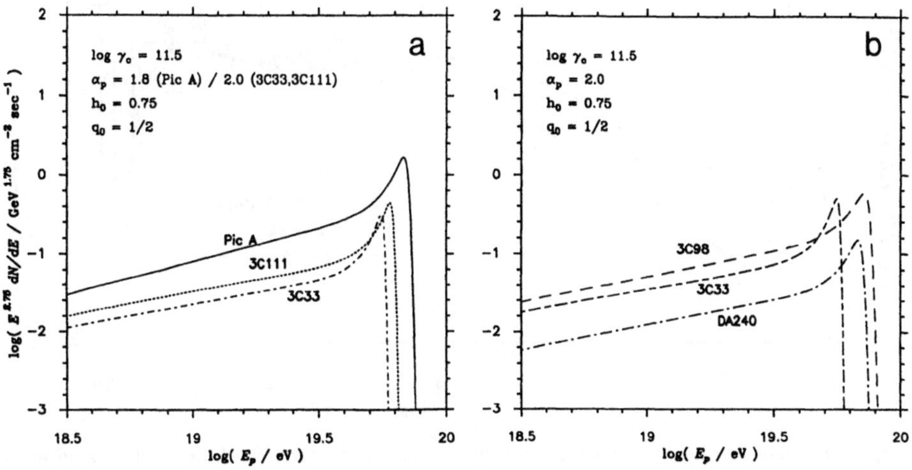

Fig. 1. a) The contribution of identified radio galaxies Pic A, 3C 111 and 3C 33; here the jet luminosities and spectral indices were taken from Meisenheimer *et al.* 1989. **b)** The contribution of identified radio galaxies 3C 98, DA 240 and, for comparison again, 3C 33; here the spectral index of $\alpha_p = 2.0$ for the particle spectra was taken in analogy to the results of Meisenheimer *et al.* 1989, and the jet luminosities are taken from Rawlings and Saunders, 1991.

not introduce errors of more than about 10%. Thus, the observed particle energy spectrum is cut off at some energy corresponding to the time that has elapsed since injection, the Greisen cutoff energy. It follows that practically no particles can reach us from any interesting distance, such as distances beyond the Virgo cluster, which have an energy upon arrival of above or even very close to 10^{11} GeV. This particle energy loss is folded in our calculation with an injected spectrum and the cosmological distribution of sources at different powers corresponding to the observed radio source evolution.

3 The Calculated High Energy Cosmic Ray Spectra

Thus, we compose total spectra of ultrahigh energy particles at earth out of the contributions from the different radio galaxies that have injected energetic particles in the past. For this calculation it is a priori not relevant whether we have straight line propagation or diffusion as long as the graininess of the sources does not become dominant.

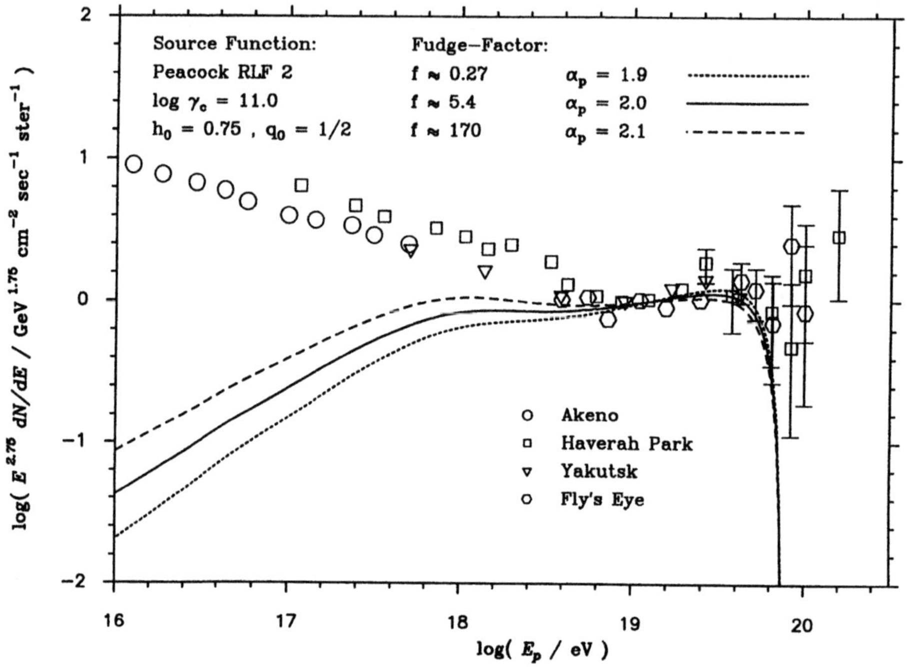

Fig. 2. The total cosmic ray spectrum for different input spectra; the fudge factor required to fit at $10^{19.1}$ eV is strongly dependent on α_p.

For instance if all energetic particles would stay close to their source and form a cosmologically small sphere filled with energetic particles, then outside the spatial sum of all such spheres there would be no contribution whatsoever. Thus, in order to be relevant as a description of real nature, our calculation imposes the condition that the path of energetic particles goes out from the sources sufficiently far, to attain a nearly homogeneous distribution. Obviously, this could be translated into a limit on the strength of the intergalactic magnetic field, given a spatial topology. Lacking such a model, we can only say that the mean free path for scattering has to be quite long, of order 100 Mpc.

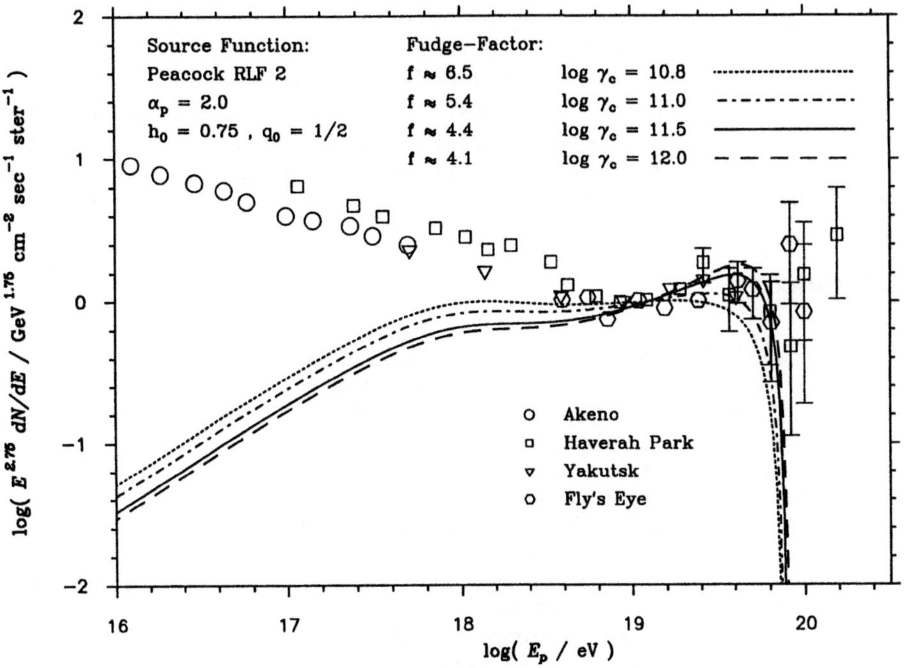

Fig. 3. The total cosmic ray spectrum for different initial cutoff energies; the fudge factor is quite reasonable for all cases considered.

The calculated spectra are displayed in Figs. 1 through 3. The first two figures give the contributions from identified radio galaxies, assuming in this case straight line propagation. Therefore, we give the spectra here as total flux without specifying which part of 4π the particles actually reach. We see that individual radio galaxies can dominate the total ultrahigh energy cosmic ray spectrum, given the simplistic propagation model. In Figure 2 and 3 we show the total spectrum using the radio source evolution model dubbed RLF2 of Peacock (1985), in Fig. 2 for a range of injection spectra, and in Fig. 3 for a range in initial cutoff energies. The fudge factor given in each case is the factor necessary to fit at $10^{10.1}$ GeV. This factor has to be compared with the function $\frac{4}{7}k_p(1+k_p)^{-3/7}$, which is the minimum energy density factor of relativistic protons relative to the case where no protons are assumed; this, however, is the case in the jet and radio lobe energies estimated by Rawlings and Saunders (1991). Keeping k_p as an unknown reduces the uncertainties involved in our estimate to questions like: a) is the fraction of energy in protons in the hot spots the same as in the entire lobe, and b) is the fraction of protons which leaves the source going into intergalactic space independent of energy? The combination of all such factors, which we assume to be near unity in their effect, with the function involving k_p has to match the fudge factor given in the figures. We see that this is easily done with the possible range for k_p; the k_p-term by itself provides a possible range of 0.42 over 7.9 to 44. for k_p from 1 over 100 to 2000. We note that spectra very close to $\alpha_p = 2.0$ or flatter are required. We also note that the cutoff energy, assumed here to be the same in all sources, could be as low as 10^{11} GeV or even a little less without changing the overall result. An early analytical version of these results was first presented at a conference in honor of M.M. Shapiro in October 1990 in Washington, D.C. (Biermann 1991). It appears possible to explain nearly all data above about 10^9 GeV within this model. We have thus combined knowledge from cosmological radio source evolution with a tested physical model for the acceleration of relativistic nuclei and have found that we can readily explain the entire ultrahigh energy spectrum of cosmic rays.

References

Berezinsky, V.S., Grigor'eva, S.I.: 1988, *Astron. Astrophys.* **199**, 1.

Biermann, P.L., Strittmatter, P.A.: 1987, *Astrophys. J.* **322**, 643.

Biermann, P.L.: 1991, in *Frontiers in Astrophysics*, eds. R. Silberberg, G. Fazio, M. Rees, Cambridge: Cambridge University Press.

Blumenthal, G.: 1970, *Phys.Rev.D* **1**, 1596.

Drury, L. O'C: 1983, *Rep.Prog.Phys.* **46**,, 973.

Fanaroff, B.L., Riley, J.M.: 1974, *Mon. Not. R. astr. Soc.* **167**, 31P.

Greisen, K.: 1966, *Phys.Rev.Letters* **16**, 748.

Meisenheimer, K., Röser, H.-J., Hiltner, P.R., Yates, M.G., Longair, M.S., Chini, R., Perley, R.A.: 1989, *Astron. Astrophys.* **219**, 63.

Miley, G.K.: 1980, *Ann. Rev. Astron. Astrophys.* **18**, 165.

Peacock, J.A.: 1985, *Mon. Not. R. astr. Soc.* **217**, 601.

Rawlings, S., Saunders, R.: 1991, *Nature* **349**, 138.

Stecker, F.W.: 1968, *Phys.Rev.Letters* **21**, 1016.

Summary of the Workshop

P.A.G. Scheuer

Mullard Radio Astronomy Observatory
Cavendish Laboratory
Madingley Road
Cambridge CB3 0HE, U.K.

Abstract: I cannot possibly do justice to all, so I will not even try to do justice to anyone, and I shall be particularly unjust in ignoring many excellent review talks. What follows is a personal selection among items that struck me most forcibly during the week.

1 M 87

John Biretta showed us wonderful pictures of the radio structure of M 87. The jet in particular is full of thought-provoking features: sharp straight boundaries, and, within the jet, shocks (?) and helical wisps that may or may not be telling us about magnetic fields. The radio and optical polarization look identical (Meisenheimer). The radio and optical intensity patterns also look identical qualitatively (HST pictures shown by Crane) though the relative intensities of features are quite different. We noted, however, that the radio and optical appearance is not at all the same in 3C 273, where the optical jet stops well short of the radio jet.

I found Meisenheimer's Magical Model of the M 87 jet extraordinarily persuasive. From the optical spectra he infers a cut-off frequency which varies by about a factor of 3 along the jet. Attributing this solely to compressions and rarefactions of a helical magnetic field (i.e. no change in the electron energy distribution) he gets not only agreement with the changes in polarization but also astonishingly good agreement with the total intensity profile along the jet. That is impressive, but also worrying: how does the jet manage to keep on re-accelerating just the high-energy electrons without also accumulating a great heap of electrons that radiate in the radio band? Standard particle acceleration recipes don' t seem to fit.

2 Jet Production

Before this meeting I wrote to Max Camenzind begging him to explain whether the many papers that make jets by twisting magnetic fields in the innermost parts of the accretion disc are (i) merely variations on a single theme, or (ii) qualitatively different in their physical assumptions and/or results, or (iii) mutually incompatible with each other and with schemes for pair production. He answered many of these questions in his admirable review, and as this is difficult stuff I shall repeat a few salient points.

- To produce jets with Lorentz factors γ above 2 or 3 one obviously needs at least special relativity in the theory; less obviously, one also needs magnetic fields.
- The maximum magnetic energy that (for example) a dynamo can make is the turbulent energy in the disc. Standard accretion disc theory gives an estimate of the maximum turbulent energy, which is a small fraction of the total energy input but nevertheless increases with the accretion rate $\dot{m}_{\mathrm{BH}}$. For a standard black hole of 10^8 $\mathrm{M}_\odot$ accreting 1 $\mathrm{M}_\odot$ per year, the maximum magnetic field is of the order of 1 T — which demonstrates the superiority of SI units.
- The ultimate speed of the jet depends on the magnetic energy per kg evaporated from the accretion disc (cf. stagnation pressure in ordinary hydrodynamics). Only Nature knows how much matter is evaporated, as that depends on complicated physics near the surface of the disc, but there is an upper bound set by the amount of angular momentum available. Taking the maximum evaporation rate as well as the maximum magnetic field, the result for relativistic flows is that the ultimate Lorentz factor of the jet is limited by

$$\begin{aligned}\gamma \leq &\ \text{Michel's (1982) magnetization parameter } \sigma \\ = &\ (\text{fudge factor of order } 10) \times \dot{m}_{\mathrm{BH}}/\dot{m}_{\mathrm{Eddington}}\end{aligned}$$

Thus, it seems likely that powerful radio sources with high-γ jets, such as 3C 273 and 3C 345, are close to the Eddington accretion rate. (Super-Eddington accretion rates don't increase γ further.)

At this point I have a question for theorists: can you set an upper limit to the γ attainable in this kind of model?

3 Variability

As Ian Robson emphasised, multi-frequency observations of variability in the mm, IR and optical bands are very important in that they probably take us closer to the nucleus than any other observations. My impression is that, at this stage, it is still a struggle to establish clear relationships between variations at different frequencies and establish time delays, so I find little that I could say as a useful summary. The short-period variations in the radio band that Witzel, Wagner and colleagues have now established beyond reasonable doubt are in a different category, because they pose an immediate challenge to the conventional synchrotron interpretation.

4 VLBI

(Zensus' review) Theorists may understand something of the nucleus, but understanding stops when we reach scales at which we can see some structure. One old and important question is whether the radio components are the bright bits in a flow down a channel. The longest and most thoroughly observed quasar with a non-linear jet, 3C 345, gives two clear answers:

Yes. Component C4 follows the path taken earlier by C3. Furthermore, when C4 gets to places previously occupied by C3, its polarization is in the same orientation (Wardle).

No. The path of C5 is quite obviously different from the path of C3 and C4.

An important observational problem is to find out whether FR I sources show superluminal motion. The work described by Giovannini is a promising start, but it will still take some time before an answer emerges.

5 The Speed of Jets

It has been known for a while that different components move with different apparent speeds across the sky, and more recent data on the graphs add conviction. In a source like 3C 273 with a straight jet it becomes highly implausible to blame these differences on different inclinations of the path to the line of sight. Now Biretta's proper motions in the jet of M 87, ranging from 0.3 c to 3 c, make the same point even more forcefully. The moral must surely be that phase velocities *in some sense* must play an important part.

While few people dispute that there is some sort of relativistic motion in jets on the parsec scale, the speed of kpc-to-Mpc scale jets in more controversial. Mitch Begelman reviewed the traditional arguments. John Wardle added a more quantitative note: statistics of the front-to-back ratios of jets, in a sample of quasars carefully mapped with the VLA, require jet speeds of 0.6 c at the very least, even if one adopts Peter Barthel's (1989) hypothesis that radio galaxies are quasars whose radio axes are further than $\approx 45°$ from the line of sight.

Begelman pointed out that the traditional formula for the front-to-back intensity ratio of relativistic jets,

$$\left(\frac{1+\beta cos\theta}{1-\beta cos\theta}\right)^{2+\alpha}$$

is not necessarily true. If the magnetic field is predominantly longitudinal, then the radiation from the approaching jet is generally emitted at a large angle to the magnetic field, but the radiation observed from the receding jet is emitted almost along the field, and the appropriate exponent may then be ≈ 4 rather than ≈ 2.5. I mention this partly for its ingenuity, and partly to say that I do not think it has much impact in practice, because the fractional polarizations observed in jets are not very large. That is relevant to another and more important point of Begelman's: in Barthel's model, the front-to-back intensity ratio for jets in radio galaxies should not exceed 15–20, provided that the exponent $2+\alpha$ is correct. That prediction can be tested by observation *now*, though we need to look at some relatively simple escape routes such as a scatter in intrinsic jet-counterjet ratios and in the critical angle of Barthel's model.

6 Geriatric Problems

What happens to a radio galaxy when the central engine switches off and jets are no more? Radio astronomers have long searched for large featureless steep-spectrum radio sources, and found relatively few (Eilek). The problem may be eased somewhat if the jet is turned off gradually. The only way we thought of that could kill a pensioned-off source quickly is to reduce the magnetic field, for that could happen through the straightening-out of fields stretched by turbulent motions, helped by the reconnection of field lines. However, such a story leads to its own difficulties:

- It is hard to understand how magnetic fields would be maintained in sources displaying synchrotron ageing in their lobes, especially as general stirring of the lobe material would destroy the observed gradients in spectral index.
- There are sources with very large wings sticking out at odd angles (several appeared in the sample shown by Adam Black), and those wings presumably do represent the relics of lobes that were made when the jets pointed along a different axis.

The question still awaits a plausible solution.

7 The Relationship Between FR I and FR II Sources

It is often said that there are lies, damned lies and statistics. In astronomy one might start the corresponding sequence with theories, scenarios – but in this case I believed the statistics! The plots which Frazer Owen showed us, the distribution of FR I and FR II radio sources in the (radio luminosity) – (optical luminosity) plane, convinced me (and, I think, most of you too) that the Fanaroff-Riley division is even sharper if one draws an oblique line across that diagram rather than dividing sources according to radio luminosity alone. From the fact that one morphological type takes over where the other leaves off, he suggested that perhaps the two classes were after all related rather than distinct phenomena.

It has generally been argued that FR II sources do not develop into FR Is (or vice versa), because FR Is are typically in the dominant galaxy in a fairly rich cluster, while FR IIs typically live in galaxies in smaller groups. Evidence is now accumulating which indicates that the more powerful FR II radio galaxies, and perhaps radio-loud quasars too, generally live in clusters (Cygnus A and 3C 295 being long-established examples). One might, therefore, be at liberty to imagine that the most powerful FR IIs decay into FR Is, while the weaker FR IIs sink into insignificance. Returning to the 'damned lies' in the proverbial sequence, let me paint you a scenario (Fig. 1).

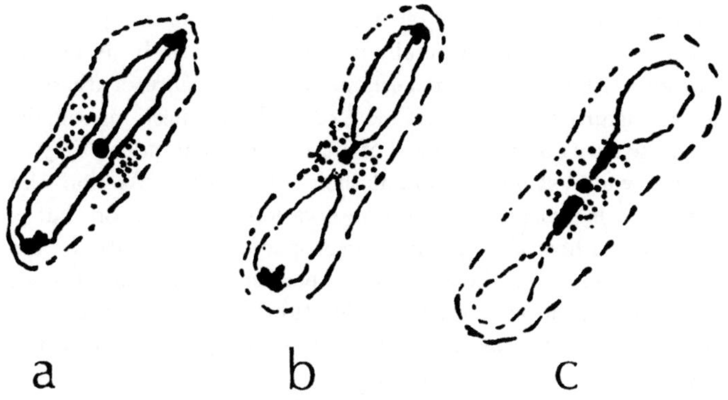

Fig. 1. **(a)** represents the standard model of a radio galaxy: the jet has blasted away some of the atmosphere surrounding the galaxy and replaced it with a 'cocoon' of very hot gas of very low density, in which the jets are fairly happy and stable. Now, the 'cocoon' can last a very long time in the outer parts of the source, but at small distances from the nucleus (less than a few kpc) the atmosphere will squeeze the 'cocoon' out in a time much less than 10^8 years (cf. Scheuer 1974, model C), as in **(b)**. The jets then become 'naked jets' in the sense of de Ruiter's talk, and inevitably start to entrain significant masses of atmosphere; they slow down, become visible on both sides of the nucleus, and we have a FR I source **(c)**.

8 Numerical Simulations (Burns, Clarke, Cox, Falle, Norman, ...)

This is a field in which there have been really big advances. A few years ago the kinds of fluid flows that were simulated were determined more by what the computer could manage than by their relevance to the astrophysics. Now one can perform the simulations one wants to do and ought to do. In particular, one is no longer restricted to axisymmetric and slab-type flows (which automatically suppress a large range of instabilities and other relevant phenomena), and we have seen many 3D hydrodynamic and MHD simulations this week which are not only beautiful to behold but bear directly upon many of the phenomena that have been invoked to explain the appearance of radio lobes.

In all this euphoria we still have to remember that the quality of the simulation ultimately depends on the quality of the physics we put into the code, and among the physics which we still don't know how to put in properly are turbulence and magnetic field line reconnection. A very interesting point that is relevant here was raised (I think) by Biermann, and certainly depends on his model of the acceleration of the cosmic ray ions of the highest energies in the hot-spots of radio sources. If indeed radio sources contain a significant quantity of ions with Larmor radii of the order of kpc, they would transport momentum with a correspondingly large 'mean free path' and thus reduce the Reynolds number to as little as 1000 or even 100.

9 Ringberg 3?

At the beginning of this week a few of us had doubts. Was this perhaps the last such 'workshop' for a long time? Had we really made any progress in understanding the jets in the last 5 or even 10 years? Was it perhaps time for a moratorium on discussing the same evidence yet again? It is true that most of the basic questions remain unanswered. We still aren't quite sure about the speed of matter in jets, if indeed they contain matter — because we know almost nothing about their composition. We don't know how they are made (and see little prospect of direct observational evidence on that point!) and we don't fully understand how they manage to propagate so far without being disrupted by instabilities. We still don't know where the particle acceleration takes place, or how (though some believe in diffusive shock acceleration), and the optical observations have deepened the mystery as well as indicating very strongly that acceleration takes place all the way along jets — *some* jets anyway.

At the same time, there has been progress in both observation and theory which has sharpened many of the questions to a point where they seem accessible. I have tried to punctuate my summary with a few questions that do look as if we could tackle them over the next few years, and I hope they may be helpful in discussing the date of the next meeting.

References:

Barthel, P.D.: 1989, *Astrophys. J.* **336**, 606.
Michel, F.C.: 1982, *Rev. Mod. Phys.* **54**, 1.
Scheuer, P.A.G.: 1974, *Mon. Not. R. astr. Soc.* **166**, 513.

Object Index

Lecture Notes in Physics

For information about Vols. 1–384
please contact your bookseller or Springer-Verlag

Vol. 385: A. Treves, G. C. Perola, L. Stella (Eds.), Iron Line Diagnostics in X-Ray Sources. Proceedings, Como, Italy 1990. IX, 312 pages. 1991.

Vol. 386: G. Pétré, A. Sanfeld (Eds.), Capillarity Today. Proceedings, Belgium 1990. XI, 384 pages. 1991.

Vol. 387: Y. Uchida, R. C. Canfield, T. Watanabe, E. Hiei (Eds.), Flare Physics in Solar Activity Maximum 22. Proceedings, 1990. X, 360 pages. 1991.

Vol. 388: D. Gough, J. Toomre (Eds.), Challenges to Theories of the Structure of Moderate-Mass Stars. Proceedings, 1990. VII, 414 pages. 1991.

Vol. 389: J. C. Miller, R. F. Haglund (Eds.), Laser Ablation-Mechanisms and Applications. Proceedings. IX, 362 pages, 1991.

Vol. 390: J. Heidmann, M. J. Klein (Eds.), Bioastronomy - The Search for Extraterrestrial Life. Proceedings, 1990. XVII, 413 pages. 1991.

Vol. 391: A. Zdziarski, M. Sikora (Eds.), Ralativistic Hadrons in Cosmic Compact Objects. Proceedings, 1990. XII, 182 pages. 1991.

Vol. 392: J.-D. Fournier, P.-L. Sulem (Eds.), Large-Scale Structures in Nonlinear Physics. Proceedings. VIII, 353 pages. 1991.

Vol. 393: M. Remoissenet, M.Peyrard (Eds.), Nonlinear Coherent Structures in Physics and Biology. Proceedings. XII, 398 pages. 1991.

Vol. 394: M. R. J. Hoch, R. H. Lemmer (Eds.), Low Temperature Physics. Proceedings. X,374 pages. 1991.

Vol. 395: H. E. Trease, M. J. Fritts, W. P. Crowley (Eds.), Advances in the Free-Lagrange Method. Proceedings, 1990. XI, 327 pages. 1991.

Vol. 396: H. Mitter, H. Gausterer (Eds.), Recent Aspects of Quantum Fields. Proceedings. XIII, 332 pages. 1991.

Vol. 398: T. M. M. Verheggen (Ed.), Numerical Methods for the Simulation of Multi-Phase and Complex Flow. Proceedings, 1990. VI, 153 pages. 1992.

Vol. 399: Z. Švestka, B. V. Jackson, M. E. Machedo (Eds.), Eruptive Solar Flares. Proceedings, 1991. XIV, 409 pages. 1992.

Vol. 400: M. Dienes, M. Month, S. Turner (Eds.), Frontiers of Particle Beams: Intensity Limitations. Proceedings, 1990. IX, 610 pages. 1992.

Vol. 401: U. Heber, C. S. Jeffery (Eds.), The Atmospheres of Early-Type Stars. Proceedings, 1991. XIX, 450 pages. 1992.

Vol. 402: L. Boi, D. Flament, J.-M. Salanskis (Eds.), 1830-1930: A Century of Geometry. VIII, 304 pages. 1992.

Vol. 403: E. Balslev (Ed.), Schrödinger Operators. Proceedings, 1991. VIII, 264 pages. 1992.

Vol. 404: R. Schmidt, H. O. Lutz, R. Dreizler (Eds.), Nuclear Physics Concepts in the Study of Atomic Cluster Physics. Proceedings, 1991. XVIII, 363 pages. 1992.

Vol. 405: W. Hollik, R. Rückl, J. Wess (Eds.), Phenomenological Aspects of Supersymmetry. VII, 329 pages. 1992.

Vol. 406: R. Kayser, T. Schramm, L. Nieser (Eds.), Gravitational Lenses. Proceedings, 1991. XXII, 399 pages. 1992.

Vol. 407: P. L. Smith, W. L. Wiese (Eds.), Atomic and Molecular Data for Space Astronomy. VII, 158 pages. 1992.

Vol. 408: V. J. Martínez, M. Portilla, D. Sàez (Eds.), New Insights into the Universe. Proceedings, 1991. XI, 298 pages. 1992.

Vol. 409: H. Gausterer, C. B. Lang (Eds.), Computational Methods in Field Theory. Proceedings, 1992. XII, 274 pages. 1992.

Vol. 410: J. Ehlers, G. Schäfer (Eds.), Relativistic Gravity Research. Proceedings, VIII, 409 pages. 1992.

Vol. 411: W. Dieter Heiss (Ed.), Chaos and Quantum Chaos. Proceedings, XIV, 330 pages. 1992.

Vol. 412: A. W. Clegg, G. E. Nedoluha (Eds.), Astrophysical Masers. Proceedings, 1992. XX, 480 pages. 1993.

Vol. 413: Aa. Sandqvist, T. P. Ray (Eds.); Central Activity in Galaxies. From Observational Data to Astrophysical Diagnostics. XIII, 235 pages. 1993.

Vol. 414: M. Napolitano, F. Sabetta (Eds.), Thirteenth International Conference on Numerical Methods in Fluid Dynamics. Proceedings, 1992. XIV, 541 pages. 1993.

Vol. 415: L. Garrido (Ed.), Complex Fluids. Proceedings, 1992. XIII, 413 pages. 1993.

Vol. 416: B. Baschek, G. Klare, J. Lequeux (Eds.), New Aspects of Magellanic Cloud Research. Proceedings, 1992. XIII, 494 pages. 1993.

Vol. 417: K. Goeke P. Kroll, H.-R. Petry (Eds.), Quark Cluster Dynamics. Proceedings, 1992. XI, 297 pages. 1993.

Vol. 418: J. van Paradijs, H. M. Maitzen (Eds.), Galactic High-Energy Astrophysics. XIII, 293 pages. 1993.

Vol. 419: K. H. Ploog, L. Tapfer (Eds.), Physics and Technology of Semiconductor Quantum Devices. Proceedings, 1992. VIII, 212 pages. 1993.

Vol. 421: H.-J. Röser, K. Meisenheimer (Eds.), Jets in Extragalactic Radio Sources. XX, 301 pages. 1993.

New Series m: Monographs

Vol. m 1: H. Hora, Plasmas at High Temperature and Density. VIII, 442 pages. 1991.

Vol. m 2: P. Busch, P. J. Lahti, P. Mittelstaedt, The Quantum Theory of Measurement. XIII, 165 pages. 1991.

Vol. m 3: A. Heck, J. M. Perdang (Eds.), Applying Fractals in Astronomy. IX, 210 pages. 1991.

Vol. m 4: R. K. Zeytounian, Mécanique des fluides fondamentale. XV, 615 pages, 1991.

Vol. m 5: R. K. Zeytounian, Meteorological Fluid Dynamics. XI, 346 pages. 1991.

Vol. m 6: N. M. J. Woodhouse, Special Relativity. VIII, 86 pages. 1992.

Vol. m 7: G. Morandi, The Role of Topology in Classical and Quantum Physics. XIII, 239 pages. 1992.

Vol. m 8: D. Funaro, Polynomial Approximation of Differential Equations. X, 305 pages. 1992.

Vol. m 9: M. Namiki, Stochastic Quantization. X, 217 pages. 1992.

Vol. m 10: J. Hoppe, Lectures on Integrable Systems. VII, 111 pages. 1992.

Vol. m 11: A. D. Yaghjian, Relativistic Dynamics of a Charged Sphere. XII, 115 pages. 1992.

Vol. m 12: G. Esposito, Quantum Gravity, Quantum Cosmology and Lorentzian Geometries. XVI, 326 pages. 1992.

Vol. m 13: M. Klein, A. Knauf, Classical Planar Scattering by Coulombic Potentials. V, 142 pages. 1992.

Vol. m 14: A. Lerda, Anyons. XI, 138 pages. 1992.

Vol. m 15: N. Peters, B. Rogg (Eds.), Reduced Kinetic Mechanisms for Applications in Combustion Systems. X, 360 pages. 1993.

Vol. m 16: P. Christe, M. Henkel, Introduction to Conformal Invariance and Its Applications to Critical Phenomena. XV, 260 pages. 1993.

Vol. m 17: M. Schoen, Computer Simulation of Condensed Phases in Complex Geometries. X, 136 pages. 1993.

GPSR Compliance

The European Union's (EU) General Product Safety Regulation (GPSR) is a set of rules that requires consumer products to be safe and our obligations to ensure this.

If you have any concerns about our products, you can contact us on ProductSafety@springernature.com

In case Publisher is established outside the EU, the EU authorized representative is:

Springer Nature Customer Service Center GmbH
Europaplatz 3
69115 Heidelberg, Germany

Batch number: 10620143

Printed by Printforce, the Netherlands